NEW DIMENSIONS IN AGRICULTURAL GEOGRAPHY

VOLUME FIVE

SPATIAL DIMENSIONS OF AGRICULTURE

Concept's International Series in Geography

Volume 1 **Historical Dimensions of Agriculture**
Volume 2 **The Ecology of Agricultural System**
Volume 3 **Socio-economic Dimensions of Agriculture**
Volume 4 **Landuse and Agricultural Planning**
Volume 5 **Spatial Dimensions of Agriculture**
Volume 6 **Human Perception and Technological Change in Agriculture**
Volume 7 **Dynamics of Agricultural Development**
Volume 8 **Anthropogenic Dimensions in Agriculture**

Concept's International Series in Geography No. 4

NEW DIMENSIONS IN AGRICULTURAL GEOGRAPHY

VOLUME FIVE

SPATIAL DIMENSIONS OF AGRICULTURE

NOOR MOHAMMAD
Editor

CONCEPT PUBLISHING COMPANY, NEW DELHI 110059

ISBN 81-7022-400-4 (Set)
ISBN 81-7022-405-5 (Vol. 5)

First Published 1992

Published by
Ashok Kumar Mittal
Concept Publishing Company
A/15-16, Commercial Block, Mohan Garden
NEW DELHI 110059 (India)

Lasertypeset by
Microtech Advance Printing Systems (Pvt.) Ltd.
H-13, Bali Nagar,
NEW DELHI 110015

Printed at
Sunil Printers
CB 386, Ring Road, Naraina
NEW DELHI 110028

PREFACE

AGRICULTURE CONTINUES to be the key industry. It is not only the dominant form of employment but also forms the basis for a way of life for, perhaps, two-thirds of the world's households. Agriculture being the function of physical, socio-institutional and techno-economic factors which are dynamic in nature, keeps on changing with the basic objective of increasing production and generation of foodgrains surplus. The traditional system of farming is in the process of transformation to modern agriculture which itself is undergoing a set of equally and generally even in more rapid changes. So it has added a new dimension in agriculture which needs to be studied and analysed.

A project was, therefore, taken up to accomplish this task. Original research papers dealing with various facets of agriculture were invited from scholars of different discipline from all over the world. The response was very encouraging and a large number of original papers were received. Besides these, some papers which have classical and everlasting value have also been picked up from my earlier publication – *Perspectives in Agricultural Geography*. The papers have been classified according to their basic themes and have been put into following eight volumes so as to give a systematic picture of various themes. The articles dealing with concepts, theories and models have been kept first followed by those dealing with the same theme at macro, meso and micro levels respectively.

The First Volume – *Historical Dimensions of Agriculture* – containing 19 articles deals with the origin, diffusion and development of agriculture, research trends in agricultural geography, world agricultural systems and methods and techniques in agricultural geography.

The Second Volume – *The Ecology of Agricultural System* – deals with the selected parameters of physical environment specially the climatic elements which affect the agricultural landscape in spite of technological advancement in agriculture.

The Third Volume – *Socio-economic Dimensions of Agriculture* –

containing 19 articles deals with socio-economic environment which have bearings on agricultural systems. The first 8 articles deal with social parameters specially labour, caste, women etc., while the last 11 articles examine the impact of techno-economic factors specially the various facets of irrigation which is a basic agricultural input.

The Fourth Volume – *Landuse and Agricultural Planning* – contains 19 articles dealing with concepts, theories, models of landuse, land capability and suitability, wastelands and dynamics of landuse.

The Fifth Volume – *Spatial Dimensions of Agriculture* – includes articles on agricultural systems of specific areas such as Himalayas, Bastar, Uttara Khand etc. It also deals with dynamics of crop landuse, cropping pattern and its diversification, crop ecology and crop-combination analysis.

The Sixth Volume – *Human Perception and Technological Change in Agriculture* – contains only 15 articles and covers four basic themes. The first set of 3 articles deals with human perception and his consequential response, while the second set includes articles on correlates of technological change and adoption of agricultural innovations. The third set of articles deals with the assessment of socio-economic transformation in rural India. The last set deals with agriculture and human settlements.

The Seventh Volume – *Dynamics of Agricultural Development* – includes articles on concept of development and its various facets. Besides assessing the trends of levels of agricultural development, it also examines and evaluates its impact on environment and ecological parameters.

The Eighth Volume – *Anthropogenic Dimensions of Agriculture* – containing 17 articles deals with regional imbalances in agricultural productivity, methods of its measurement, and also its correlates and determinant. It also focuses on food availability and food security at regional and continental level.

The collection of papers in eight volumes of *New Dimensions in Agricultural Geography* is designed to encapsule the major themes and recent trends and concepts in the field of agricultural geography with particular reference to conceptual revolution in the discipline. All the volumes contain a brief introduction which gives a glimpse into the topics dealt with in each of them.

The variety of articles contributed by the renowned scholars of different disciplines has made the character of these volumes multi-disciplinary and has enhanced their value. These volumes have been designed to help not only the students of a particular discipline but also to general students who wish to keep abreast of the latest developments in the field of agricultural geography. From these points of view these volumes hold a promise of rich fare and in addition to being heuristic, will provide a mine of information for discerning scholars. It is hoped that this unique collection of research papers will not only be valuable to the researchers in the fields of agricultural geography, economics, sociology, history, and agricultural science, but will also provide guidelines to decision-makers and planners.

The completion of this project has been possible due to the constant inspiration and unstinting co-operation of a host of people too numerous to acknowledge individually. However, I owe a debt of gratitude to my mentor — Dr. Mohammad Shafi, Professor Emeritus, Department of Geography, A.M.U., Aligarh who gave me valuable comments and suggestions ever since the inception of this project. Prof. R. Ramachandran and my other colleagues who offered sincere comments on the title of volumes and the collection of articles deserve special thanks. I would also like to thank my father, wife and children who provided a favourable working atmosphere to me for successful completion of this project. Last but not the least, I would like to put a special word of appreciation, love and affection to Mohammad Imran Ali Khan who is a source of hope and inspiration.

Department of Geography
Delhi School of Economics
University of Delhi
Delhi-110007

NOOR MOHAMMAD

January 26, 1992

PUBLISHER'S NOTE

When Concept's International Series in Geography No. 1 — *Perspectives in Agricultural Geography* edited by Dr. Noor Mohammad in five volumes was issued over a decade back, we expressed our apprehensions that 'Posterity will be the best judge of our efforts in this direction. The response to this set of volumes will provide the impetus for hastening the publication of further series that are already on the anvil'.

Perspectives in Agricultural Geography received such an overwhelming response that it became out of print soon after its publication. Its success provided the necessary encouragement to issue thereafter Concept's International Series in Geography No. 2 — *Perspectives in Urban Geography* edited by Dr. C.S. Yadav in 16 volumes and No. 3 — *Perspectives in Geomorphology* by Dr. H.S. Sharma in four volumes.

We are glad to offer now Concept's International Series in Geography No. 4 — *New Dimensions in Agricultural Geography* edited by Dr. Noor Mohammad in eight volumes. In view of the new emerging trends, concept, theories and models in the field of agricultural geography, the set presents a wide panoramic view of the latest advances in this field. Hope this set too would be equally welcomed by the students, researchers, scholars and planners in various disciplines.

February 11, 1992 NAURANG RAI

CONTENT

LIST OF FIGURES

LIST OF TABLES

LIST OF CONTRIBUTORS

Bishnoi, O. P.
Department of Agronomy, Haryana Agricultural University, Hisar

Bryant, Christopher R.
Department de Geographic, Universtic' de Montrial, Montrial, Que'bec, Canada

Burman, Savitri G.
Professor of Geography, (Retd), Department of Geography, Delhi School of Economics, University of Delhi, Delhi

Coutinho, O.
Department of Geography, Rani Parvati Devi College, Belgaum

Das, K. K. L.
Department of Geography, University of Bhagalpur, Bhagalpur

Das, K. N.
Department of Geography, University of Bhagalpur, Bhagalpur

Deslauriers, Pierre
Department de Geographic, Universtic' de Montrial, Montrial, Que'bec, Canada

Gupta, H. S.
Department of Geography, Ravishankar University, Raipur

Hill, R. D.
Department of Geography, University of Hong Kong, Hong Kong

Jain, C. K.
Department of Geography, Dr. Hari Singh Gour University, Sagar

Jalal, D. S.
Department of Geography, Kumaon University, Nainital

Joshi, R. C.
Department of Geography, Kumaon University, Nainital

Khan, A. Q.
Department of Plant Breeding, G.B. Pant University of Agriculture and Technology, Pantnagar

Majeed, Abdul
Department of Geography, S.S. School, Jamia Millia Islamia, Jamia Nagar, New Delhi

Marois, Claude
Department de Geographic, Universtic' de Montrial, Montrial, Que'bec, Canada

Mohammad, Noor
Department of Geography, The Delhi School of Economics, University of Delhi, Delhi

Pal, Indra
Department of Geography, University of Rajasthan, Jaipur

Pant, B. R.
Department of Geography, Kumaon University, Nainital

Ramanaiah, Y. V.
Department of Geography S.K. University, Anantapur

Reddy, M. V.
Department of Geography, S.V. University, Tirupati

Reddy, N. B. K.
Professor of Geography (Retd), Department of Geography, S.V. University, Tirupati

Sharma, Pankaj
Department of Genetics, University of Delhi, South Delhi Campus, Benito Juarez Road, New Delhi

Sharma, T. C.
Department of Geography, Rani Parvati Devi College, Belgaum

Shukla, Lakshmi
Department of Geography, University of Rajasthan, Jaipur

Singh, J. P.
Agricultural Economics Research Centre, University of Delhi, Delhi

Singh, Surender
Department of Geography, Shivaji College, University of Delhi, Delhi

Tuteja, Usha
Agricultural Economics Research Centre, University of Delhi, Delhi

Verma, T. L.
Department of Geography, Government College, Kankar, Bastar

INTRODUCTION

THE CONCEPT of region which has been derived from the Latin word *rigio* referring to a straight line delimiting the regions of heaven and drawn by the augers in their search of omens, is still highly controversial in geography, although it has been a central theme of geographical studies for at least hundred years (Symons, 1972), and has indeed almost theological status in the discipline (Haggett, 1977). Various concepts of regions such as uniform or homogeneous areas or as a unitary entity or coherent whole, have been put forth, discussed and criticised (Kimble, 1951). Here it would be out of place to recount the history of regional thinking in geography (Hartshorne, 1939) but the common denominator of all regional studies is that the region is part of earth's surface having certain characteristics which enable it to be recognised as a unit distinguishable from other units which surround it, and which may themselves be identified by similar or different characteristic features (Symons, 1972). Hence, regions may be defined in several terms e.g. physical regions — relief, climatic, vegetation, soil, etc.; economic regions — agricultural, industrial etc.; and socio-cultural regions — religious, linguistic, racial, etc.; or in terms of Whittlesey (1954) single feature, multifeature and total regions. Thus a single point in space may be included in several regions differently defined, the boundary of which will rarely coincide unless marked physical features such as sea coast or steep mountain wall sets a limit. A region, however, cannot be defined with sharply demarcated boundary line as the political boundary line of a nation. A region generally has a transitional zone in which a change in regional pattern occurs. Not only the limits of a region differ according to the criteria by which it is defined but at least in some cases, these limits will change with time. Thus the definition, delimitation, and mapping of region is very difficult and agricultural regions are no exception to it.

Agricultural regionalisation is a comparatively recent phenomenon in its strict sense, although different forms of agriculture belonging to particular phase of agricultural development has been taken as agricultural regions e.g. shifting cultivation, intensive cultivation, commercial agriculture (Thomas, Conkling and Yeats, 1968; Jones and Darkenwald, 1965; and Alexander, 1977). The agricultural regions of U.S.A. according to the scheme of Baker (1926-32) is, perhaps, the classical example of agricultural regionalisation which has been done purely on the basis of qualitative agricultural attributes. Whittlesey (1936) has made an improvement upon it and has used few sets of criteria related to crop and livestock to divide the world into agricultural regions. Although Whittlesey's scheme is logically imperfect, after certain modifications and improvements, major agricultural regions of world are delineated and discussed (Thoman, 1962; Kawachi, 1959; Von Royen and Bengston, 1954; Timmons, 1944). But Kostrowicki (1968), contrary to Bunge (1966) who believe that classification (type) and regionalisation are synonymous and all sciences have in their history a classificatory or taxonomic phase and geography is no exception, pointed out that types and regions are two different sort are two different sort of notions. Type being based on similarities between the object with respect to some attributes, may recur over space given similar environment, whereas region is a unique area characterised by a particular set of characteristics different from all others and unlike type, is spatially continuous.

However, based on different criteria or set of criteria, agricultural regions are formed which differ from one another (Wood, 1931; Stamp and Beaver, 1971; and Vince, 1951) but four major views emerge: firstly, agricultural regions are not different from natural regions or physical divisions of a land (Ahmad, 1979; and Sengupta and Sdysuk, 1968); secondly, agricultural regions as an area where a particular assemblage of crops/livestock is found (Weaver, 1954); thirdly, agricultural region as a unique set of interaction of different attributes of variables; and finally, agricultural regions formed on the basis of results of the rigorous quantitative analysis (Sharma, 1968; and Weaver, 1954).

Hence, the existence of so many ideas regarding agricultural regions is undoubtedly a stumbling block for any attempt to define agricultural region precisely, and it leads to the question whether an agricultural region in the proper sense of the term exists or not. None of the ideas regarding agricultural region is adequate. However, agricultural regionalisation requires the consideration of all the agricultural variables, which do not usually co-vary exactly (Knowles and Waring, 1976). It is, therefore, very difficult to conceive an agricultural region as a cluster of points which stand for the unit of agricultural landscape in a 'n' dimensional space where the co-ordinate denote the values of 'n' numbers of variables, like the idea of Berry and Ray (1966) in the context of socio-economic regionalisation. It is true that the unit of agricultural landscape can be clustered on the basis of their similarity co-efficient, but the clusters are less likely to be defined in terms of all the relevant variables. In such cases clusters should be named only according to the involved attributes and they should be called as special purpose agricultural regions.

However, in this age of specialisation regions and especially agricultural regions are worked out by keeping certain specific theme in view. It could be the measurement of level of agricultural development or diversification in agriculture or agricultural resource/planning regions.

The Himalaya is an young and vulnerable mountain region with an extremely fragile eco-system. Man has lived here for centuries and has utilised the least vulnerable lands – broad mature valley, flat alluvial terraces, and the lower slopes so as to maintain ecological balance. But during this century, specially in the past three decades, commercial economy has made inroad in the region which necessitates the removal of the natural cover consciously or unconsciously resulting into acceleration of erosional processes leading to degradation of soil and land here as well as creation of problems in the plains below. **Burman** has given a systematic account of agriculture and its impact on environment in the Kumaon Himalaya. The authoress has described and explained the traditional system of agriculture in which type of agricultural utilisation varies along a slope, farming is characterised according to its location signifying certain environmental characteristics. Typical farming systems developed according to locations are recognised as *talaon* – lower land, *upraon* – upper lands, and *katil* – highest land, running upward from the basal cavity of the river valley of the upper part of the slope. These systems derive their support from buffaloes in lower altitudes and cattle in higher — largely in terms of manure and ploughing.

However, in the present century commercial economy in the form of commercial forestry, dairy farming and commercial crops has been introduced in the system which has already reached the maximum level of its land capability and it has resulted into excessive pressure on land and other resources leading to serious environmental problems. The authoress has assessed and explained the ecological problems and consequences of agricultural development, specially land degradation, deforestation, accelerated soil erosion, silting of water bodies, drying up of springs, replacement and disappearance of species and increased ratio of energy expanded in direct agricultural activity and out-migration of labour force together with their health problem. She has finally suggested suitable strategies for sustainable agricultural development.

In spite of target area and target group approach of planning coupled with bottom top approach at micro-level, the regional disparities in the levels of agricultural development could not be removed. The tribal areas remained backward and the agriculture being still in backward stage, is a low productive enterprise. Since the agriculture fails to meet the food requirement of tribals, they use to supplement it with forest products which are dwindling and the prohibitive laws of forest also deprive them from this support. Therefore, there is a need to develop tribal agricultural which is only possible after its thorough study and understanding. **Gupta** and **Verma** have studied it and have exposed the weaknesses of the tribal development schemes which, in spite of devoting millions of rupees, have not been able to bring desired improvement in the tribal situation in north-east Bastar. The study is based on 334 respondents (322 land-owners and 12 landless) chosen from 8 randomly selected villages located in Kanker and Kondagaon tahsils of Bastar district. The authors have critically examined the quality of land, nature and characteristics of soil, size of holdings, cropping pattern, agricultural inputs, labour, bullock power, seed, manure, fertiliser and irrigation, and crop production. They have also assessed the agricultural input-output relationship in value-added terms, and have worked out the carrying capacity and food sufficiency by converting the available food in calorific value. The study concludes that the physical(undulating topography, infertile soil, low rainfall) and human (illiteracy, government negligence, low level of innovativeness etc.) factors restrict the scope of agricultural development. It seems unlikely that the pace of development will be so fast to make up the present shortage and then cope with growing population in meeting their food requirements. Therefore, an integrated developed programme is needed to improve the tribal situation.

Uttarakhand is another underdeveloped and backward region where besides the threatening problem of soil erosion, indiscriminate cutting of forests, overgrazing of pasture lands and mismanagement of land

and water resources, undernutrition, malnutrition and associated problems are posing severe threat to the local population. **Pant** et al., have described and explained the agricultural scenario, nutritional status and related diseases at meso and micro levels in this region. The study is based on both secondary and primary data. The former is used to analyse landuse, cropping pattern, man-land ratio, occupational diversities, holding size etc., while the latter is intuned towards the assessment of nutritional status and prevalent deficiency diseases among the rural people. On the basis of stratified purposive sampling 3 villages located in different altitudinal zones have been selected and Food Balance Sheets of each village have been prepared indicating intake of various foodstuff and their nutritive value per head per day. Deficiencies were marked and correlated with prevalent nutritional diseases. The study reveals that the net sown area is very low ranging from about one per cent in Uttarkashi to 39 per cent in Naini Tal with a small percentage of irrigated land (30%), and very small size of holding. Rice and wheat are the predominant crops occupying more than two-thirds of the cropped area. Food Balance Sheet indicates that 'dunfloor' villagers get more calories as compared to the inhabitants of the upland. The diet of the people is deficient in 8 out of 11 elements which leads to the occurrence of as many as 20 diseases such as night-blindness, kuwashiorkor, rickets etc. The upland village is having more diseases than the 'dunfloor' village, and similar is the position with respect to mortality and morbidity. The authors have suggested several suitable measures for the development of agriculture which will in turn increase the nutritional status of the inhabitants.

In the post-war period the restructuring of farm-business has been done in many western countries, most which have adopted farm diversification. The phenomenon is most prominent in the rural-urban fringe and comprises of three broad types: degeneration of farm structure, agricultural adaptation and innovation, and continued agricultural development along traditional lines. The paper of **Bryant** et al., after giving a critical review of various studies done in this field with special reference to conceptual framework and its evolution, discusses and examines the concept of farm diversification, method of selection of study area, general characteristics of farm structure and change and diversification strategies, and offers concluding comments on future research direction. The authors elaborate the mode and type of diversification at individual farm and regional levels. For indepth study of this issue they have selected Montreal region located on the south bank of St. Lawrence in Canada. Primary data for the year 1988-89 from 139 farmer respondents chosen from three different zones viz. most urbanised having definite urban fringe; semi-rural with low population densities, and rural, were collected. The analysis of farm size and the changes in it, the extent of farmland renting, and farm types in term of enterprises structure shows considerable interzonal variations — Zone I recording the higher degree of change with diversified structure. In case of diversification strategies — crop and livestock enterprises and marketing enterprises, the first zone comes on the forefront. This study, however, provides an enhanced opportunity to advance our understanding of how the mosaic of farming structure in the rural-urban fringe evolves, particularly by permitting us to trace the changes in the interaction pattern that link farmers, their families and farms into different system of exchange at different scales of geographic analysis bringing them into contact with different set of forces.

Cropping pattern is a reflection of the interplay of the complex physical, social and economic factors which are dynamic in nature. Consequently cropping pattern also gets changed and modified with the passage of time. It is more so around a metropolitan city like Delhi which is expanding very fast encroaching on immediate surrounding agricultural land. **Singh** in his paper has studied two problems simultaneously: the dynamics of cropping pattern in the union territory of Delhi and the relevance of Von-Thunen theory in a region of subsistence economy with entirely different physical, social and economic background. The author has selected six villages on the basis of their connectivity and distance from the core of the city i.e. villages near the city, away from the city, near the road and away from the road. Year-wise data of agricultural landuse of kharif and rabi seasons are taken, processed, mapped and analysed for a time span of thirteen years (1965-78). In order to get the clear picture of the dynamics of land use, the crops have been put into five groups viz. cereals, millets and pulses, fruits and vegetables, cash crops, and fodder crops. Finally the changes in all these groups have been analysed and discussed in detail. The study reveals that the total cropped area and area under cereals and vegetables has increased, while it has decreased in case of millets and cash crops. The intensity of cropping has also increased. The author has also examined the spatial organisation of landuse, cropping intensity zones and the intensity zones in terms of value and nature of crops in order to test the relevance of Von-Thunen theory.

The study concludes that in spite of the subsistence economy, different socio-economic conditions, rapid urbanisation and technological change, the age old ideas of Von-Thunen and his followers have stood the test of the time and still hold good.

Change in cropping pattern overtime is a continuous and dynamic process which with reference to its pace and patterns depends on a set of dynamic physical, techno-economic and socio-institutional factors, which play differential role over space and time. **Singh** examines the extent of change in cropping pattern together with its determinants and rational in different agro-climatic zones of Haryana during the period 1969-77 at five years interval with three years moving averages. The state is divided into 4 agro-climatic zones and changes in net sown area, irrigation and its intensity, crop landuse, and cropping intensity is discussed and explained for each region with an aggregate analysis for the state as a whole, which indicates considerable inter-regional variations in it. The author, being an economist, has explained these changes and differential levels in them in terms of prices, productivity, profitability, technology and irrigation. The study has come up with several revealing results. The production performance of crops is quite good showing annual compound growth rate of 3.2, 0.65 and 2.5 per cent respectively but more remunerative and fine crops—wheat, rice, cotton etc. have attained a higher gain, while less remuncrative crops—bajra, gram etc. has lost heavily. Although the net sown area has reached to its maximum levels in almost all regions, it has shown a little gain. But intensity of irrigation has considerably increased which has brought substantial change in cropping pattern and intensity of cropping. However, the study concludes that value productivity and profitability of crops are determining factors of change.

In India the major portion of natural resources consists of land and, by far, the largest proportion of her inhabitants are tiller of land, but unfortunately more than 60 per cent of her population is underfed. Perhaps, the reason is that the considerable carrying capacity of land has never been coaxed. The need now is to raise uniformly the food production efficiency of the land which, if used properly, can feed as many as five times India's population (Shafi, 1969). Hence, in any scheme of planned development of the country, agricultural landuse planning is of basic importance which requires a scientific investigation of the current use of land. In any plan which aims at utilising the available resources in a scientific, rational and economic manner, landuse surveys are of fundamental importance, as they provide detailed information with regard to existing use and misuse of land and possibilities of its further development. But such type of surveys cannot be carried out by an individual for a vast country with a diversified pattern of landuse. Moreover, each region has its own peculiarities, and hence, a plan formulated for one region may not suit to the other region. Therefore, **Mohammad** has selected a small region (Ghaghara-Rapti Doab) in order to study the patterns of landuse. The author after giving a balanced account of all the five major categories of general landuse, has given for the improvement, wherever it was necessary. The study reveals that the region is endowed with potential agricultural land but it is not properly coaxed. Hence, the region is economically backward.

As a result of technological change and introduction of agricultural innovations, the agricultural landuse pattern has changed at all levels. **Pal** and **Shukla** have studied and analysed changing pattern of agricultural landuse at micro level i.e. in Chittaurgarh district lying in the hilly tract of Rajasthan. The regional imbalance in agriculture, future prospects and potentialities, and problems that confront the development are the vital issues discussed in the paper. The authors have studied at Tahsil level data, the changes that have taken place during 1954-74 in forest, barren land, land not available for cultivation, net area sown, current fallow, old fallow, arable and non-arable land. Besides this, they have also worked out land use combination regions after Kostrowicki's percentage method of land use combination for 1954 and 1974, which shows hardly any change in the combination which is surprising to me. Finally cropping intensity is worked out at tahsil level and the district is divided into high (> 130 per cent), medium (130-110 per cent) and low (< 110 per cent) intensity zones. The authors have also worked out the standard nutrition unit (S.N.U.) by applying Shafi and Jasbir's technique in order to assess the productivity and carrying capacity of land in term of population. The area is divided into high, moderate, low and very low efficiency zones. The study reveals that the change is very slow and insignificant in almost all aspects of land use. Although cereals hold a place of pride, the area under gram, cotton and spices has increased during latter period of the study.

After Independence concerted efforts have been made to increase agricultural production by inducting modern technology into it. It has brought a major change in landuse, irrigation, cropping pattern, crop-combination, productivity and overall agricultural system which may collectively be termed as transformation of agriculture. But the pace and pattern of transformation is not uniform both in space and time. **Jain** in his paper has described and explained the transformation of agriculture in the 15 tribal districts of Madhya Pradesh during 1955-56 and 1984-85. Because of the restrictive physical base, there has been very little increase in net sown area and cropped area. However, districts like Bastar, Surguja, Sidhi, Jhabua have recorded higher increase as compared to state average. Intensity of cropping has also not shown remarkable increase over the base year. However, there has been a change in farming system which has turned to be irrigation-intensive farming. The author has given a descriptive account of use of various modern farm technologies — farm implements, chemical fertilisers, improved irrigation system, use of power and high yielding varieties of seed which have been instrumental in transformation of agriculture in this region. Finally the study concludes that owing to low level of adoption of modern agricultural technology due to variety of reasons, the pace and trend of agricultural transformation is very slow in this region.

In order to minimise the regional disparities in the levels of agricultural development, target groups and target area approach of planning was adopted in which problem/weakers areas such as hilly, tribal, canal command, drought prone areas were selected and massive modern technology was inducted into it for development. As a result agricultural transformation has taken place. But all such areas are not covered under this programme. However, **Reddy** and **Ramanaiah** have described and explained the changing pattern of agricultural landscape in Anantpur district — a drought prone area, specially with reference to dynamics and trends of crop land use, irrigated land and land under fallow, mutual shift of area between net sown area and fallow land, spatial distribution of HYV cultivation, agricultural landuse efficiency, and spatio-temporal variations into it. The study is based on secondary data collected for three points of line — 1962-65, 1972-75 and 1982-85 at Taluka level. Statistical techniques such linear regression equation, correlation co-efficient and modified standard co-efficient method have been used. The analysis of data shows a decrease in net area sown where as fallow land indicates an average gain of almost same percentage. It is mainly because of the occurrence of drought in several consecutive years. Intensity of cropping also shows decreasing trend but there is an average increase of 9 per cent in irrigated area. There has been considerable variation in all these parametres at taluka level. Consequently cropping pattern has undergone change. Among the four crops studied (groundnut, millets, pulses and paddy), all the crops except groundnut have shown a declining trend which indicates increasing detrimental conditions. HYV have mark inroad in the area showing limited increase. The study, however, concludes that the erratic behaviour of monsoon and severe drought conditions have a debilitaling effect on agricultural economy. The moisture deficiency caused by low rainfall with high variability, meagre irrigation facilities, poor red soil undulating terrain with rocky exposures, low technology, as well as poor socio-economic condition of the people are the detrimental factors which came in the way of development of agriculture. However, a welcome feature in the cropping system of Anantpur is the shift from inferior millets crop to superior variety of crops such as groundnut, sericulture and fruit farming.

Among the various strategies inducted for maximising agricultural production, multiple cropping i.e. raising more than one crop in a sequence in an agricultural year on a unit of crop land, holds a vast potential in most rural areas of India in view of prevailing abundance of farm labour, favourable climate and only a small proportion of net sown area being sown more than once. **Coutinho** and **Sharma** have analysed the growth patterns in multiple cropping and the role of some important factors responsible for regional disparity in Uttar Pradesh for the period from 1966-67 to 1980-81 with the presumption that rise in yield of most crops has served as an added incentive and there has been considerable rise in it since the launching of HYV programme in the earlier year and that the trend has stabilised by the latter. The study covers all the five agro-economic regions of Uttar Pradesh. The authors have used a simple correlation matrix, step-wise regression model with 8 explanatory (independent) and cropping intensity as dependent variable to assess their role in the spatial disparity in the index value of the latter. The independent variables are: sugarcane cropped area; gross irrigated area; irrigation intensity; tube well irrigation; fertilizers consumption; density of rural population; size of land holdings; and rainfall. The first six variables have shown high degree of positive correlation in descending hierarchical order, while the last two have revealed negative correlation. The study reveals that since there is a lot of spatio-temporal variations in all the independent variables, cropping intensity also varies to a great extent. The irrigation has emerged as the most important determinant of cropping intensity which is lowest in Bundelkhand region and highest in western region. The study suggests to focus immediate attention on neglected regions and to take measures for reclamation of ravines, badlands, waterlogged and saline-alkaline affected soils and to prevent excess of irrigation which is the root cause of degradation of last two types of land.

The changing pattern of landuse as a result of increasing pressure of population could be realised more clearly and vividly at a micro level in a predominantly agricultural tract. **Das** and **Das** have studied this problem in a backward region which is beset with a number of natural hazards. The paper makes an attempt to portray conditions of agricultural landuse that have fostered settlements and population expansion in a backward region such as the study area, which is beset with a number of natural factors. These, in harmony with historical, social and economic interactions have given rise to high density of population. Natural hazards like floods and occasional droughts act as push factors but natural fertility of soils coupled with monsoonal rhythm favoured with plentiful rainfall, have made the region a veritable landscape for agricultural use as a predominant feature, which together with high natural increase and controlling effects of preventive checks on death, have given rise to staggering explosion of population. Despite efforts for economic development and thrust for improvement of the agricultural situation, the agriculture still remains backward and majority of the people are poor. The region very much needs face lift and landuse planning to meet the challenge of population pressure. In the absence of other avenues to ward off pressure of population on land, secondary and tertiary sectors are not as dominating as the primary. It becomes abundantly clear that the pressure goes on increasing and there is inverse relationship between population increase and depletion in per capita available land. The farmland remains heavily taxed almost throughout the year as a result of desperate efforts on the parts of the farmers to eke out living for a great majority among the rural population. The pressure on the land has been examined with the help of various parametres and the results of the study have been tested at the yardstick of sound statistical principles. The problems have impelled the farmers and the planners alike to devise ways and means to promote a number of remedial measures to make sure fairly good living conditions for the people inhabiting the region. In order to corroborate facts, possibilities about the potential carrying capacity of land have been explored and planning proposals have been suggested to counter the ill-effects of the pressure of population on land.

Among the various modern technological inputs in agriculture, improved varieties of seed, in terms of yield, quality and adjustment to various detrimental environmental conditions, have contributed quite a lot. These varieties are evolved and developed at various research organisations from where they are diffused in the potential areas after being tested on experimental farms. **Sharma** and **Khan** in their paper have analysed the status and prospects of somaclonal variation in sugarcane and some other selected crops. The authors have described tissue culture in sugarcane and have examined the somaclonal variation highlighting the origin of somaclonal variation which may be accounted to pre-existing and cultural induced variation, chromosomal abnormalities, nuclear gene changes, cytoplasmic gene changes, gene amplification and diminution and activation of transposable elements. The study examines somaclonal variation in sugarcane in detail and discusses the variation in chromosome nuclear, disease, resistance and increase in sucrose content.

Most of the strategies of agricultural development adopted after Independence in India were geared towards increasing the production of foodgrains in which the country was largely deficient. As a result the acreage and production of pulses — the major source of protein, have shown a marginal increase of 0.5 and 0.16 per cent between trienium ending 1971-72 and 1988-89 respectively. The situation is even more grave in northern states. The government has, therefore, given special attention by implementing Pulse Production Programme which has not yielded expected results. **Tuteja** in her paper has critically examined — area under pulse crop, production and productivity, profitability of pulse, crop *vis-a-vis* competing crops, and constraints to pulse growth in the state of Haryana. The authoress has analysed the areal strength of pulses and other important crops during 1964-65 to 1983-84 and has worked out the compound growth rate which shows that area under pulses has continuously declined whereas most of the other crops have gained in it. The study examines this issue at district — Jind and Mahendergarh, and farm level which also shows the similar results. The relative importance of factor influencing acreage of pulses is statistically ascertained. Production and productivity of pulses in Haryana at the same levels for the same time periods has been statistically analysed which shows almost similar results, It is mainly due to low level of agricultural input, uncertainty of rainfall and lack of inadequate irrigation facilities, lack of plant protection measures, and weed control practices. The analysis of profitability of pulses *vis-a-vis* competing crops shows that it is not only low in case of pulses but is negative in some cases. However, the major constraints to pulses growth in Haryana are poor physical environment, low level of adoption of improved technology and marketing problems. The study concludes by giving suitable suggestion and highlighting policy implications.

The common notion is that the agricultural production is not keeping pace with the ever increasing pressure of population on land resources in Southeast Asia as a whole. **Hill** in his paper has endeavoured to show that in most of the countries of Southeast Asia agricultural production has more than kept pace with population growth through horizontal and vertical expansion of agriculture. A dramatic rise in cereals production in tropical and sub-tropical Asia since the mid-1960s has been heralded widely as a green revolution. The crore of the revolution was the development and the propagation of high yielding varieties of food cereals—rice and wheat, adaptable to local ecologies. The new varieties are designed to be high yielding in response to high level of input (Hayame, 1973). The paper sets out an analysis of national crop production and area data of six crops — rice, maize, soyabean, groundnut, coconut and oil palm for the last three decades and has compared with the growth of population in the corresponding period. The author has examined in detail the spatial distribution, production and spatio-temporal changes in all these crops together with growth and distribution of population along with their dietary habits. The study reveals that the increase in both production and area over the last three decades have been sustained and large. The analysis tend to support Boserup's (1965) notion that present economies contain feedback mechanism which lead to intensification of agriculture in response to population growth. Moreover, the addition to the economy of maize, annual oil crops and specially export crops such as oil palm and rubber, together with vigorous industrialisation employing labour-intensive technology and involving further processing of agricultural raw materials than now the case, do offer some prospects of escape from the 'the low level productivity trap', though, in the capitalist oriented countries, at the cost of vulnerability to the vagaries of the international market place.

During the past two decades there has been shift in the acreage strength of crops in the southern part of the Andhra Pradesh which is specially in favour of groundnut cultivation. **Reddy** in his paper has presented the area and production of groundnut and industries-based on it in Chittoor district which is characterised by gentle slope, red soil, semi-arid climate, low industrial base and predominantly agricultural economy. The study analyses the locational aspects of cultivation of groundnut and groundnut-based industries and suggests suitable measures to strengthen the economy of the district. Administration unit *firka* is taken as unit of study and analysis is done for the year 1982-83. Locational pattern of groundnut and groundnut-based industries in studied applying Florence Location Quotients. Groundnut ranks first in 44 *firkas* and has high concentration in 34 *firkas* which speaks about importance of its cultivation. As a result there are as many as 171 groundnut based industries here spreading over 37 *firkas*. Out of which 68 are groundnut decorticating mills and 105 are oil mills. *Firka*-wise location quotients for groundnut cultivation and industries based on it shows that the former is widely spread, whereas the latter is concentrated in only 37 *firkas*. The study reveals that the district is surplus producer of groundnut and the groundnut industrial by-products, being put to uneconomic use, are sent out of the region for processing. Hence, there is potentially for further development of groundnut-based industries and the author has suggested a large number of industries for the district.

Crop ecology is the study of relationship between crop plant and their environment. The environmental factors especially climatic, edaphic, biotic and anthrogonic are highly influential in determining the natural distribution of plants. Hence, the appraisal of crop ecological conditions over a worldwide picture is essential for the transfer of plant and animal material and technological experience from one area to another. According to Good (1931) who has formulated certain principles of plant distribution, climatic conditions primarily control the distribution of plant; it is secondarily controlled by edaphic factors; the great movements of floras have taken place in the past and are still continuing; the movement of species is brought about by the transport of individual plants during their mobile dispersal phases; there have been considerable change in climatic conditions especially at higher latitudes; and the relative distribution of land and sea has considerably changed. However, **Bishnoi** has examined in detail the influence of ecological factors — temperature, moisture, light, edaphic and biotic on the distribution of plants. In order to make it more comprehensive he has described the distribution of important crops in tropics, sub-tropics and in the intermediate climatic regions on the ecological basis. The major crops like rice, wheat, cotton, maize, millet, barley, groundnut, potato, coffee, bananas, cacao and vegetable and fruits have been discussed in detail in relation to their distribution with the climatic, edaphic and biotic factors.

The fertility of soil deteriorates with the continuous cropping, and that too under the same crop. Different types of crops absorb different quantities of plant nutrients from the soil. Hence, in order to maintain the soil fertility and productivity, it is necessary to have good soil management practices including the supply of organic matter, the adoption of proper rotation of crops and the adoption of other recommended package of practices. Out of several factors that play a vital role in maintaining production and the productivity of the soil, suitable crop rotation has a definite place. **Bishnoi** has dealt with this theme in his paper. Crop rotation in fact is growing of different crops in the same field in the proper sequence for getting the maximum crop production (Singh, 1975). He has thoroughly discussed the feasibility of various crop rotations in different soil-climatic zones of India. The various crop rotations have been presented in 38 important soil-climatic zones of India. Besides this, the crop rotation in some specialised farming areas like fodder, vegetables, sugarcane, rainfed conditions have been presented in order to get maximum benefit by managing the available natural and input resources to best possible way without deteriorating the existing situations in situ.

During the third quarter of this century, the importance of combinational analysis of different elements in geographical studies has greatly increased, as such studies, by an large, minimise the chance of oversimplified generalisations. Whether it is a study of towns, industries, or the study of agricultural production and landuse, it is more necessary to highlight or to focus the most important element or elements that govern the other smaller elements in the whole complex system. However, various methods have been devised to identify the set of elements which play a significant role in particular system by all of them have their own merits and limitations. **Majeed** in his paper has made a judicious attempt to review some of the important methods. The author has critically examined with suitable illustrations, wherever necessary, the merits and limitations of the Quartile method, Mean Positive Deviation method, Standard Deviation method, Minimum Deviation method, Least Squared Deviation method, Maximum Positive Deviation method, Modified Minimum Deviation method, Maximum Distance methods, Lower Limit method, and Successive

Quoptient method. The author has come to the conclusion that none of the methods is perfect and their use and utility depends on the nature and character of the elements.

REFERENCES

Ahmed, E., 1980 : *Regional Planning with Particular Reference to India*, New Delhi.

Alexander, J. W., 1977 : *Economic Geography*, New Delhi.

Baker, O. E., 1926-32 : 'Agricultural Regions of North America' *Economic Geography*.

Berry, B. J. L. and Ray, M., 1966 : Multivariate Socio-Economic Regionalisation: A Pilot Study in Central Canada (Unpublished).

Bunge, W., 1966 L 'Theoretical Geography', *Lund Studies in Geography Series C — General and Mathematical Geography*, No. 1, Lund.

Haggett, P., 1977 : *Locational Analysis in Human Geography*, p. 450.

Hartshorne, R., 1939 : *The Nature of Geography: A Critical Survey of Current Thought.*

Jones, C. F. and Darkenwald, G. G., 1965 : *Economic Geography,* New York.

Kawachi, K., 1959 : 'On a Method of Classifying World Agricultural Regions', *Proceedings of the I.G.U. Regional Conference in Japan 1957*, pp. 355-56.

Kimble, G. H. T., 1951 : 'The Inadequacy of the Regional Concept', in Stamp and Woolridge, *Essays in Geography*, London, pp. 151-74.

Knowles, R. and Waring, J., 1976 : *Economic and Social Geography*, London, pp. 34-38.

Kostrowicki, J., 1968 : 'Agricultural Typology, Agricultural Regionalisation, Agricultural Development', *Geographica Polinica* (Special issue), Warsaw, pp. 265-74.

Sengupta, P. and Sdysuk, G., 1968 : *Economic Regionalisation in India*, Census Monograph.

Shafi. M., 1969 : Can India Feed Five Times of her Population, *Science Today*, Vol. III, No. 9, New Delhi.

Sharma, P. S., 1968 : *Agricultural Regionalisation of India*, New Delhi.

Stamp, L. D., and Beaver, S. H., 1971 : *The British Isles*, 6th Ed., London, pp. 229-53.

Symon, L., 1972 : *Agricultural Geography*.

Thomas, R. S., Conkling, E. C. and Yeats, M. H., 1968 : *The Geography of Economic Activity*, pp. 126-27.

Timmons, J. G., 1944 : 'Distribution of World Land Resources' — *Land Policy Review*, (Winter), pp. 8-14.

Thoman, R. S., 1962 : *The Geography of Economic Activity,* pp. 124-37.

Vince, S. W. E., 1951 : 'The Agricultural Regions of Belgium' in Stamp and Woolridge, *Essays in Geography*, Memorial Volume, pp. 255-87.

Von Royen, W. and Bengston, N. A., 1954 : *Fundamentals of Economic Geography*.

Weaver, J. C., 1954 : 'Crop Combination Regions in the Middle East', *Geographical Review*, Vol. XLIV, pp. 175-200.

Whittlesey, D., 1936 : 'Agricultural Regions of the Earth', *Annals of the Association of American Geographers,* Vol. 26, pp. 199-240.

Whittlesey, D., 1954 : 'The Regional Concept and the Regional Method', in James and Jones, *American Geography: Inventory and Prospects.*

Wood, H. J., 1931 : *An Agricultural Atlas of Scotland,* London.

1

HIMALAYAN AGRICULTURE AND ITS IMPACT ON THE ENVIRONMENT

SAVITRI G. BURMAN

THE HIMALAYAS are a young, vulnerable mountain system where there are several geological weaknesses like unconsolidated deposits, abundance of soluble rocks like limestones, highly faulted regions and earthquake zones, all being highly erosion prone. The height of the Himalayas, the copious monsoonal rainfall and the sub-tropical temperatures all favour weathering and active erosion. Nature's regulatory mechanisms to slow this process are the vegetation cover and healthy humus rich soils. In many places especially in the Central and Eastern Himalayas, these highly weathered materials are held together along valley sides by the massive root systems of the tropical/sub-tropical forest systems only. The valley sides are likely to 'flow' out as soon as this 'binder' is removed. The present mode of human use of the Himalayas, which necessitates the removal of the natural cover consciously or unconsciously has led to the acceleration of the erosional processes, leading to the degradation of soils and land in the region, as well as creation of problems in the plains below. Both the High Himalayas and the Siwaliks are ecologically more vulnerable than the Middle Himalayan Zone, the former because of altitude, lower temperatures and drier climate and the latter because of hot, wet conditions, loose deposits and seismicity.

Humans have lived in and used this region for at least 7,000-8,000 years, but as far as known, they have lived on a sustenance basis and utilised the least vulnerable lands, i.e. the broad mature valleys, flat alluvial terraces and the lower slopes of the Middle Himalayan Zone for year round utilisation. Their economy had always been of a subsistence type and trade consisted of sharing of surpluses only, both within and outside the region. For a little over a century and particularly in the last two and a half decades commercial economy has made inroads in the region whereby the region still has to sustain itself for its basic needs (emanating from the mountainous character of the region) and produce commercial products as well — timber and forest produce, fruits, vegetables, and other agricultural products including those from livestock for the increasing urban areas and tourist industry in the region as well as the plains below. The above has led to undue pressure on the use of fragile, i.e. steeper upper slopes, and the highly vulnerable Siwaliks and Higher Himalayas, because of overall lack of direction and planning suitable to the inherent character of the area.

Agricultural Systems

Traditional

A major fact that is often overlooked is that agriculture in the mountains is very different from that in the plains, not only in terms of crops but also in its different components. Resulting from the paucity of land suitable for cropping, agriculture is not a pure pursuit but is combined with a variety of economic activities that are interlinked, catering to basic human needs and utilising resources from different types of physical environments in the locality, as well as available labour and skills. The resulting mix of activities may be followed in different parts of the year, dependent on seasons; or by different members of the family; and may also be different in the different ecological zones.

Basically, agriculture here is subsistent, and therefore, the basic human needs have to be met with from the local area and community. The core activities of this system are based on land (together with the plant cover — natural and cultivated), people and livestock. This core provides all basic necessities of food, fodder, fuel, fibre and fertiliser or what is often called the 5 Fs. The peripheral part of this system is collecting, processing and in some cases trading of products, as well as providing seasonal labour in ancillary activities and neighbouring regions — mostly in latter. The predominance of one activity over the other varies basically with the location, especially as regards altitude, aspect and slope/landform; and to a minor degree with history and ethnic communities.

The web of energy-flow between the people, land and livestock in this system is intricately connected (Fig. 1.1). There is a continuous feedback from each of the constituents of the system, i.e. land, people and livestock. Not only crops but the surrounding forests and pastures form an integral part of this agricultural system and the area involved is often termed the Support Area for farming. The cultivated lands, therefore, form only a small proportion; the ratio between them and a healthy Support Area is generally 1:4. Not only does the Support Area, i.e. public lands with their natural vegetation service the agricultural lands with inputs indirectly via the animal chain, but the forests also provide organic matter and plant nutrients via the leaf litter used as manure and the nutrient-rich water flowing downslope from the forest soils rich with litter. This Support Area also provides the community with fibres, fuel and vitamin-rich food. This subsistent system is not necessarily primitive in nature as is often felt, but is environment adaptive for difficulties of movement and transport make this system a more feasible one.

Location

Location of agricultural lands varies a great deal in different situations of landforms and population pressures. On the whole cropping seeks areas of 'deposits' whether they be alluvial in river terraces and old lake basins or fluvio-glacial terraces in the higher altitudes, for all deposited materials have a relatively gentler slope with thick soils. In cases of necessity, like in the Higher Himalayan Zone, agriculture also occupies morainic material of lateral and ground moraines left behind by glaciers. Increase in population numbers and corresponding dearth of suitable agricultural lands have pushed agriculture higher up into areas of steeper, erosion slopes and thinner soils. This handicap to some extent has been overcome by terracing of hillsides. This, however, cannot wholly provide the natural advantages of a gentler slope and thick deposit of soil, which has a far better retentivity of water and nutrients. Majority of agricultural lands are located on slopes varying from 10°–40° with 20°–30° slopes being the most favoured in the Kumaon region. The higher slopes, however, are not bereft of cultivation.

Altitude and aspect in a mountainous zone are closely linked with insolation and temperature and they also form a major influence on the location of cultivated lands. In the region, cultivated land is generally

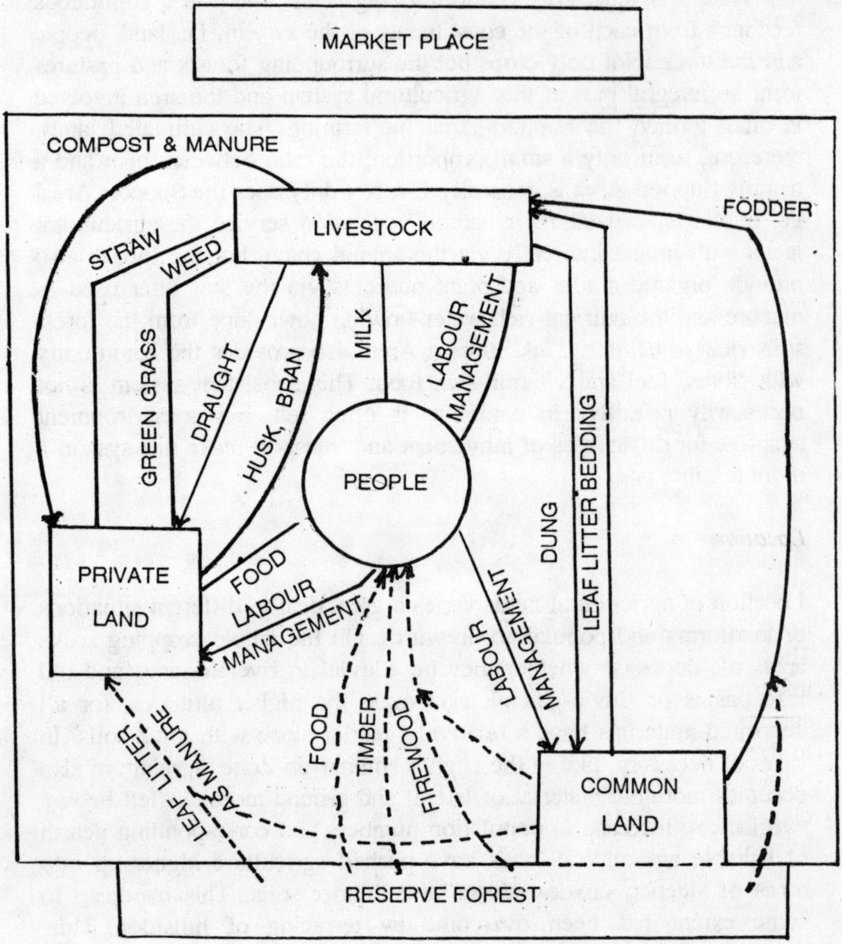

FIG. 1.1: Land-people-livestock system (Subsistence forming).

spread between 1,000–4,000 metres. The generally favoured altitudes however lie between 1,100 to 1,700 metres. Aspectwise, the most favoured in descending order are, Southeast, South, East and Northeast. Aspect influences the altitudinal limit as well, being highest on the Southern face. Relative relief most favoured is 200–500 metres, which makes it possible to grow a range of crops and some pasture as well on the same slope. Therefore, many villages are made up of a series of clusters along a slope to utilise the land for different aspects of agriculture in different seasons. Quite often a single village is made up of 2-3 hamlets located on different parts of the slope, with each area being utilised for different crops and for animal grazing according to seasons. Some members of the family move up and down the hamlets accordingly.

Ultimately it is the combination of the above discussed attributes together with soil suitability and water availability that makes for cultivated land in this region. Most of the favourable or near favourable characteristics are largely concentrated in the Middle Himalayan Zone and this is the Zone of extensive cultivation where 74 per cent of the population of the region makes its home.

In this traditional system where the type of agricultural utilisation varies along a slope, farming is categorised according to its location which in turn signifies certain environmental characteristics. Typical systems of farming have developed according to these locations and are recognised as 'talaon', 'upraon' and 'katil' running upwards from the basal cavity of the river valley to the upper parts of the slope.

The *talaon* or *lower lands* are relatively flat and are situated on the lower terraces near the rivers or their tributaries which are perennial. They are low-lying and have fine alluvial deposits of considerable thickness and a perennial water supply from streams, rivulets or springs, making irrigation possible. They are further divided according to their location relative to the slope, called 'seras' or 'panchars'. The 'seras' are low-lying and flat (abandoned river channels) with fine clayey soils and can be flooded whenever required. They are often located on the beds of rivers and old lakes. The 'panchars' lying slightly higher, are also irrigated but not so easily as the 'seras' and may have to go without irrigation in the years of drought or dwindled water supply. Their soils are somewhat coarser. In some parts of the Siwaliks and other low-lying valleys, a third type of 'talaon' lands are found, i.e. 'shimar' which are waterlogged for most parts of the year and can only produce a crop in the dry season.

The *upraon* or *upper lands* which are situated above the *talaon* are at a higher level and have very little or no means of irrigation. These are lands of terraced agriculture as here the slopes are steeper and soils are dry, thin and somewhat stoney. They are often the upper parts of alluvial fans and cones — materials deposited by the minor streams in the above named shapes. The lower parts of these lands, i.e. a transition zone between 'talaon' and 'upraon' are called *talliya* or first class 'upraon', having alluvial silt and being free or first class 'upraon', having alluvial silt and being free from stones. Old river terraces now lying above the river level also form this category. They have better water supply and are quite productive.

The 'katil' also known as 'khil' or 'irjan' in Kumaon are the highest places on steep 'erosion' slopes and are cultivated intermittently, i.e. once in 4, 5, 7 or 12 years, in order to allow the soil to recover its fertility. Here the soils are thin, stoney and dry. Terracing or ploughing is traditionally not done here but they are often terraced to provide additional cultivated land as the pressure of population and its requirements build up from below.

Types of Farming

Three distinct types of farming known by the same names are practiced on these areas, closely conforming to the physical character in terms of crops, livestock and cultivation methods.

The 'talaon' lands are usually double cropped and are capable of producing tropical and sub-tropical crops. Here, rice is the dominant crop and 'kharif' the major cropping season. Rabi season is generally slack. Terracing is nominal and irrigation predominant. The 'upraon' farming which is based on terracing is also divisible into two in terms of location, i.e., first class 'upraon' — 'talliya' lower, and second class 'upraon' — upper. The former are relatively flat alluvial silts that are capable of broad terraces, which produce rice and wheat. The second class 'upraon', capable only of dry farming, produces barley and coarse grains like 'madua'. Different types of crop rotation systems based on two or three year cycles are practiced on the basis of altitudes. 'Upraon' lands are by far the major category of cultivated land in this region.

The 'katil' culture is the least productive of the three and is an intermittent farming system where generally terracing is not done. Here land is cleared of natural vegetation and tilled by hoeing. These lands are cultivated by migratory harders or those under pressure of population. They are often located next to 'kharaks' or animal shelters which supply animal manure. The 'madua', a millet is the main crop here, though barley, 'jhangora', turmeric and potatoes are also grown. They are left fallow for a number of years for the soil to recover their fertility.

All the three farming systems derive their support from buffaloes in lower altitudes and cattle in higher, largely in terms of manure and ploughing. They in turn graze or are fed from the surrounding forests and pastures. Quite often, the communities involved in the three systems of agriculture are also different. Even their past movements have been traced along similar altitudinal zones and types of farming.

On the whole, the agricultural system is characterised by small and scattered *land holdings*, the average size being 0.53 hectare with a variation between 0.04 and 0.14 hectare per capita in the Kali river watershed, in 1988. In the same watershed, 19.6 per cent of the total land was under cultivation in 1984-85. A decade earlier, however, it was under 16 per cent (According to the slope factor, only 12.5 per cent of the land was cultivable). Here, 67 per cent of the cultivated land was under seasonal crops, 32 per cent under tree crops and only 1 per cent under fallow and with almost 45 per cent of the gross cropped land, being double cropped. The latter also shows a rise of 5 per cent in last ten years.

The agricultural pursuit here is highly *labour* intensive as almost two-third or more of the cultivated land lies in the 'upraon' and 'katil' areas which means unsuitable slopes that need to be terraced. This manufacture, maintenance and preparation of fields requires hard manual labour. Each new terrace requires a period of 3-4 years to get ready for an adequate harvest. Further, anywhere from 25 per cent to 50 per cent of the terraces break in the course of a single monsoon; their repair also requires a lot of hard labour. Practice of irrigation wherever possible is labour intensive too, as in the traditional systems, most is got out of every drop of water. Both irrigation reservoirs and ducts also require a lot of maintenance as the seasonal floods destroy the structures during the monsoon season. The farmer of the area has often to contend with two crops simultaneously at different altitudes where one is ready to be harvested and the other ready to go in the field. Sometimes a crop is sown in a particular field even before another is harvested. Different members of the farmer's family are involved in different localities; and sometimes in different trades altogether. In the case of Bhotias some young men are out with flocks for breeding; the women and children are out for cropping in the summer villages, and the older people and tenants tilling the winter villages.

Livestock, a very major and essential part of the entire agricultural system as said earlier, are kept in considerable numbers and closely interlinked with the rest of the system. The animal numbers per household in the Kali river watershed in 1987-88 (based on a household survey) was 9, with a zonal variation of 5 in the Siwaliks and 11 in the middle Himalayan zone. There are in fact, three streams of livestock, one for agricultural operations i.e. cattle for ploughing and manure; second for protein requirements of milk and meat, i.e. buffaloes and goats, and third for wool trade and transport i.e. sheep that migrate from the lowest foothills to the glaciers. Grazing system of livestock are interlinked with the agricultural operations and physical environment of climate and natural foods. Manuring of the fields is done by a complex of systems where transport of manure is avoided by transporting animals to the fields for different periods of time.

The outstanding character of this traditional system of mountain agriculture is the adaptation of the humans to get the most out of each situation in terms of locality, product, time and technology available where cropping, livestock raising and other economic practices are closely intermeshed.

Introduction of Commercial Products

In the above described type of farming system which has already reached the maximum level of its land capacity, there is a very little or no room for a commercial product which is not a surplus, but is implanted with the sole purpose of cash generation and sale outside the region. Any given product for the above purpose makes undesirable inroads into the traditional system in the absence of suitable replacements. The first commercial product that made its presence felt in this region was the export and planting of commercial timber, almost a hundred years ago. This not only encroached on the 'Support Area' of agriculture, but it also reduced and eliminated some of the species that formed a part of agricultural operations. Although there are grazing and fodder rights in the reserved forests in Kumaon even then because of the changes in plant species, the pressure on suitable 'Support Area' has been increasing, leading to degradation therein. On the other hand, the loss of space to commercial forests brought hardly any returns to the locals, for the revenue on these was not spent in the locality except for the meagre employment in resin tapping or carrying loads, and did nothing for the improvement of agriculture in the 'Support Area'.

In the post-independence period, several measures have sought to be introduced in the field of agriculture, all of them meant to alleviate the poverty of the farmer, largely to generate a cash income for him by introducing new commercial crops that would be cashable in urban areas in and outside the region; and by bringing in high yielding varieties of basic foods like grains and milk, as well as improved technologies and equipment. However, the new commercial cash crops that are being encouraged are temperate vegetables, soyabeans to be processed in the factories, and fruits — all to be marketed in the plains. This is supplemented by higher milk yielding animals like jersey cows. However, all the above 'improvements' are confined to the broad irrigated valleys and the medium to large farmers. The small farmers and the 'upraon' lands have hardly been touched by these innovations. Needless to say that these introductions have not reduced the land under basic crops, nor the overall cultivated land. In fact, the land under cultivation has increased into still more unsuitably lands. Nor have the numbers of local animals decreased by the introduction of newer breeds for the purposes of the two are different. They have kept pace with the growth of human population and cultivated land. On the other hand, land under irrigation has decreased and so have the yields per hectare. With the introduction of the new crop varieties and techniques, a number of new problems and pressures have developed, for the new systems have hardly replaced anything; they have just been added on, pressurising the land further.

Pressure on Land

The advent of commercialisation in the form of forest and its products and the change in revenue system with the British rule started it on its way to degradation by pressures on land, which have demanded more than the land can give. The process not only continued but escalated in the post-independence period when the pressures have increased due to higher demands, both in quality and quantity, in agricultural as well as forest products, which here is a part and parcel of agriculture.

The pressures on land emanate not only from a larger population that the region has to support but also the intensity and quality of the demands of that population in the context of reduction of Support Area by degradation. Population growth, i.e. growth in numbers is not only the result of the higher natural growth of the area but also the sizeable number of incomers in urban areas and other services, as well as pilgrims and holiday makers. With the growth of population and a static land area, and with agriculture being the only possible source of livelihood for the local people, the demand on agricultural and attendant lands have obviously increased. The result is a continuous division of agricultural land, with half an hectare being the average holding size per family in the 1980's. In a survey of village Chausli an average holding of 0.38 hectare was found spread over five separate locations, had 32 fields with an average area of 0.01 hectare (Shah, 1981).

On the other hand, the requirements of an average family (of 5) calculated at about 3 kgs of food grains per day requires 2 acres of land at the rate of a yield of 500 kg per year for two crops. Another half acre for spices and such would mean that family could subsist on $2^1/_2$ acres or approximately one hectare. It is said that on an average a healthy Support Area, i.e. pastures and forests, required would be 4 hectares, totalling the land required per family to be 5 hectares minimum as against 0.5 hectares available on the average. In fact, the Support Area is so degraded that according to some experts like Madhav Ashish, land required would be 10 times more or 40 hectares (Madhav, 1983). In such a situation, the majority of the people have barely enough to eat, and of course no purchasing power for other necessities or for any kind of inputs. 'Land to the landless' scheme or gift of a Jersey cow hardly helps anyone; it only augments the agricultural pressure.

Resulting from this system of agriculture, livestock population is 1.5 per capita and a growth of 2 per cent per annum (same proportion as the population). With the increase in agricultural area and consequent increase in the animal population the 'Support Area keeps on reducing further leading to a stress situation or a pressure syndrome which perpetuates itself towards a gradual degradation of land and ultimately ends up in abandonment of agricultural land. Reduced yields and reducing agricultural holding are another reasons which have led to moving away of the younger people of the family, leading to abandonment of agricultural land by the older people who find it impossible to cultivate it. Abandonment is also perpetuated by women cultivators where young males have migrated, by not being able to repair the agricultural terraces. (Also refer Overburdened Women). The abandoned land, however, is not left alone but is continuously grazed, degrading it to the point of no return. The quality of pastures reduced by overgrazing leads to perpetuation of inferior grasses, leading to further pressure, rendering some lands totally useless.

Degradation and disappearance of some natural species of plants have led to the transference of their pests to cultivated crops. For example, 'kurumulla' is a major plant pest currently. It is active in the *Kharif* season and attacks crops of that season. In the *Rabi* season it goes underground. In areas of heavy snowfall, it is absent because it dies in the snows. Earlier, it lived off the oak forests, and with their disappearance, the pest has come to the agricultural fields where pine needles are its main food. The latter are used as manure along with cattle dung.

Economic Pressures

A mountainous terrain has definite limitations in the expansion of cultivation. The concept of 'Support Area' helps to understand why food output cannot be increased beyond a point in any geographical area by a continued extension of cultivation, for it means the reduction of Support Area. This reduction leads to the fall of average output per hectare. In other words, the ratio between Support Area and cultivated area varies in a given environment. In a hilly terrain, with poor or thin soils and lack of irrigation, this Support Area has to be larger. The reduction of the Support Area, which not only includes animal foods but also several human foods and necessities as well, leads to economical pressures.

Whereas the overall growth in numbers of population escalates the volume of demands; urbanisation and incoming communities introduce higher demands per capita, newer types of demands, as well as influence an attitude change in the agricultural communities. For example, the demand for milk and its products leads to more animals near the urban centres. The demand for fuelwood is another example. It is said that in the Kumaon region, trees are felled for firewood only for the use of town dwellers. For the rural folk, branches and waste from other operations is the usual fuel and trees are not felled for this purpose.

According to one field officer, there is a definite change in working attitudes of the younger generation of women who are in a comparative hurry for they would like to finish their chores and go listen to the radio, etc. Some of the examples given were scattering fresh cowdung in the fields rather than composting it; cutting the top of one entire tree for fodder rather than cutting the lower branches of several of them, cutting away trees and bushes along *nallah* or stream banks and extending their agricultural land towards the nallah, thereby getting more land as well as water. This obviously interferes with the flow of water and reduces it for the lower people, and leads to water erosion. The above kinds of changes pressurise certain lands, leading to spoilage and poor yields.

The current economic pressure on agricultural lands in the Region results from a number of causes, some of whom escalate each other. The outstanding ones are population growth, changing and higher demands per capita, decreasing yields, increase in degraded lands, and inability of the majority of the farming population to improve its plight by cash inputs. This has led to acute pressure of subsistence crops on the one hand, and relatively unsuccessful attempts at alleviating the situation on the other by introduction of cash crops and high yielding varieties. In other words, the demands are both for subsistence and for cash on a land that has already over reached its capacity or the current agriculture is drawing on the capital resources of natural environment, which in the words of Madhav Ashish is "a highway to environmental and economic bankruptcy".

Environmental Impact of Agriculture and Allied Activities

All ecosystems in their natural state change extremely gradually so that the different parametres of the system are continuously adjusting themselves to these changes. Vegetation and soils have a very vital role to play in this dynamism of a mountain system, for the type, size and thickness of vegetation adjusts itself with the amount and type of precipitation which is the main agent of erosion. A vegetation cover (natural or planted) with a healthy soil regulates the run-off of water thereby regulating the rate of erosion and the detritus that the water will carry away. These two also act like a sponge forming a water reservoir which releases the water gradually during the non-rainy periods, thus ensuring a continuous water supply throughout the year and preventing the seasonal 'flood and drought' situation. Furthermore, a healthy vegetation and water soaked soil also reduces the variation of temperature, thereby reducing the rate of weathering as well.

Inherent or physical *vulnerability* of the Himalayas tends to enhance environmental instability because of mode and intensity of usage. The geologic and geomorphic youth of the Himalayan landscape renders the Region highly vulnerable to degradation. The 'youngness' implies that the Area is seeking its own balance in terms of its relative relief, structure and morphology, particularly altitude and slope. Great amounts of tectonic activity accompanied by folding and faulting of the strata has led to large scale crushing and crumbling, which has made the bedrock very susceptible to weathering and erosion. Many landslides and slumps are attributable to the above, together with recurring tectonic activities. As the bedrock in many places is composed to weak sedimentary rocks, heavy downpour of rain results in accelerated erosion. The rate and degree of erosion is accentuated by the availability of tools for scouring. These are supplied by the immense amounts of debris left by the last glaciation in the upper reaches, as well as fresh one being generated by extremes of temperatures and freeze-thaw action.

In the Kumaon Region, there is great deal of erosional susceptibility, not only from the above discussed factors but also from its slope and relative relief characteristics. Nearly three quarter of the slopes are steep to very steep, being nearly vertical at places. As most of the rivers have carved deep valleys, the relative relief is also extremely large. At some places within a horizontal distance of less than a kilometre, the elevation of land changes by more than a thousand metres. An approximate relationship between the degree of slope and erosion intensity can be seen from the figures below: (Pandey, 1981).

Degree of Slope		*Erosion Intensity*
15 – 20°	Gentle	Low
20 – 25°	Semi-gentle	Moderate
25 – 30°	Moderate	Moderately high
30 – 35°	Moderately high	High
35 – 40°	High slope	Very high
Above 40°	Very high	Maximum

Another factor responsible for the removal of materials and enhanced rate of runoff is the low infilteration rate here. This results from the shallowness and texture of the soil. Clay as well as stoney soils, abundant here, reduce the rate of infilteration and enhance the rate of runoff. Shallow soils, especially when bare, are unable to absorb and hold moisture for hydrologically, the vegetation, particularly trees represent a sub-surface flow system in shallow soils — where much of the rain water is transferred laterally to the channels through a sub-surface flow process. On the removal of vegetation, the underground system begins to deteriorate, with the water running downhill unchecked as a fast moving torrent, taking along soil and other debris. In time, the slopes lose their stability and even a small rain storm can lead to landslides and cataclysmic floods downstream.

The location, altitude and physical character of the Region itself is prone to intense geomorphic action. The removal of the cover or activities that lead to it only enhance the rate and volume of this erosional activity. The earlier economic systems of utilising the resources according to their capabilities and regenerative capacities is breaking down in the face of modernisation, commercialisation and cash generation or from the colonial type exploitation for extra regional uses. This relentless exploitation of forests, expansion of agriculture on steep slopes, construction of heavy structures, and above all, the all pervading network of roads necessary for commercial production have led to increased geomorphic hazards in the form of landslips and landslides, accelerated erosion of soils, as well as changes in water supply and climatic phenomena.

Impact of Land Utilisation

Different modes of landuses affect the rate of erosion by influencing the rate and amount of water runoff quite profoundly. A study at Khairna in the Kosi valley revealed that there was considerable difference in the surface runoff of total rainfall according to the manner in which the land was used, with geographical, climatic and topographical conditions being the same. Forest land, being the least disturbed, produced the minimum runoff, i.e. 3.2 per cent of the total rainfall; barren land produced 10.38 per cent and agricultural land 20 per cent of the total rainfall. (Rawat, 1988).

Studies also reveal that there is a significant correlation between landuse and silt discharge in a mountainous area. Silt discharge is a good indicator of rate of erosion. A study conducted by Joshi and Dani (1984) in the lower catchment of the Ramganga watershed revealed that both forests and pasture-land had a negative correlation, with values being –0.54 and –0.15. Cultivated land has a positive correlation, being +0.53, cultivable wasteland having an r value of +0.87. The latter, which forms the community grazing land and belongs to no one in particular is utilised ruthlessly, with no concern or care and soil depth and low moisture holding capacity.

As seen above, *agricultural landuse* is one of the predominant uses adversely affecting the rate of runoff as well as silt discharge. The expansion of agriculture, especially on moderate to steep slopes replacing the natural vegetation has caused severe erosion of soils, contributing to increased frequency and severity of flooding. Cutting of vegetation along *nallahs* and expansion of agricultural land along their banks, especially for the facility of irrigation, is another cause of erosion.

The depletion of the forest system is indirectly responsible for the adverse effects on agriculture, for the hill agricultural system is largely dependent on the forests for energy supply as explained earlier. According to Pandey and Singh (1984), the actual ratio of forest to agricultural land currently is 1.33 : 1.00, well stocked forest ratio being only 0.84 : 1.00 (Melkania) et. àl, (1988), while according to others like Valdiya (1988), it is 1.66: 1.00. With the actual requirement being about 4.00 :- 1.00, the above is too meagre to support the required agricultural production. It not only leads to expansion of agriculture on undesirable slopes but also exhaustion and destruction of soils leading to the degradation spiral.

It is not only the increase in the extent of cultivation, but crops unsuitable to the environment, but coaxed out of the soil for commercial gains that also cause environmental problems. The cultivation methods for certain crops like maize and potatoes on steep slopes require drainage of excess water thus promoting soil erosion. It is calculated that up and down cultivation of maize as practised by farmers on 8 per cent slopes gives a soil loss of 28.5 tonnes per hectare. In another study of farming on 4 per cent slopes, soil losses were 4.9, 4.8, 4.5 and 1.5 tonnes per hectare under maize, paddy, mandua and jhingora respectively, while napier grass showed a loss of 0.5 tonnes per hectare only (Gupta, 1981). This results in the removal of top soil and loss of nutrients, leading to severe soil erosion. The outward sloping terraces on steep slopes in areas of heavy rains promote the same situation.

The use of *chemical fertiliser* in order to coax more yields are also leading to damaged soils by making them more compact and dry. This in turn inhibits the growth of life in the soil and makes them susceptible to greater runoff, with compactness leading to the same results. The chemicals that flow out in adjoining water bodies lead to greater weed proliferation and consequent loss of water in transpiration.

Large scale *irrigation* by modern methods has its own drawbacks, for it has to be remembered that the water for irrigation is subject to considerable losses due to evaporation, percolation, surface waste, and non-productive consumption. This loss is estimated to be nearly 50 per cent for hilly regions and 50 to 70 per cent in the Bhabbar area at the foot of the Siwaliks (Pathak, 1983). In other words, increase of production by irrigation would be at the expenses of overall water resources. A very substantial effect of current irrigation methods, especially under government management, is waterlogging of the soils because of excess water input and lack of adequate drainage facilities. The traditional methods of irrigation did not suffer from such a situation, for both application of water and its drainage was meticulously done, as it was based on community labour. One of the common causes of erosion is the neglect and non-repair of irrigation channels, which has become more frequent, resulting from government role in the irrigation facilities.

Food for livestock is a major burden of the natural environment. As agricultural in the Region is based on animal population with each household on the average possessing more than 4-6 animals and animal food being obtained largely from natural vegetation or the Support Area. According to a study conducted by G.B. Pant University in the Dwarahat Block, Almora district in 1978-81 the following 'mix' of fodder was obtained per household per day from different sources (Table 1.1):

TABLE 1.1: Fodder requirement (on dry matter basis) per household*

Source/Type	*Kg per Day*
*Common Land***	
Dry grass	7.1
Green grass and weeds	11.5
Green leaves	0.9
Grazing	4.7
*Private Land****	
Dry grass	2.8
Green grass and weeds	2.7
Straw	12.6
Husk and Bran (from house milling)	0.3
Total	42.6
Bedding material	1.2

* From Jackson, M.G., 1985, p. 137-38.

** Includes village common land and wasteland including private and reserved forest with grazing rights.

*** Includes cultivated, abandoned, and current fallow lends.

Note: Based on a Household Survey in Kali R. Watershed, 1988. It was found that 64 per cent of the livestock grazing was in the Reserved forests, 10 per cent in the village grazing grounds, 18 per cent in the harvested fields and only 8 per cent of the food was from other sources.

Even in 1961, on the basis of the various categories of forest lands available and the units of land required per animal (1/2 hectare = 1 unit) it worked out that the district of Pithoragarh had a carrying capacity of 5.7 lakhs units and the district of Almora 8 lakh units against the required grazing of 15 lakh and 31 lakh units respectively (Working Plan 1971, p. 46). In 1981, the carrying capacity in terms of grazing units in the Kali R. Watershed had reduced to 5.34 lakh units and the requirement had increased to 16.92 lakh grazing units. In the Middle Himalaya, the situation was even worse; the requirement was 13.18 as against 2.95 lakh units available for grazing. Grazing pressure is high everywhere. In the villages, the available grazing land was symbolically low in some areas. For example, in Dharchula area, the villages near the ridge tops just west of the Kali River had only 9.5 per cent of the required grazing land, while those at the lower levels had 10 per cent. The above mentioned over-grazing clearly leads to several situations like interference in the development of raising of forest plantations and fruit orchards. It has also led to constant degradation of the ground flora. Many fodder species are vanishing at a fast rate. Natural regeneration of the forest species is also affected by reducing moisture due to reduced undergrowth, especially on northern slopes. Over-grazing also results in a denudation, leading to erosion of the soil and reduction in the carrying capacity of pastures. Because of loss of nutrients, the grasslands are infested with scrub vegetation, inferior grasses and weeds, which lead to further pressure on the pasture grasses, ultimately leading to bare areas and erosion of the soil and land.

Generally, the gentler slopes are set aside for grazing and steeper parts for grass cutting. Grass cutting in 'chir' forests outside regeneration areas is not very damaging, except for the repeated firing. Extensive damage is caused by grass cutting in 'chir' regeneration areas as seedlings get cut along with the grass. This is particularly so near the villages and at the edges of the reserved forests where whole areas fail to regenerate. Planned firing to induce the growth of grass is particularly responsible for continuous shrinkage of the area under oak because of desiccation and for removing miscellaneous broad-leaved understorey from 'chir' forests, but villagers continue to do so for firing results in better grass growth because of instant nitrogen availability.

Lopping to leaves and stripping of small branches is another human activity that surpasses the capacity of the forests Oak is lopped for leaf fodder and 'chir' branches for fuel. 'Chir' does not suffer much damage from this activity but oaks and other broad-leaved species like hornbeam (which is popular) are harmed by excessive lopping and stripping to the point of ruination. Many of these, especially near settlements, are reduced to oak scrub though some of them a few decades ago were such thick forests that people were afraid to pass through them. The above described grazing, grass cutting and regeneration methods lead to continuous poverty of the soil, with reduced production setting up a degradation spiral.

Finally removal of wrong trees for the sake of convenience, especially near habitations, has become a common practice, creating large gaps in the forest canopies. Another menace that has cropped up in recent years (especially from 1953 onwards) and is on the increase in *encroachment* on forest lands for agricultural purposes. It is much more rampant in the lower zones of Terai-Bhabbar as compared to the hills, where it is reaching serious proportions. Unwittingly, the Government itself has been responsible for this situation by legalising the poachers' land in the early stages. A Government order dated November 1, 1975 authorised regularising encroachment up to June 30, 1966 and in case of harijans, tribals and landless, up to November 8, 1973. An example of the outcome of this policy was that at the end of 1987, more than 8,000 hectares of land had been encroached upon in one working circle of National district alone. Organised groups of unsocial elements continue this activity here with Government as a helpless spectator (Joshi, 1988). These negative impacts on natural vegetation result from two major sets of causes. One, as discussed above, are the agricultural and allied activities dependent on forests which are not as much commercial as they are subsistent in nature. The availability of these have consistently reduced in time, for there has been practically no management and no augmentation of the required natural vegetation resources. As this leads to lack of regeneration, degradation, and even change of species, the availability of required species like oak have consistently reduced, leading to further chain reactions. The higher demand because of growth of population is a minor cause in this case.

Ecological Problems and Consequences

Resulting from the negative interactions between the inherent characteristics of the physical and human systems and the agricultural and the allied uses of land as highlighted in the earlier pages, are a number of ecological maladies that are now quite conspicuous. There are apparent both among the physical and the human systems. The major effects are on land, soils, water and biotic resources, the degradation of which results in out-migration of the labour force. Pollutions, climate changes, and adversely affected human health are further outcomes of overall degradation. So far, there are hardly any specific studies available which analyse these problems, but several authors have pinpointed them on the basis of observation and inference. The ecological problems/situations highlighted below are based on the above writings; personal observations and those of several persons in the Region; a survey of the perception of ecological changes in the Kali R. Watershed and inferences from indirect evidences.

Physical System

Symptoms of environmental ill health in general are degraded lands, reduced forest cover, accelerated soil erosion, increased silting of water bodies, drying up of springs, replacement and disappearance of species, and increased ratio of energy expended in direct agricultural activity and out-migration of labour force to augment resources for daily living.

Land Degradation

Land in this area is subject to a number of ecological problems, based on its inherent characteristics as well as the manner and intensity of usage. Movement of loose materials downslope and erosion are the major types of land degradation. In Himalayan areas a lot of loose debris is generated by the process of weathering especially mechanical, and glaciation both past and present. This natural debris is supplemented by materials loosened by several economic activities. The foremost among them are road construction and mining which resort to blasting, and loosening more rocks than utilise. Bare surfaces formed by excessive grazing, deforestation and abandoned agricultural land also produce a lot of loose material which may even slide down by vibrations caused by heavy traffic. All this results in *movement of materials* downslope by gravity flow, both on the surface as well as in the sub-surface strata. The surface movements are generally known as 'rockfalls' 'landslips' 'landslides', or 'slope failure', and can be anywhere from a few feet, both horizontally and vertically, to hundreds of yards or even miles in the case of the former i.e. horizontal; and a few thousand feet in the case of the latter i.e. vertical flows. They may be intermittent or continue for years. These movements are most frequent during the monsoon season when the gravity flow is aided by water. The sub-surface movements called *mudflows* are common where sub-surface material is clayey and where moisture seeps underground from bare surfaces above. Here a whole mass of materials flow out from underneath, causing *land slumping* and *subsidence*. Both the above described movements are fast and sudden movements. These same movements can also take place at a slow or imperceptible speed, with similar results. Such a surface movement is known as *Soil Creep* and is perceptible in bent trunks of trees while the underground movement is perceived in *cracks* on surface areas and human structures.

Land slumping and subsidence are quite important problems on the river terraces which are generally cultivated and thus unprotected because of a shallow root system. Here 'toe' erosion by the rivers causes a part of the terrace to slump into the river. This is very frequent and causes a great amount of loss to agricultural land and property. Towards the upper margin of terraces and other debris slopes, subsidence is also common because of similar reasons causing much damage to land and houses. This is a very common phenomena in the entire region.

Several *landslides* and their scars are frequently visible along the arterial roads of the Region being attributable to bare surfaces, road cutting and use of explosives. Abandoned agricultural terraces also show scars of varying sizes which are fore-runners of current and potential slides. These slides and flows are very frequent in a wide calcareous belt running N.E. — S.W. along the foot of the High Himalayas from Khela on the confluence of Kali and Dhauli Ganga rivers, westwards. Innumerable scars of these slides are visible in the entire zone especially along the Eastern Ramganga where the valley sides abound in abandoned agricultural terraces. Here, overcut Panchayati forests are also not immune to these running land slides. In addition to the above, another major landslide prone zone, also calcareous in nature, stretches along the southern part of the Middle Himalaya, i.e. from Kosi to Sera. Sometimes these slides can also cause considerable loss of life. More than 20 persons were killed in a massive landslide on August 15, 1977, at Tawaghat, below Khela, where huge boulders still lie in testimony to the disaster. Large scale landslides have occurred in the ten years since early 1970's in the Bageshwar, Dharchula and Munsiari tehsils. (Bisht, 1982). According to the same source, some villages in the above areas have been noticed to be sinking at a steady rate. The author saw cracks in several houses on the bottom of the slope at Tawaghat in 1984, possibly resulting from a slow sub-surface 'mudflow' from the fields above. Landslides, mudflows and subsidence are also common phenomena along the Siwalik zone, especially in the ranges on both sides of the Ladhiya river — a tributary of the Kali River.

Land erosion is compounded by the scouring effect of loose materials that the water carries downslope and its rate is based on the volume and velocity of water and the load it carries. It is estimated that in many catchment areas, the rate of erosion has increased five times as compared with that in the past, the present rate being one mm per year. The amount of eroded material the rivers in the Kumaon Himalayas are transporting is at the rate of 16.5 hectare –m/100 km^2/year. (Melkania et al, 1988). Though the major contributor of the eroded material are construction activities, i.e. roads, dams and the like but all bare surfaces put in their share as well. Both sheetwash and rill/gully erosion is rampant and are most conspicuous in the Calcareous zone talked of above where the landslides are also rampant. It is conspicuous along nallahs where vegetation is being removed for agricultural usage and in the abandoned agricultural land, as well as in community lands. Rill and gully erosion scars are visible all over the Region in the above two types of land, especially in the Middle Himalaya.

There is abundant hydrological evidence to indicate that *water resources* of Kumaon are diminishing and altering in character. Some of the major indicators are retreating glaciers, changes in trends of stream discharges, diminishing ground water in the form of discharges of springs and even their disappearance, enhanced siltation and the dwindling capacities of water bodies and channel network. Much of this altering is resultant from overall land degradation.

Accelerated Silt Discharge

Another corollary of faster run-off apart from floods and reduced supply in the dry season is the enhanced amount of silt carried by running waters by enhanced rate of erosion. Studies on two small plots on a 6 per cent over a period of 10 years showed that the bare soils yield more than 100 times the sediment as compared to a well vegetated catchment (Gupta, 1981). The increasing silt load is leading to rapid filling of lakes and reservoirs as is evident in the Bhim Tal and Naini Tal lakes. The former has been reduced by 46 per cent by developing depositional cones and fans around its fringes, and the latter has become shallower by about 6 metres in 83 years or by 8 cms per year (Melkania et al, 1988).

The load piling in the channels and valleys is also bringing changes in the channel network capacity. An experimental study conducted in the micro-catchments of the Nana Kosi basin (District Almora) reveals that about 50 per cent changes are man-induced in the water-flow capacity of the channels (Rawat, 1987). In the month of heaviest rainfall, streams of the agricultural land (i.e. greatly disturbed land) have maximum channel capacity, carrying water at the rate of 8,734.3 m^3/km^2/day, while the streams of the forest land (i.e least disturbed land) have the minimum channel capacity, 8,478.2 m^3/km^2/day. During the driest month, the situation of flow capacity is opposite. The agricultural land has minimum channel capacity of 180 m^3/km^2/day while the forest land has the maximum channel in the capacity of 307.7 m^3/km^2/day. It is evident from the experimental study that the type of land usage has a direct impact on the channel capacity as manifest in increase of the flood water and decrease of the base flow water. (Rawat, 1988).

Soil Degradation

Several ecological problems can beset soils which very often occur in linear/cyclical stages and are brought about by the interaction of intensity and mode of usage versus the inherent characteristics. Soils are extremely closely linked with biota which, dead of alive, not only furnish them with a continuous supply of nutrients, but also organic colloids that increase the water holding capacity and form a building force in the soil body. Thus lack of a vegetation cover, natural or cultivated, and non-replenishment of organic matter sets in motion a series of changes like reduction in the moisture holding capacity, changes in texture and structure, soil exhaustion, erosion, and finally total loss.

Structural and textural changes in the Region's soils are largely towards increased looseness, coarseness and stoniness. Looseness is brought about by reduced water holding capacity. Coarseness and stoniness are brought about by two methods, i.e. removal of the finer materials by wind and water and/or an overlay of coarse materials. Daily cycles of strong, adiabetic winds (anabatic and katabatic) in deep valleys and higher altitudes, as well as seasonal winds are powerful enough to carry away all fine particles of exposed soils, making them steadily coarser and more vulnerable to being washed away. In cultivated areas, especially where potato harvesting in autumn has loosened soils, veritable dust storms are created by the evening winds. Similar phenomenon is observed where different kinds of shrubs and herbs are uprooted for commercial purposes and soil loosened and left bare. Soils on morainic materials in the High Himalayas are also subject to the same phenomenon. Tons of fine soil particles are continuously removed in this fashion.

Another menace to cultivated soils in many areas is the overburden of coarse materials and stones from the upper slopes. This results not only from natural phenomena of slips and slides occurring in the steeper slopes above, but also from many economic activities like cultivation on steep glaciated slopes; blasting of rocks for quarrying and road making and pushing of unwanted debris downslope to save removal costs. This process continues to plague soils of the downslope locality for years after the activity is over. A good example is that of Ladhiya River leaving behind thick deposits on fields along its banks in the monsoon season as it carries large quantities of gravelly material that has moved downslope in its upper catchment.

Soil exhaustion is brought about by nutritional losses and structural changes, including moisture holding capacity. Of the three major plant nutrients — Nitrogen, Phosphorus and Potash — Phosphorus is generally low to medium everywhere, while Nitrogen content is low in the Siwaliks. Indirect evidences of soil exhaustion are lowering of yields and abandonment of agricultural lands and out-migration of the male labour force. The latter is measurable by high female ratio of the population and lower percentage of males in the working age groups. There are enough records to show that yields of the major crops have been steadily going down in the last one hundred years. A comparison of the yields of the four major crops of the region — rice, wheat, 'madua', 'urad' — with those of mid-ninteenth century yields show a definite decline. This decline is to the tune of 6 qtls/hectare in the case of irrigated rice, 3 qtl/ha in wheat, 1 qtl/ha in 'madua' and 5 qtl/ha in 'urad' (Calculated from Atkinson, E.T., Vol. III, 1882 and official records, 1984-85, p. 482).

Soil erosion is quite rampant in the above mentioned areas, resulting from usage and exposure, including abandoned agricultural terraces. About half of the bare slopes above and around the villages show traces of abandoned terraces, from where soils are being washed off layer by layer, leaving scars ranging from rills to deep gullies as described earlier.

Based on a perception survey of 10 persons in 1988 in Kali R. Watershed, some interesting inferences are also possible as to changes occurring in the soils of the area as regards their structure, texture, moisture, exhaustion, erosion, and fertiliser effect.

Changes in Soil Characteristics	Change Perceived No.	Increase No.	Since when No. of Years
Structure Texture:	6	*	3 - 10
Stoniness	8	5	3 - 7
Coarseness	7	7	3 - 5
Moisture	8	0	3 - 8
Exhaustion	5	5	3 - 10
Erosion	7	7	3 - 8

* Structural changes perceived vary from spoilage of soils to drier soils, affecting the entire fields.

The above figures do show that several changes in the character of soils are perceived in the Study Area. Most changes which are inter-related are perceived over a relatively short span of time, i.e. 3-10 years. As to the location of these changes, most of them, i.e. changes in structure, moisture and erosion, entire fields have been affected. In some cases, especially stoniness, exhaustion and erosion, the fields near river banks have been particularly affected. In one case, 2-3 feet of top soil has been eroded and in another, a landslide has been the cause of erosion. In the case of chemical fertiliser effect on soils, the perception was mixed. Of the 8 people that perceived changes, only 5 felt that there was increase in productivity; others perceived a decrease. At the same time, the majority perceived that the fertiliser was making the soils poorer, drier and more compact.

Not only land, water and soil changes have come about by agriculture and allied activities but several *biotic changes* are also taking place, resulting from overall changes in the habitat. This change in manifest in two major spheres among the plant communities. One is the change in type, i.e. in species and formations of plants, and the other is in the character, i.e. changes in density, on the ground, canopy, as the body of the formations. The above habitat changes show their repercussions on the wild life as well, in numbers as well as habits. The latter is often visible in their raids on agricultural lands, and lifting of livestock.

From the above discussions, observations and inferences, it becomes quite clear that in spite of relative lack of concrete evidence of ecological problems in the Kumaon Region, there is ample scope to infer that most of the maladies have already begun and have also started to show their impact on the human system as seen in the following pages.

The Human System

The ecological degradation of the natural resources in the form of reduction or near disappearance of wild gathered food, fodder and medicinal plants, reduced and erratic water supply, exhausted soils and eroded land are showing their effects on the human and animal system of the area. In the case of animals, it is reduced milk supply and poor health nutrient-wise, leading to proneness to disease. In the case of humans two major effects are visible. One is health, making them prone to a number of vitamins and other nutrients deficiency and water-borne diseases. The second major effect is lack of assurance of adequate agricultural production resulting from exhausted soils and erratic water supply, leading to out-migration of the labour force to earn cash for their requirements. Both of these in turn affect the Region in different ways in the ecological sphere, setting up what may be called a 'Degradation Syndrome'.

Out-Migration

The exodus of working age males in sizeable numbers seems to have started after 1941, and has been steadily increasing as seen below (Table 1.2) from the ratio of females to males in the Region.

TABLE 1.2: Females per 1,000 males, 1901-81

District/State	1901*	1921*	1941*	1961**	1971**	1981**
Almora	978	1,013	1,027	1,091	1,088	1,099
Pithoragarh	976	999	1,010	1,044	1,026	1,055
U.P.	937	999	907	909	879	886

* Kamlesh Kumar, 1983, p. 129 ** Khanka, S.S., 1988, p. 144

On the whole, the long settled mountainous parts of Middle Himalaya show a higher exodus especially those villages which have a southward and eastward aspect, as revealed by a survey in the Kali River Watershed. Another survey of Vishwa Yuvak Kendra, New Delhi of migrant employees revealed that 63 per cent of the out-migrants have lands on the higher ridge flanks and another 33 per cent on both higher and middle locations. By inference, the southern and eastern aspects means longstanding cultivation and higher locations means shallow soils. As most of the Vishwa Yuvak Kendra migrant employees gave lack of assurance of water supply and not enough returns as one of the major reasons of migration, it implied lack of moisture holding capacity of the soils.

On the human front, this kind of male migration increases the *intensity of work for women*, for men are either away (working age group), or are being trained for jobs outside (i.e. in schools), or are an older age group (and often would not form a part of domestic labour force). The women, therefore, are engaged in-farm activities, off-farm activities, in addition to child bearing and rearing. "The farm activities comprise cultivation, levelling, sowing, hoeing, weeding, harvesting, threshing, winnowing, drying and storage of grains and their bye-products. The off-farm activities include relating to rearing of animals and fetching fodder and fuel wood from forests. The domestic work is primarily related to upkeep of the household — cleaning, washing, cooking, handicraft, processing fruits, drying vegetables and grains, etc." (Pandey, 1988). A study conducted by Nirmala Khetwal in 1980 illustrates this burden as shown in the Table 1.3 below:

TABLE 1.3: **Working Days of Rural Women in Farm, Off-Farm and Domestic and other Activities, in Kumaon**

Size of Land-Holding (hectares)	Economics Activities of Women		Domestic and other Activities	Total Working Days in a Year
	Farm	Off-Farm		
0.25 or less	79.8	55.7	195	330.5
0.25 — 0.5	108.8	49.5	177	335.3
0.5 — 1.0	156.2	47.0	139	342.2
1.0 — 2.0	187.1	42.6	126	355.9
More than 2.0	212.7	39.7	105	357.4
Total	128.9	49.1	154	344.2

Source: Nirmala Khetwal, 1980, p. 463.

According to another Survey in 1988, near Munsiari (Pithoragarh district), the average number of hours spent daily on different chores by the women folk worked out as follows:

Food preparation and eating	3.6 hours
Water collection	1.0 hour
Fuel collection	4.0 hours
Agriculture and grazing	2.0 hours
Weaving woollen goods	6.0 hours

Grazing animals was generally the job of younger girls, below 14 years, for which they travelled up to 2 km; collecting fuel wood was generally assigned to the age group of 16-25 years who went from 5-6 km distance for the purpose and spent 4-5 hours on the job. In this group, fodder collection for milch cattle did not enter the scene; it is also a job of younger women who may go as far for collection of fire wood and spend as much time. According to the above survey, a woman on the average worked from 6 a.m. to 9 p.m.

Health Problems

The greatest impact of this over burden, however, is in the health field where both mental and physical diseases are rampant. Most of the physical diseases in the area result from malnutrition and bad hygiene. A study in Dwarahat Block in Western Ramganga Watershed, in 1979-80, on the food consumed by rural households, found that it was not only deficient in nutrients but the people were also underfed. It was found that energy (calorie) deficiency on the whole was more than 20 per cent. Largest deficiencies were vitamin A and C, followed by vitamins B_2, calcium and protein. Malnutrition results from not only lack of enough quantity and the right kind of food, but also lack of time to prepare and administer it to the children and to themselves. The latter is directly connected to the lack of time and shortage of fuel and water supply. In addition, the divided family at two places leads to social unhappiness at both ends, i.e. the migrants and the families they left behind. This situation has led to a number of mental diseases among women and a high suicidal rate. The men folk, generally living in urban slums, are also not spared and are prone to alcohol intake and diseases like tuberculosis.

Added to the above problems, is the pattern of marriage and child bearing. Early marriage is the prevalent pattern. A study by Srivastava (1979) showed that in Pithoragarh district 57 per cent of the girls were married by the age of 15 years, in 1961; and 41 per cent in 1971. (Pandey, 1988). The Vishwa Yuvak Kendra survey revealed that the average marriage age of the male migrant employee was 17. Here marriage provides the benefit of an utterly loyal female worker without wage, for all kinds of chores in the field and at home and that too unregulated by time or hours of work. Thus, marriage is a typical economic institution under the social cover.

Several economic deterrents also get set in the Region emanating from lack of labour force for performing heavy manual work. This has affected the region's economy in more ways than one. Such labour is either absent or is at a premium. Upkeep of agricultural terraces has suffered considerably because of the above and has often led to their abandonment. Ploughing agricultural land also requires male labour, the general practice being that most migrants in civil and private services return home at ploughing time, but that is costly and not always possible. Thus many migrants' families are reducing cultivation and increasing livestock instead, especially as there is more cash available for their purchase as well as a growing demand for livestock products like meat and milk in the expanding urban areas of the region. Several migrants' families have abandoned agriculture altogether, but let their lands be used for grazing of the village livestock with the result that the agricultural production is substantially reduced together with an increase in the number of animals. There is, however, no attempt to increase pasturage and they are dependent on the dwindling forests. Many of the new schemes for economic development cannot also be fully operative in these areas because of lack of adequate labour supply. The long term *ecological consequences* of this type of out-migration are quite serious and are unfortunately on the land itself. The earnings away from the land are never invested back in it and no attempts made to revatilise it. This situation actually pressurises the land further by firstly, trying to get as much out of it as before by increased population of the area; secondly, by its wrong usage, i.e. increasing animal population without making any attempts to increase fodder supply; thirdly, by its neglect, i.e. leaving it bare, not checking erosion, not taking interest in it as the income emanates from elsewhere.

The increasing rates of erosion and landslides puts an ever increasing amount of land out of production directly and reduces the water supply indirectly. A typical sequence that can be worked out is: land and soil degradation→reduction in soil productivity→less yields and reduced water supply→necessity to earn away from land→working age male migrates→lack of repair of agricultural terraces→less returns→increased animals→grazing on agricultural terraces→erosion scars→further grazing→further erosion→final abandonment→more grazing→landslides and enhanced erosion→ further loss of water supply→overall desiccation.

Strategies for Sustainable Agricultural Development

Sustainable agricultural development of the Region is a corollary of rehabilitating the catchment and keeping it in shape. Being a mountainous area, it is largely suitable for agriculture-forestry combinations, together with development of specialised products and in utilising its special potentials like water and forests.

A mountain farming system has to be a largely self-contained system because of the peculiarities of terrain and soil that retard large scale cropping and its movement. In this system land, livestock and people have to be tightly integrated. Land is required not only for cultivation, but also for pastures to feed the livestock and forests to fulfil the needs of both in terms of not only extra food, but to maintain the nutrients and water supply as well. Therefore, it becomes imperative to strike a balance between economic requirements and land capability, with suitable techniques for sustainable development in order to achieve a reasonably adequate standard of living for the people of the area. The power to buy imported necessities has to be built up gradually by surpluses and special products of the area as well as generating secondary level employment from the local potentials. Reducing the numbers of people by reducing family size is a commendable strategy but not very likely to take place in a hurry; besides too small a family in mountain farming can prove to be a disaster to farming itself where a machine is not easily applicable.

A sound, ecologically oriented sustainable agricultural system would not only have to be an integrated system of cultivation animal husbandry, horticulture and forestry, it would have to be devised according to the type of land. This does not involve only the altitude but the entire environmental milieu. In the Region, the major types have already been defined traditionally as the 'talaon' or lower lands, the middle level lands or the lower parts of the 'upraon', lands and the higher 'upraon' lands and 'katil' lands.

The agricultural strategies for the *lower valley land (talaon)* could gradually be steered towards modern technologies for grain production as well as for commercial products like vegetables. Here, suitable contour terracing with efficient mechanisms of water disposal and irrigation systems that are manageable at local levels, from hydrums, pumps or ponded water and gravity flow, can be encouraged. The terrace risers, if used for grass production would be beneficial as they would take up the extra water and nutrients, produce fodder in the form of hay and have a protective influence. The middle level or lower slopes ('talliya') lands could largely be given over to grain and legumes production, i.e., wheat, barley, and mixtures of millets and legumes.

The 'upraon', mid-slope or slightly higher lands would do best under agro-forestry practices or dairying economy with specialised agro-forestry products like seed production, honey, medicinal plants and sericulture, among others depending on the suitability of the location. The above products should be processable in the area and used locally as well as for export. Most community lands being in these locations, when restored, should fulfil the local biomass requirements.

The *upper slopes,* Class VII and VIII lands, used for grazing and *katil* cultivation currently, should be treated primarily as environment protection lands, with no cultivation but with controlled pasturage and forestry usages.

The products within each of the major farming categories would have to be locality-suitable to fulfil the local requirements first. There would be need of an integrated agricultural development programme in terms of policy, planning and management. The core areas of this programme would have to be agriculture, horticulture, animal husbandry and forestry, with close integration. If this integration is lacking, the above strategies would obviously not be successful.

The above suggestions in the field of agriculture would provide the basic requirements to a large extent, as well as generate income from specialised products as well as from secondary occupations. If implemented suitably, it would break the degradation syndrome to a great extent by holding back adequate amounts of soil, water and labour force within the Region.

REFERENCES

Atkinson, E. T., 1973 : *The Himalayan Gazetteer* (The Himalayan Districts of the North Western Province of India, 1882), Vol. I, II and III, Cosmo Publications, Delhi (Reprint).

Bisht, D. S., 1982 : *Guide to Garhwal and Kumaon Hills,* Trishul Publications, Dehra Dun.

Burman, S. G. : *Study of Integrated Resource Management and Development* (Kali River Watershed) India, Parts I & II, (CSCD), October 1990 (Unpublished).

Gupta, R. K., 1981 : "Ecological Consequences of Irrational Landuses on Loss of Productive Soil" in *Proceedings of the South-East Asian Regional Symposium of Problems of Soil Erosion & Sedimentation,* Bangkok, Thailand, Jan. p. 219-30.

Jackson, M. G., 1985 : "A Strategy for Improving the Productivity of Livestock in the Hills of Uttar Pradesh", in Singh, J. S. (Ed.) *Environmental Regeneration in Himalaya:* Concepts and Strategies, p. 130-52.

Joshi, D. P., 1988 : "Forest Resources and Wild Life of Kumaon", in Valdiya, K.S. (Ed.) *Kumaon: Land and People,* p. 223-41.

Joshi, S. C. and Dani, D. D., 1984 : "Erosion Intensity and Landuse in a Himalayan Watershed: Some Hydrographic Implications", in Joshi, S. C., *Rural Development in the Himalaya: Problems and Prospects,* Gyanodaya Prakashan, Nainital, p. 249-63.

Kamlesh Kumar, 1983 : "Population Structure of the U.P. Himalaya", in Singh, O. P. (Ed.), *The Himalaya: Nature, Man and Culture,* Rajesh Publications, Delhi, p. 117-48.

Khanka, S. S., 1988: "Demographic Profile of Kumaon" in Valdiya, K. S. (Ed.), *Kumaon: Land and People,* Gyanodaya Prakashan, Nainital, p. 137-60.

Melkania, U., Melkania, N. P. and Johri, B. N., 1988 : "State of Forests in Kumaon and their Regeneration", in Valdiya, K. S. (Ed.), *Kumaon : Land and People,* p. 223-25.

Pandey, G. C., 1988 : "Role and Status of Women in Kumaon", in Valdiya, K. S. (Ed.), *Kumaon : Land and People,* p. 161-78.

Pandey, R. K., 1981 : "Influence of Geomorphology and Land Utilisation: A Case Study of Gagas Basin (Almora District)", in Singh, J. S. et al, *Science and Rural Development in Mountains,* Gyanodaya Prakashan, Nainital, p. 309-17.

Pathak, H. G., 1983 : "Water Resources of the U.P. Himalaya: Their Uses and Potentials" in Singh, O. P. (Ed.), *The Himalayas: Nature, Man and Culture,* Rajesh Publications, New Delhi, p. 161-73.

Ram Ganga Watershed Project (RAPRAP), 1984 : *Integrated Natural and Human Resource Planning and Management in the Hills of U.P. — Ram Ganga Watershed,* Ford Foundation Project, Oct. 1978 — Dec. 1984, New Delhi, G.B. Pant University (Unpublished).

Rawat, J. S., 1988 : "Water Resources and Burning Hydrological Problems of Kumaon", in Valdiya, K.S. (Ed.) *Kumaon: Land and People,* Gyanodaya Prakashan, Nainital. p. 201-12.

Shah, S. L., 1981 : "Ecological Degradation and Future of Agriculture in the Himalayas", Presidential Address at the 41st Annual Conference of the Indian Society of Agricultural Economists, Karnataka University, Dharwar, Dec. 25.

Valdiya, K. S. (Ed.), 1988 : *Kumaon : Land and People,* Gyanodaya Prakashan, Nainital.

———, 1971 : Working Plan for the Pithoragarh Forest Division (Kumaon Circle, U.P. 1968-69 — 1977-78, Working Plans Circle, Nainital, U.P.

2

STATE OF TRIBAL AGRICULTURE IN NORTH-EAST BASTAR

H. S. GUPTA AND T. L. VERMA

THE TRIBALS, by tradition, are cultivators. But the tribal agriculture being still in a backward stage is a low productive enterprise. Therefore, the tribals are unable to meet their food requirement from agriculture. They used to supplement their food with the forest products, but the dwindling forest resources and prohibitive forest laws have deprived them of this support. The present paper reveals the poor state of tribal agriculture and exposes the weaknesses of the tribal development schemes which in spite of devoting millions of rupees have not been able to bring desired improvement in the tribal situation. The study discusses agricultural production in relation to the population and the quality and quantity of different inputs.

The Region

Two north-eastern Tahsils of Bastar district, viz. Kanker and Kondagaon constitute the region under study. The region extends over an area of 9132 km^2 from 19°12′ to 20°35′ N Lat. and 81°13′ to 82°6′ E long. Physiographically it is a plateau, but a precipitous scarp, named Keshkal Ghat, divides it into two parts. The northern part, i.e. Kanker tahsil has a lower elevation and is drained by the river Mahanadi. The southern part, i.e. Kondagaon tahsil is more elevated and dissected, and is drained by the tributaries of the river Indravati which itself is a tributary of the river Godavari. Non-tribal influences are less marked in this part than in the northern part of the region. The tribals numbering 3,45,920 constitute 64.9 per cent of the total population of the region which is 5,32,741. The share of tribals in the total population is 57.5 per cent in Kanker Tahsil and 70.1 per cent in Kondagaon Tahsil.

Data and Method

The study is based on primary data collected through a survey of tribal households of eight villages. The households sampled out for the study numbered 334 out of which 322 own land and 12 are landless. There are eight development blocks in the region — three in Kanker tahsil and five in Kondagaon tahsil. The villages were selected randomly in such a way that each block is represented by one village (Table 2.1). The data were collected by administering to each household a schedule which included questions pertaining to population structure of the family, land holding, soil types of land, nature and quantity of different agricultural inputs, and production of crops.

Quality of Land

The undulating terrain has restricted the net sown area in the region to 2,13,751 hectare which is only 39.3 per cent of the total area according to village papers and only 23.4 per cent of the total geographical area which includes the reserved and the protected forests also. The forests occupy 62.5 per cent of the total geographical area of the region. The agricultural land is not only limited in area but its productivity is also low due to undulating terrain and presence of gravels and pebbles in the soils.

TABLE 2.1: Location of sample households

Tahsil	Development Block	Village	No. of Sample House-holds
Kanker	Charama	Kilepar	49 (+6 landless)
	Kanker	Keotintola	50
	Narharpur	Kapaspoti	48 (+2 landless)
Kondagaon	Keshkal	Uparbedi	16
	Bare Rajpur	Lihagaon	55 (+3 landless)
	Farasgaon	Korrabargaon	51 (+1 landless)
	Makadi	Deoharduli	18
	Kondagaon	Chimdi	35
Total			322 (+12 landless)

According to the author's survey, the agricultural land of the sample households extends over 840 hectare. About 254 hectare, i.e 29 per cent of the total land, occupying positions on the plateau tops, have gravelly and lateritic soils locally known as Marhan. The Marhan land has very low fertility status and is cultivated with least care. It is more extensive in the Kondagaon tahsil where it occupies one-third of the total land owned by the sample households than in Kanker tahsil where about one-fifth of the total land belongs to this category. About 84 hectare of land, i.e. 10 per cent of the total, has sandy and gravelly soils on the slopes which are also very infertile.

The soils of low lying areas and valley bottoms are dark brown in colour and heavy in texture. These clay loan soils, locally known as Kanhar, are the most fertile soils of region. They are found in 196 hectare, i.e. 22.3 per cent of the total land. About 346 hectare land, i.e. 39.3 per cent of the total, occupies positions between Kanhar and Marhan and has sandy loam soils known as Matasi. Kanhar and Matasi are more extensive in Kanker than in Kondagaon tahsil. It may be noted that the Kanhar soils are used exclusively for paddy cultivation, the Matasi soils for paddy and to some extent for the millets, pulses and oilseeds, and the Marhan soils mainly used for the millets, pulses and oilseeds and to some extent for paddy. The average per hectare yield of paddy is 7.5 quintals in the Kanhar, 6.0 quintals in the Matasi and 2.7 quintals in the Marhan and sandy soils.

Size of Landholdings

The landholdings of 322 sample households are 840.34 hectare. Thus the average size of landholdings is 2.61 hectare. As Table 2.2 reveals large tribal farmers are few. Small to medium farmers constitute more than 75 per cent of the total farmers which indicate that disparity in the distribution of land among the tribal households is not as great as in the non-tribal regions of the state.

TABLE 2.2: Categories of tribal landholdings

Size category	No. of landholdings	Per cent of the total landholdings
Marginal (< 1 ha)	68	21.1
Small (1 - 2 ha)	77	23.9
Semi-medium (2 - 4 ha)	105	32.6
Medium (4 - 10 ha)	67	20.8
Large (> 10 ha)	5	1.6
Total	322	100.0

Cropping Pattern

The total cropped area of the sample households is 851.6 hectare. Paddy is the most important crop occupying 511.43 hectare, i.e 60.1 per cent of total cropped area (TCA). Its share in TCA is 72.4 per cent in the sample villages of Kanker tahsil and 52.1 per cent in those of Kondagaon tehsil. The millets, like Kodo, Kutki and Madia cover 20.2 per cent of TCA (15% in Kanker and 23.5% in Kondagaon). Maize is grown on 10.3 per cent of TCA in the sample villages of Kondagaon tehsil, but its cultivation is not common in Kanker tehsil, but its cultivation is not common in Kanker tehsil. Pulses occupy 7.4 per cent of TCA. Kulthi is the main pulse. Urad finds some place in Kondagaon tehsil, while Teora and Gram are grown in some area in Kanker tehsil. Oilseeds are grown over 5.5 per cent of TCA. Mustard and Ramtil are the main oilseeds of Kondagaon tehsil, while linseed is important in Kanker tehsil.

Input of Labour and Bullock Power

All the agricultural operations are performed by human labour and bullock power. Since paddy is the main crop and occupies the best agricultural land, it receives more attention of the cultivators and its production processes are more elaborate than those of other crops. As Table 2.3 indicates paddy cultivation in one hectare of land engages 660 man hours and 270 bullock hours in Kanker tehsil, and 529 man hours and 210 bullock hours in Kondagaon tehsil. As regards the cultivation of millets, the input of labour and bullock power amounts to 216 man hours and 150 bullock hours in Kanker, and 201 man hours and 140 bullock hours in Kondagaon tehsil.

There are three types of labour in the tribal region : (i) family labour, (ii) exchange labour, and (iii) wage labour. Family labour constitutes about 75 per cent of the total labour inputs of the tribal agriculture. At the time of weeding and harvesting, usually more labour is required than what is available in the family. To meet the shortage, exchange and wage labour is employed. By exchange labour we mean reciprocal barter of labour by households. In this system, the cultivator does not pay the labourer anywage either in cash or in kind, but he has to provide something to eat and drink. On such days pork and liquor are the most preferred items. With the loosening of traditional ties of reciprocity, the system of exchange labour is giving way to the employment of wage labour, but the latter is still only a small constituent (only about 5%) of the total labour input. Only medium and big farmers employ wage labour and that too only in paddy cultivation.

TABLE 2.3: Per hectare input of labour and animal power

Agricultural Operation		Paddy		Kodo/Kutki/Kulthi	
		Kanker	Kondagaon	Kanker	Kondagaon
1. Mamuring		24 man hrs. 24 bullock hrs.	—	—	—
2. Sowing		35 man hrs. 70 bullock hrs.	35 man hrs. 70 bullock hrs.	50 man hrs. 100 bullock hrs.	45 man hrs. 90 bullock hrs.
3. Levelling		60 man hrs.	—	—	—
4. Biasi (Ploughing)		25 man hrs.	25 man hrs.	—	—
5. Weeding		270 man hrs.	252 man hrs.		
6. Watering		15 man hrs.	8 man hrs.		
7. Harvesting		115 man hrs.	115 man hrs.		
8. Fetching	(A)	25 man hrs. 25 bullock hrs. or	—	150 man hrs.	140 man hrs.
	(B)	80 man hrs.	66 man hrs.		
9. Threshing and Winnowing		36 man hrs. 126 bullock hrs.	28 man hrs. 90 bullock hrs.	16 man hrs. 50 bullock hrs.	16 man hrs. 50 bullock hrs.
Total	(A)	605 man hrs. and 295 bullock hrs. or	529 man hrs. and 210 bullock hrs.	216 man hrs. and 150 bullock hrs.	201 man hrs. and 140 bullock hrs.
	(B)	660 man hrs. and 270 bullock hrs.	—	—	—

Seed

The tribals use only traditional seeds of the crops. The use of HYV seeds is negligible. During 1988-89 the HYV seeds of rice were used by some farmers in 7.4 hectare in Kilapar, 4.3 hectare in Keotintola and 3.6 hectare in Lihagaon. But this area of HYV seeds is only 3 per cent of the total area under paddy. In the rest of the villages there has been no use of HYV seeds.

Manure

Use of cowdung as manure in paddy fields is quite common in Kanker tehsil, where 645 cart load of cowdung manure was used in 140.12 hectare, i.e. 58 per cent of the total area under paddy in three villages of Kanker tehsil. As regards Kondagaon tehsil, the condition is more primitive and paddy is grown without use of manure. Only in Lihagaon and Deoharduli villages, manure was used in 49 hectare of paddy land. However, the tribals of Kondagaon tehsil use cowdung manure in the fields of maize which are mostly attached with their dwellings. In 1988-89, 580 cart load of manure was used in 84 hectare of maize fields and 49 hectare of paddy fields in Kondagaon tehsil.

Fertilisers

While there has never been any use of chemical fertilisers in Uparbedi, Deoharduli and Chimdi villages of Kondagaon tehsil, some tribal households have started the use of fertilisers in the rest five villages. The use of fertilisers started less than 5 years ago in the villages of Kondagaon tehsil and less than 10 years ago in those of Kanker tehsil. In all 120, i.e. 37.3 per cent, households used 129.2 quintals of fertilisers in 134.37 hectare, i.e. 15.8 per cent, of the total cropped area (Table 2.4). Urea and superphosphate are used in the paddy fields and urea in the maize fields. Ignorance, mainly among the tribals of Kondagaon tehsil, poverty and lack of irrigation are the main reasons of low use of fertilisers.

Irrigation

The tribal agriculture is totally rainfed. Undulating terrain and rocky surface have prevented the development of irrigation in the region. Only in Kanker tehsil, about 8 per cent of net sown area receives irrigation from canals and tanks. As regards the village under study, only 17.16 hectare of area under paddy was irrigated in village Kilepar. In the rest of the village no facility of irrigation is available.

TABLE 2.4: Use of chemical fertilisers

Village	Paddy				Maize		
	No. of house-holds	Area (ha)	Urea (Q.)	Super-phos-phate (Q.)	No. of house-holds	Area (ha)	Urea (Q.)
Kilepar	17	24.15	21.5	—	—	—	—
Keotintola	33	63.65	34.5	38.0	—	—	—
Kapaspoti	3	8.90	5.5	2.0	—	—	—
Lihagaon	20	18.41	7.8	—	9	2.47	3.0
Korrabargaon	—	—	—	—	47	16.79	16.9
Total	73	115.11	69.3	40.0	56	19.26	19.9

Production of Crops

Due to undulating terrain, poor soils, lack of irrigation, little use of fertilisers and low level of innovativeness of the tribals, the productivity of tribal agriculture is very low. Per hectare yield of rice in the sample villages ranges from 2.53 quintals in Uparbedi to 6.0 quintals in Kilepar, the average being 4.9 quintals for Kanker tehsil, 3.76 quintals for Kondagaon tehsil and 4.3 quintals for the total region. The yield rate of other crops is extremely low — 2.8 quintals per hectare in the case of maize and 2 or much less than 2 quintals in the case of other crops (Table 2.5).

Input-Output Relationship

The tribals spend a very small amount of cash on agriculture. Wage labour and fertilisers are the only inputs involving some expenditure on the production of paddy and maize. No cash expenditure is incurred on any other crop. Per hectare expenditure on the production of paddy ranges from a small amount of Rs. 14 in Uparbedi to Rs. 150 in Keotintola, and on the production of maize from Rs. 25 in Lihagaon to Rs. 110 in Korrabaragaon.

However, if the costs of family and exchange labour, bullock work, seed and manure are also added to the cash expenditure, the total cost of input becomes much higher. Examples of Keotintola (Kanker tehsil) and Uparbedi (Kondagaon tehsil) villages indicate that the total cost of inputs is greater than the value of yield in case of all the crops except oilseeds and maize. Table 2.6 depicts the cost of inputs involved at different stages of paddy cultivation. It reveals that the agriculture of Kondagaon tehsil being more primitive involve less inputs. Table 2.7 makes it clear that the tribal agriculture is an enterprise of loss. But since the cash expenditure is small, the tribals do not feel the loss.

TABLE 2.5: **Production and yield rate of crops**

Crop	Area (ha)	Production (Q)	Yield per (ha)	Crop	Area (ha)	production (Q)	Yield per ha (Q)
Cereals				*Pulses*			
Rice	511.43	2197.9	4.3	Kulthi	43.49	35.9	0.8
Kodo	62.70	104.2	1.7	Urad	14.75	19.4	1.3
Kutki	79.24	106.8	1.3	Teora	2.00	2.3	1.1
Madia	30.19	29.5	1.0	*Oilseeds*			
Maize	54.57	155.2	2.8	Linseed	10.85	8.1	0.7
Wheat	2.04	4.0	2.0	Mustard	23.22	32.4	1.4
Jowar	1.00	1.7	1.7	Ramtil	12.90	12.4	1.0

TABLE 2.6: **Per hectare cost of inputs of paddy cultivation**

Stages	Inputs			Keotintola (Rs.)	Uparbedi (Rs.)
1. Manuring	Labour	Bullock Power	Manure	115	Nil
2. Sowing	"	—	Seed	275	310
3. Levelling	"	—	—	50	Nil
4. Biasi	"	Bullock Power	—	90	100
5. Weeding	"	—	—	240	22(
6. Fertiliser	"	Fertilisers	—	100	Nil
7. Watering	"	—	—	10	Nil
8. Harvesting	"	—	—	105	100
9. Fetching	"	Bullock Power	—	80	25
10. Threshing & Winnowing	"	"	—	65	45
Total				1130	800

TABLE 2.7: Per hectare cost of inputs, actual expenditure and value of yield

Crop	Keotintola			Uparbedi		
	Cost of Inputs (Rs.)	Actual Expen-diture (Rs.)	Value of Yield (Rs.)	Cost of Inputs (Rs.)	Actual Expen-diture (Rs.)	Value of Yield (Rs.)
Paddy	1130	150	1070	800	14	600
Kodo/ Kutki	396	Nil	140	374	Nil	190
Madia	—	—	—	357	Nil	158
Maize	—	—	—	282	Nil	408
Kulthi	—	—	—	365	Nil	195
Teora	410	Nil	338	—	—	—
Linseed	254	Nil	466	—	—	—
Mustard	—	—	—	319	Nil	460
Ramtil	—	—	—	153	Nil	271

Carrying Capacity of Land and Sufficient of Food

The total production of crops by the sample tribal households amounts to 2197.9 qnt. of rice, 401.4 qnt. of other cereals, 64.1 qnt. of pulses and 52.8 qnt. of oilseeds. The quantities remaining after deducting the allowance of 12.5 per cent for seed, animal feed, wastage, etc. provide 8282.4 lakh calories. Thus on an average one hectare of cropped land produces crops having 9.726 lakh calories which can feed 1.11 adult units at the rate of 2400 calories per adult unit per day. The age and sex structure of the tribal population is such that one person is equal to 0.8547 adult unit. Hence, one hectare of cropped land can support 1.3 persons. The carrying capacity of land varies from 0.69 person per hectare in Uparbedi to 1.8 persons per hectare in Kilepar. On the whole the tribal land of Kondagaon tehsil has lower carrying capacity (1.158 persons per hectare) than that of Kanker tehsil (1.52 persons per hectare).

As Table 2.8 reveals, the actual pressure of population is 2.457 persons per hectare of cropped land which is 1.89 times the carrying capacity of land. Thus the tribal land can support only 52.9 per cent of the present tribal population. As regards the state of food sufficiency at the household level, only 44 households are able to produce surplus food, therest 278 households are deficit food producers of food and 12 households are landless.

TABLE 2.8: Carrying capacity of land and food sufficiency

Village	No. of House-holds incl. land-less	Popula-tion	No.of Adult Units	Cropp-ed Land (Ha)	Available Food (12.5% deducted from Production					Per Ha Kcalo-ries (lakh)	Per Ha Carry-ing capa-city		Actual Press-ure person per Ha	Surplus produc-er House-holds	Deficit Produ-cer & landless House holds
					Rice (Q)	Other (Q)	Pulses (Q)	Oil-Seeds (Q)	calo-ries (lakh)		AU.	pers.			
1. Kilepar	55	299	262	102.7	381.8	23.0	6.5	—	1420.5	13.83	1.597	1.802	2.911	13	42
2. Keotintola	50	334	291	114.7	371.6	11.8	3.3	6.9	1373.6	11.98	1.368	1.570	2.912	8	42
3. Kapaspoti	50	344	304	115.6	279.4	25.4	12.3	0.1	1094.0	9.46	1.080	1.222	2.976	3	47
4. Uparbedi	16	75	65	59.8	39.8	33.7	1.2	10.8	313.3	5.24	0.598	0.690	1.254	2	14
5. Lihagaon	58	383	321	174.9	301.0	114.0	10.2	4.8	1488.9	8.51	0.971	1.559	2.190	7	51
6. Korrabar-gaon	52	328	270	108.9	260.8	73.8	3.2	—	1164.5	10.69	1.220	1.482	3.012	4	48
7. Deoharduli	18	119	100	47.3	89.3	20.1	10.6	0.7	416.3	8.80	1.005	1.196	2.516	3	15
8. Chimdi	35	210	175	127.7	199.5	49.4	8.8	22.9	1011.3	7.92	0.904	1.085	1.644	4	31
Total Region	344	2092	1788	851.6	1923.2	351.3	56.1	46.2	8282.4	9.37	1.111	1.300	2.457	44	290

Conclusions

The tribal agriculture is an extremely low productive enterprise. The factors responsible for this state of affair are physical and human both. No doubt, the difficult physical conditions, such as undulating topography and infertile soils, restrict the scope of agricultural development, but no sincere effort has been made either to improve the situation even what is possible. Irrigation is still negligible. Although it is not possible to launch any major irrigation scheme, yet there is much scope of undertaking small scale measures of irrigation development throughout the region. The tribals have very low level of innovativeness. Therefore, it is essential to make efforts with missionary zeal to motivate the tribals to adopt agricultural innovations and to provide them with necessary facilities.

However, it seems unlikely that the pace of development will be so fast that the tribal agriculture is able to make up the present shortage and then cope with the growing numbers in meeting the food requirement of the tribal population. Therefore, an integrated development is called for the improvement of the tribal situation.

3

AGRICULTURAL PANORAMA OF UTTARAKHAND

B.R. PANT, R.C. JOSHI AND D.S. JALAL

BESIDES THE threatening problems of soil erosion, indiscriminate cutting of forests, over-grazing of pastire lands, merciless lopping of trees and mismanagement of water and land resources, Uttarakhand is facing the problem of malnutrition and related diseases which are posing a severe threat to the local population. Unscientific use of land, consumption of less nutritive foodstuffs and ignorance about the nutritional importance for health are responsible to accelerate this fatal problem.

Per capita daily consumption of food is far below the optimum requirement in mountaineous regions of developing countries. To overcome this problem, the landuse pattern should be changed according to the nutritional requirement of the society. The nutritional study in present context particularly to the regions where population increases day by day, is a prime necessity for understanding the present nutritional requirements and the potentiality of their upliftment from easily available resources.

The nutritional status of the people in Uttarakhand vary from one place to another according to physical environment and consequent cropping pattern. Most of the Himalayan people do not get sufficient amount of food but the diet is deficient in nutritive elements of protein, fat, minerals and vitamins. In the absence of adequate nutritive elements, people suffer from several deficiency diseases. Present study attempts to describe the agrarian scene of Uttarakhand and finding out the nutritional status and related diseases in the rural people.

The Geographical Profile of Uttarakhand

Uttarakhand consists of eight districts, namely, Nainital, Almora and Pithoragarh of Kumaun and Chamoli, Uttarkashi, Tehri, Pauri and Dehradun of Garhwal Divisions of Uttar Pradesh. Physiographically, it exhibits in four longitudinal sub-divisions of Himalayas namely, the Outer Himalaya, the Lesser Himalaya, the Great Himalaya and the Trans Himalaya which are separated by several thrusts and faults (Pant et. al. 1990). Lying between 28° 43′ 24" and 31° 27′ 30" N. lat. and 77° 34′ 27" and 80° 7′ 22" E long. It encompasses an area of 54,18,108 ha (54,181 km^2) and a population of 48,35,712 (Census 1981). It accounts for 18.15 per cent of the state's land area and 4.36 per cent of its population.

Uttarakhand is a natural unit bounded by Kali river in the east, Tibet in the north, Himachal Pradesh in the West and fluvial Gangetic plain in the south. In general the altitude ranges below 300m to more than 7000m. It is mainly drained by Yumuna, Ganga, Western Ramganga and Kali or Sarda rivers. These generally take their rise in the snowfed mountains and provide huge volume of water for drinking, irrigation and power generation. A number of hydroelectric power plants are constructed on these rivers. Variations in altitude, topography, slope and vegetation has offered various micro-climatic features in Uttarakhand. There is tropical climate in lower elevations (below 600m) and sub-tropical climate in between 600 and 1200m. As one moves to uplands upto 1800m, the climatic become cool to cold temperate. Above 2400m in elevation, the climate is almost alpine to polar.

The complex environment with its dramatic differences in altitude and climate, supports a great variety of forest eco-system. Below 1200m in elevation there lie tropical and sub-tropical broad leave forests. Between 1000 and 2000m it manifests in pine forests. Moist temperate evergreen broad leave and mixed conifers forests are formed from 1500 to 3500m. Sub-alpine forests occur around 3500m under these circumstances. The area supports livestock based farming system. The system is responsive to lead geo-ecological conditions. The majority of cultivation is rainfed (69.19%) producing three crops in two years while irrigated area (30.89%) yields three harvests a year. It is an important area for growing temperate fruits and off season vegetables. Its forest resources supplies timber, firewood, medicinal plants and other minor products. Its abundant water resources has potential for productive use for hills as well as for plains.

Seven soil groups — conglomeratic, loamous, stony hilly, crushed slaty, hilly sandy, regolithic and podzolic soils (Joshi, 1987 and Pant, 1988) have been identifies in Kotadun area of Uttarakhand. The foot hill soils of Dehradun are loam to clayee loam in texture and acidic in nature. Four broad soil groups — redloams, brown forest, podzolic and meadow soil (Ghildyal, 1981) are identified in Ranikhet area.

Out of the total population of 48,35,721 of Uttarakhand 11,36,523 are in Nainital, 7,57,373 in Almora, 4,89,267 in Pithoragarh, 3,64,346 in Chamoli, 1,90,267 in Uttarkashi, 4,97,710 in Tehri, 6,37,877 in Pauri and 7,61,668 in Dehradun districts.

Data — Base and Methodology

An attempt has been made here to analyse the landuse, cropping pattern, per capita land, occupational diversification, holding size etc. of Uttarakhand. It is based on secondary source of data. The basic aim of this study is to find out the nutritional status and related deficiency diseases of rural people which is based on primary information collected from a few sample villages. Keeping in mind the environmental diversities, the area is divided into three homogeneous zones with respect to elevation, slope, landuse, cropping pattern, animal husbandry, crop production, institutional infrastructure etc. Selecting one representative village from each zone of below 800m, 800–1800m and above 1800m altitudes, three case studies have been made. Patkot represents the zone below 800m elevation or valley (dun), Amgarhi represents 800–1800m elevation or mid slope zone and Sigarhi represents above 1800m elevation or upland zone. These sample villages are purposely selected as they fairly represent the nutritional status and related diseases of their respective physical strata. Average Food Balance Sheet (FBS) of these villages are prepared on the basis of per acre yields and external supplies of various foodstuffs. Each nutritive element is converted into per head daily share and its percentage departure is compared with the optimum requirement (Gopalan et al, 1985). An attempt is made to correlate the deficiency of a particular food constituent and the prevailing diseases in the people.

Patterns of Agriculture

The economy of Uttarakhand depends on agriculture which engages about 70 per cent population of the region. But there is scarcity of agricultural land and the share of per capita land is far below from the standard requirement. Only 0.18 ha land goes to the per head share of the cultivated land in Uttarakhand while Madhav Ashish (1983) estimated it about 0.20 ha as the optimum requirement in this region. The population pressure on the cultivated, forest and pasture land is considerably high. About 80 per cent of fodder and 100 per cent of fuel requirement is fulfilled from the forest by the inhabitants of some village in this region (Singh et al, 1984).

Irrigation is the most significant constraint to determine the use of land and cropping choice. Table 3.1 indicates the district-wise extent of irrigation in Uttarakhand. It shows that 30.89 per cent of net area sown is irrigated in the region. The extent of irrigated area is 69.06 per cent of the total area in Nainital district as against 75 per cent irrigated area in the district of plains — Tarai and Bhabar. It is followed by Dehradun district (10.02%). There are numerous streams 'gad' 'gadhera' in the region, but very little use of them is made for irrigation because of diverse topographic features and high costs of lift irrigation schemes on the one hand, and silting and sliding of the gules constructed by villagers on the other (Pant et al., 1989).

TABLE 3.1: Irrigated area — 1984-85

Districts	Net Irrigated area (ha)	Irrigated area as percentage of net area sown	Percentage of the total irrigated area
Nainital	1,49,982	73.17	69.06
Almora	11,828	11.02	5.45
Pithoragarh	6,678	7.52	3.07
Chamoli	2,758	6.46	1.27
Uttarkashi	4,328	13.68	1.99
Tehri Garhwal	11,068	15.02	5.10
Pauri Garhwal	8,780	8.88	4.04
Dehradun	21,753	39.51	10.02
Uttarakhand	2,17,175	30.89	100.00

The size of holding and the field in the region is very small. Table 3.2 shows that 53 per cent holdings are less than one hectare in size and cover only 10.2 per cent of the cultivated area. In all about 93 per cent holdings are in below 3 ha size and account 66.8 per cent of the cultivated area. The study reflects that small and marginal holdings are largest in number in Uttarakhand.

TABLE 3.2: Operational holdings of various size groups — 1976-77

Size group (ha)	Number	Operational Holdings		Percentage
		Percentage	Area (ha)	
Less than 0.5	3,93,544	53.3	71,942	10.2
0.5 – 1.0	1,33,852	18.2	96,321	13.6
1.0 – 2.0	1,15,195	15.6	1,59,226	22.5
2.0 – 3.0	45,444	6.2	1,08,869	15.5
3.0 – 4.0	20,868	2.8	69,311	9.8
4.0 – 5.0	11,211	1.5	48,247	6.8
5.0 – 10.0	14,741	2.0	96,513	13.6
10.0 – Above	3,058	0.4	5,783	8.00
Uttarakhand	7,37,913	100.00	7,08,262	100.00

However, reporting area of Uttarakhand is 54,18,108 ha of which 63.76 per cent area is under forest, 5.22 per cent is not suitable for cultivation, 2.28 per cent is area under other than cultivation, 12.97 per cent is net area sown. Remaining 5.88 per cent is culturable waste, 5.03 per cent permanent pasture and other grazing land, 8.84 per cent is under trees, bushes, gardens excluding cultivation area, 0.22 per cent under current fallow and 0.82 per cent under other fallows (Table 3.3). It is evident that the maximum area is classed as forest but it is not under full tree canopy, since it is the area owned by State Forest Department. A study conducted by Singh and his colleague (1984) indicates that only 28.71 per cent area of the Indian Central Himalaya is now forested and only 4.4 per cent of the area has a forest cover with greater than 60 per cent crown density while 60 per cent forest cover the minimum requirement for environmental balance in the Himalaya. The net area sown is only 3.95 per cent in Uttarkashi district, while it is 29.36 per cent in Nainital district. In this respect Dehradun stands second (17.45%). Remaining districts possess below 15 per cent cultivated area.

TABLE 3.3: Major landuse — 1984-85 (percentage of reporting area)

Landuse class	Nainital	Almora	Pithoragarh	Chamoli	Uttarkashi	Tehri	Pauri	Dehradun	Uttarakhand
1. Forest	57.69	54.52	50.39	58.18	88.69	68.97	61.31	69.76	63.44
2. Usar and not suitable for cultivation	0.67	4.11	3.85	19.81	2.24	1.88	3.93	0.56	5.52
3. Area under other uses excluding cultivation	4.41	2.33	2.17	1.95	0.51	1.30	2.22	5.12	2.28
4. Culturable Waste land	4.39	8.69	9.34	4.07	1.52	11.70	4.87	3.60	5.88
5. Permanent Pasture and other grazing land	0.20	8.09	12.38	6.86	1.79	2.28	5.59	0.02	5.03
6. Area under trees, bushes, gardens, excluding cultivated area	1.99	6.52	6.74	4.23	0.94	0.04	7.02	1.40	3.84
7. Current fallow	0.52	0.09	0.15	0.01	0.04	0.16	0.24	1.13	0.22
8. Other Fallow	0.77	0.82	1.43	0.18	0.32	0.88	1.52	0.96	0.82
9. Net cultivated area	29.36	14.83	13.55	4.71	3.95	12.79	13.30	17.45	12.97

Source: Agricultural Data of Uttar Pradesh, Director, Agricultural Statistics and Crop Insurance, Krishi Bhawan, Uttar Pradesh, March 1987.

Cropping Pattern

Table 3.4 shows the districtwise cropping pattern. Paddy and madua are two major Kharif foodgrains, each respectively accounting for 41.43 and 22.95 per cent of the cultivated area. Net 17.1 per cent area is devoted to other uses. Corn and sawa are other millets. Urad is the main pulse in this region. Madua is usually intercropped with legumes-bhat (black soyabean), gahat and urd (black gram).

In the rabi season wheat is the major foodgrain grown in 81.38 per cent of the net cultivated area. The second crop of this season is barley accounting for 7.27 per cent of the cultivated area. Only 6.86 per cent area is under other uses. It is evident from Table 3.4. that the cropping pattern of all the districts of Uttarakhand is almost similar.

Food and Nutrition

Dietary systems of the villagers are almost similar in Uttarakhand. People take food four times a gastronomic day. Between the breakfast in the morning and tiffin in the evening, people take heavy lunch because they engage themselves in strainuous agricultural or other activities as ploughing, manuring, lumbering, weeding, harvesting, threshing, collection of fuel and fodder. Other principal meal is taken in early night (dinner). Overwhelming proportion of the rural people take two meals only except during agricultural peak seasons.

It is clear from Table 3.5 that more than 50 per cent of the energy is obtained from cereal and pulses, while about 20 per cent is taken from milk, meat and sweets. Remaining energy is gained from spices, vegetables, hydrogenetic oil, ghee and fruits. Table 3.6 shows the average per head per day annual food intake for 1986-87. Table 3.7 indicates that about eight elements i.e. fat, protein, carbohydrates, calcium, iron, carotene, riboflavin and ascorbic acid fall short from the optimum requirement while the proportion of phosphorus, thiamine and niacin are sufficient in the diet of people (Pant, 1988).

Owing to the shortage of nutrition, several deficiency diseases occur in the people. Approximately twenty diseases are observed in the sample villages (Table 3.7). The deficiency of a particular nutrient causes to generate a specific nutritional deficiency diseases in the people. Nutrition intake is high in dun floor or valley villages which infrastructurally are developed than the mid slope and upland villages.

TABLE 3.4: Cropping pattern, 1984-85

(percentage of cultivated area)

Crop	Nainital	Almora	Pithoragarh	Chamoli	Uttarkashi	Tehri	Pauri	Dehradun	Uttarakhand
*Rice	60.39	35.05	47.54	41.28	32.50	22.97	27.41	30.79	41.43
*Corn	5.82	2.12	6.03	0.69	2.40	2.51	2.58	26.53	5.46
*Madua	2.57	37.15	30.32	34.70	24.70	29.57	37.66	8.45	22.95
+Wheat	81.61	82.89	76.28	88.92	84.85	86.08	81.48	76.55	81.38
+Barley	1.16	10.38	12.05	6.83	4.63	9.45	14.59	6.34	7.27
+Lentil	3.18	3.43	8.50	0.32	2.77	1.88	1.26	2.72	3.45

* Percentage to kharif cropped land.
\+ Percentage to rabi cropped land.

TABLE 3.5: Actual per head per day food intake — 1986-87

Food Stuff	Sample villages								
	Dunfloor (Patkot)			Mid Slope (Amgarhi)			Uplands (Sigarhi)		
	kg/year	gm/day	Kcal/day	kg/year	gm/day	Kcal/day	kg/year	gm/day	Kcal/day
Rice	57.79	158.32	552.54	60.51	165.78	587.57	52.38	143.50	500.80
Wheat (Flour)	49.50	135.62	462.42	48.47	132.81	452.89	52.40	143.55	489.51
Madua (Flour)	22.78	62.41	204.70	27.93	76.53	251.03	14.33	39.27	128.81
Barley (Flour)	—	—	—	5.26	14.41	48.43	5.25	14.39	48.30
Gram	1.71	4.68	16.85	4.59	12.57	45.23	—	—	—
Corn	—	—	—	—	—	—	10.76	29.52	101.00
Lentil	3.95	10.82	37.11	1.22	3.35	11.56	—	—	—
Other pulses	6.57	18.00	53.82	9.20	25.20	75.35	18.81	51.54	154.11
Soyabean	1.08	2.96	12.78	—	—	—	—	—	—
Vegetable	37.10	101.64	381.15	34.82	95.39	357.71	17.70	48.48	24.00
Ginger	1.23	3.36	2.25	—	—	—	—	—	—
Spices	1.48	4.05	19.27	1.34	3.68	17.52	1.07	2.93	13.95
Lahi	6.75	18.49	166.44	7.46	20.45	184.05	3.11	8.52	76.91
Peas	—	—	—	—	—	—	5.22	14.30	48.62
Potato	—	—	—	—	—	—	17.45	47.81	46.38
Milk Product	54.33	148.85	174.15	50.05	137.12	160.43	36.92	101.16	118.36
Flesh foods	8.65	23.69	35.98	5.26	15.24	23.01	8.13	22.28	33.64
Fruits	26.89	73.67	81.80	13.09	25.85	39.79	6.35	17.39	10.26
Sweets	19.48	53.37	212.40	13.49	36.96	147.10	12.00	32.68	130.80
Total	—	—	2413.66	—	—	2392.67	—	—	1925.61

TABLE 3.6: Average per head per day supply, 1986-89

Elements of diet	Sample Villages									
	Optimum Dunfloor (Patkot)				Midslope (Amgarhi)			Uplands (Sigarhi)		
	Optimum requirement	Actual intake	—Deficiency + Surplus	Percentage Departure	Actual intake	—Deficiency + Surplus	Percentage Departure	Actual intake	—Deficiency + Surplus	Percentage Departure
Energy Kcal	2430.0	2413.66	—16.34	—0.67	2392.67	—37.33	—1.54	1925.61	—504.39	—20.67
Protein gm	66.0	57.94	—8.06	—17.46	58.12	—1.88	—3.13	63.95	—2.05	—3.11
Fat gm	50.0	41.27	—8.73	—11.76	39.22	—10.78	—21.56	24.51	—25.49	—50.98
Carbohydrate gm	430.0	379.46	—50.54	—11.75	382.40	—47.60	—11.07	364.49	—65.51	—15.23
Calcium mg	800.0	689.54	—110.46	—13.81	722.12	—77.88	—9.74	533.44	—266.56	—33.32
Phosphorus mg	1400.0	1556.18	+156.18	+11.16	1607.78	—207.78	+14.84	1654.18	+254.18	+18.16
Iron mg	40.0	28.63	—11.37	—28.43	29.43	—10.57	—26.43	27.74	—12.26	—30.65
Carotene μg	960.0	900.51	—59.49	—6.20	625.54	—334.46	—34.84	263.63	—696.17	—72.52
Thiamine mg	1.8	1.98	+0.18	+10.00	1.87	+0.07	+3.89	1.96	+0.16	+8.89
Riboflobin mg	1.5	0.85	—0.65	—43.33	0.86	—0.64	—42.67	0.78	—0.72	—48.00
Niacin mg	16.0	21.74	+5.74	+35.88	16.50	+0.50	+3.13	16.15	+0.15	+0.94
Ascorbic acid mg	50.0	30.90	—19.10	—38.20	23.71	—26.29	—52.58	32.63	—17.37	—39.74

TABLE 3.7: Nutritional deficiency diseases in the sample villages — 1986-87. (in % total patients)

Sample village	Total population	Total patients	Percentage of patients to the total population	Kwashiorkor	Hypoprotonamia	Dental Decay	Osteomalacia	Rickets	Haemophilia	Diabetes
1	2	3	4	5	6	7	8	9	10	11
Dun Floor (Patkot)	1127	96	8.52	3 (3.13)	2 (2.08)	5 (5.21)	4 (4.17)	10 (10.42)	2 (2.08)	4 (4.17)
Mid slope (Amgarhi)	326	37	11.35	1 (2.70)	2 (5.41)	3 (8.11)	1 (2.70)	5 (13.98)	1 (2.70)	2 (5.41)
Uplands (Sigarhi)	117	30	25.64	—	—	3 (10.10)	—	7 (23.32)	2 (6.67)	2 (6.67)
Total	1570	163	10.38	4 (2.45)	4 (2.45)	11 (6.75)	5 (3.07)	22 (13.50)	5 (3.07)	8 (4.91)

(Contd.)

(Contd.)

Anaemia	Night blindness	Xerophat-halmia	Kerotom-alacia	Pyorrhaea	Scurvy	Diarrhoea	Angular-stomatics	Dyspep-sia	Consti-pation	Pallegra	Goitre	Glossitis
12	13	14	15	16	17	18	19	20	21	22	23	24
12 (12.05)	1 (1.04)	4 (1.04)	1 (6.25)	6 (6.25)	6 (6.25)	2 (2.08)	10 (8.32)	8 (8.32)	7 (7.29)	1 (1.04)	3 (3.13)	5 (5.21)
2 (5.41)	—	3 (8.11)	—	3 (8.11)	2 (5.41)	2 (5.41)	2 (5.41)	3 (8.11)	2 (5.41)	—	2 (5.41)	1 (2.70)
5 (16.67	1 (3.33)	1 (3.33)	—	2 (6.67)	2 (6.67)	— —	— —	2 (6.67)	2 (6.67)	—	— —	1 (3.33)
19 (11.67)	2 (1.23)	8 (4.91)	1 (0.61)	11 (6.75)	10 (6.13)	4 (2.45)	12 (7.36)	13 (7.97)	11 (6.75)	1 (0.61)	5 (3.07)	87 (4.29)

Notes: Figure in Parentheses, refer to percentage.

Table 3.8 illustrates the mortality and morbidity rates in the villages. It is evident from it that dun floor or valley village shows highest caloric intake while mid slope and upland villages indicate a decreasing trend. Variation in production and consumption results in prevalence of various deficiency diseases. It is worth to mention that the mortality and morbidity scores are maximum in low caloric intake village (Sigarhi) while maximum caloric intake village (Patkot) reflects minimum mortality and morbidity number (Pant, 1989). It is because that Patkot is situated in favourable environmental conditions — level land, irrigation facility, infrastructurally developed, high yield etc. The supply of energy decreases while the percentage of patients increases from lower to higher elevations. As a result of this deficiency, 163 persons (10.38%) of the surveyed population suffer from various deficiency diseases (Table 3.7).

TABLE 3.8: Mortality and morbidity, 1986-87

Sample Village	Energy perhead per day (Kcal.)	Percentage of morbidity to the total population	Percentage of mortality to the total patients
Dunfloor (Patkot)	2413.66	8.52	10.42
Mid Slope (Amgarhi)	2392.67	11.35	13.51
Uplands (Sigarhi)	1925.61	25.64	20.00

Recommendations

On the basis of these observations following recommendations are made for quantitative and qualitative enhancement of food supplies in Uttarakhand :

1. Cropping pattern should be changed according to the needs of the society and capability of the land.
2. Total cultivated level should be brought under double cropping pattern particularly in favourable topographic areas.
3. Irrigational facilities should be provided in mid slope and upland areas.
4. Villagers should be educated to consume leafy and root vegetables, pulses and millets regularly.

5. New agricultural techniques particularly use of HYV seeds, chemical fertilisers, insecticides, pesticides, ploughing means, reclamation of wastelands, land consolidation and soil conservation programme should be introduced in the region.

On the basis of above measures the economic and nutritional status of the rural people of Uttarakhand may be improved to a certain extent.

REFERENCES

Anonymous, 1980 : The Sixth Five Year Plan, U.P. Hill Region Govt. of U.P., Lucknow, pp. 62-63.

———, 1981 : *Census of India*, Uttar Pradesh, Series 22 Primary Census Abstract, pp. 28-55.

———, 1987 : *Agricultural Data of Uttar Pradesh,* Director Agricultural Statistics and Crop Insurance, Krishi Bhawan, Uttar Pradesh, Lucknow, p. 165.

Ashish, M., 1983 : Agricultural Economy of Kumaun Hills : Threat of Ecological Disaster, in O. P. Singh (ed) *The Himalaya : Man and Nature,* Rajesh Publication, New Delhi, pp. 184-97.

Ghildyal, B. P., 1981 : Soils of the Garhwal and Kumaun Himalaya, in J. S. Lall and A. D. Moddie (eds) *The Himalaya : Aspects of Change,* Oxford University Press, Delhi, pp. 120-37.

Gopalan, C., B. V. Ramashastri and S. C. Balasubramaniam, 1985 : *Nutritive Value of Indian Foods,* Indian Council of Medical Research, Hyderabad, p. 204.

Joshi, R. C., 1987 : Soils of Kotadun, in D. S. Jalal (ed.), *Optimal Landuse for Environmental Restoration in Kotadun, Kumaun Himalaya,* Technical Report II, submitted to the Deptt. of Environment, Govt. of India, New Delhi, pp. 11-20. (mimeo).

Pant, B. R., 1987 : Land Capability Classification in D. S. Jalal (ed.), *Optimal Landuse for Environmental Restoration in Kotadun, Kumaun Himalaya,* Technical Report — II, submitted to the Deptt. of Environment, Govt. of India, New Delhi : 28-37, (mimeo).

———, 1988 : Land Utilisation in Kotadun, Kumaun Himalaya, (Unpublished Ph.D. Thesis, Kumaun University), Nainital, p. 270.

———, 1989 : Food, Nutrition and Their Impact in D. S. Jalal (ed.), '*Optimal Landuse For Environmental Restoration in Kotadun, Kumaun Himalaya,* Final Technical Report submitted to the Deptt. of Environment, Govt. of India, New Delhi, pp. 190-210, (mimeo).

Pant, B. R., R. C. Joshi and D. S. Jalal, 1988 : Land Capability Classification Model for Kotadun, Kumaun Himalaya, *Geographical Review of India,* Vol. 50(1), pp. 47-52.

———, 1989 : Water Management in Kotadun, Kumaun Himalaya in R.D. Khulbe (ed) *Perspectives in Aquatic Biology,* Papyrus Publishing House, New Delhi, pp. 509-14.

———, 1990 : Cropping Pattern and its Impact on Nutrition and Health in Kumaun Himalaya with a special reference to Kotadun, M. P. Joshi, A. C. Fangar and C. W. Brown (eds), *Himalaya : Past and Present,* Shree Almora Book Depot, Almora, pp. 365-86.

Singh, J. S., Uma Pandey and A. K. Tiwari, 1984 : Man and Forest : A Central Himalayan Case Study, *Ambio* : 13 (2), pp. 80-87.

4

DIVERSIFICATION STRATEGIES IN AGRICULTURE IN THE RURAL-URBAN FRINGE

CHRISTOPHER R. BRYANT, PIERRE DESLAURIERS AND CLAUDE MAROIS

AGRICULTURAL STRUCTURE and change in the rural-urban fringe has been recognised increasingly as complex phenomena during the 1970s and 1980s. The stereotyped view of agriculture in the rural-urban fringe as unambiguously deteriorating because of urban pressures has been giving way to a perspective that recognises the role of factors unrelated to urban development (Munton, 1974; Bryant, 1976, 1984, 1986), the multiple ways in which they combine with urban development forces and the role of the individual farmer as a mediating agent between the different factors influencing agricultural change.

This changing perspective has been parallel and supported by the development of an increasingly more complex conceptual framework which has helped bring order to what otherwise appears to be a bewildering array of situations. This conceptual framework has recognised the existence of metropolitan (urban) and non-metropolitan (non-urban) forces affecting agriculture (Bryant et al., 1982), the different ways in which these forces combine to produce different patterns of evolution of farming in the rural-urban fringe (Bryant, 1976, 1984; Bryant et al., 1984; Munton et al., 1988) and the role of the individual farmer in integrating and developing adaptive and innovating strategies on the farm (Bryant 1981, 1986; Ilberry, 1988; Johnston, 1989).

The complexity of patterns of change in farming in the rural-urban fringe was suggested by Bryant's (1984) framework, in which three types of agricultural change were outlined and which represented different types of interaction between the factors and forces of change in farming as well as different patterns of decisions taken at the farm level. These three broad types of farming landscape change comprise:

1. *Degeneration of the farm structure*
   Here, the various forces impinging on farming combine to create conditions that thwart agricultural development. Frequently, urban development forces with a negative impact on farming, such as speculative pressures and land fragmentation, are reinforced by other negative forces such as poor market conditions for the local enterprises and poor quality land. The negative influences outweight any positive urban influences, such as the market factor, although it is still possible to find the occasional farmer who has managed to survive and adapt successfully to the overall negative conditions.

2. *Agricultural adaptation and innovation*
   In these situations, there are still forces with negative impacts on farming, e.g. some uncertainty because of possible urban development or scattered non-farm development. However, they are not sufficient to counteract the positive influences of a near urban environment, e.g. direct marketing possibilities to the nearby non-farm population, and other positive influences contribute to this, e.g. a sound farming structure, a fertile environment and favourable market conditions. Under these conditions, some farmers will be able to cope relatively easily with the negative conditions present [adopting a *management* mode of operation (Bryant, 1989)] while others are able to develop more innovative approaches and capitalise on the advantages of a near urban location [adopting, therefore, a more *entrepreneurial* mode of operation (Bryant, 1989)].

3. *Continued agricultural development along "traditional" lines*
   This situation recognises that some form of agricultural change has been occurring in near urban zones unrelated to urban and non-farm development pressures. Technological change in farming, competitive forces in the international market place and the pull of non-farm employment opportunities in metropolitan zones are among the powerful forces that have led to substantial changes in farm structure throughout the Western world. Farm consolidation, mechanisation and farm

specialisation are some of the indicators of the changes that have occurred (Wong, 1983), frequently labelled collectively as "agricultural modernisation" or "agricultural progress". Where, for instance, conditions for urban or non-farm development are lacking, where opportunities for innovative marketing of farm produce are small (or perceived to be small), and where a good physical environment for agricultural exists, we can expect such areas to be characterised by such "traditional" or "normal" agricultural development.

It should be emphasised immediately, however, that these three broad types of agricultural landscape change are not to be equated with homogeneous patterns. The key in the way this conceptual framework has been developed is the individual farm decision-taking unit (the farmer and farm family). The management and entrepreneurial capabilities of this decision-taking unit are of major importance in understanding how individual farms evolve, as are the particular influences that affect decision-taking at the individual farm level, e.g. family wealth, farm family size and children's career intentions. While the broad parameters of the farm's decision-taking environment may vary from one major zone (broadly defined) to another within the rural-urban fringe as the result, for instance, of variation in non-farm development pressure and soil capabilities, there is room for variation in farm-level responses within any broad zone as Bryant (1981) has already argued in an earlier paper. What this suggests that the three broad zones or types of situation (agricultural degeneration, agricultural adaptation and innovation, and agricultural development) are, in reality, comprised of farms which could be typified as farms experiencing degeneration, those engaging in adaptation and innovation and those following the more general lines of traditional agricultural development. The difference between broad zones is, therefore, in the relative mix of these farm-level types, rather than in terms of homogenous landuse patterns and tendencies. In fact, each of the three broad zones (and remember, the threefold categorisation is only a schematised representation; the empirical reality could likely reveal a greater number of types of situation, depending upon the number of discriminatory variables selected) is in itself a *mosaic* of types of agricultural change.

Considerable research remains to be done, both on the types of agricultural change and responses and on the mechanisms and paper processes by which farm-level change occurs. This paper is aimed at expanding the discussion of the types of agricultural change involved in adaptive and innovative modes of response in the rural-urban fringe, drawing upon recent research undertaken in the Montreal region, Canada. The main focus is on diversification versus specialisation in the farm enterprise as a means of adaptation to the special market conditions frequently found in metropolitan regions as well as to the additional pressures encountered there. First, some brief comments are offered on the concept of diversification; second, the study area and methodology are reviewed; third, the general characteristics of farm structure and change in the study are described; fourth, diversification strategies in the sample are discussed for the period 1981 to 1988; and finally, some concluding comments are offered on future research directions.

Diversification

Given that the basic decision-taking unit of interest to us is the farm family, diversification at the farm family level can be defined as a decrease in the dependency of the farm family upon a particular set of income sources. Diversification generally involves either:

1. an increase in the number of enterprises (products and services) produced on or off the farm with the resources accessible to the farm family, and/or
2. an increase in the number of different methods of marketing one or more products or services produced with those resources.

Diversification, therefore, involves both enterprise allocation decision as well as marketing decisions. Marketing decisions involving the decision to sell a particular product through more than one marketing channel (e.g. fresh vegetables to a processing plant, wholesalers or an a pick-your-own basis) are as much decisions to diversify as are decisions to increase the number of products or services produced because different marketing channels can be associated with different market segments which can behave quite differently from each other.

The definition offered above the diversification includes the introduction of unconventional enterprises and marketing methods as well as the "traditional" enterprises and marketing methods. It, therefore, differs from Gilg's (1990) and Ilberry's (1990) definitions which focused exclusively on non-traditional enterprises. The notion of diversification used here covers a net increase in the number of "traditional" enterprises, non-traditional enterprises (e.g.

"horticulture") and marketing channels (e.g. various forms of direct marketing of farm produce), the addition of non-farm enterprises on the farm that use redundant or underused farm resources (e.g. farm vacations, using farm building for long term storage, camp-grounds), adding value to farm produce through processing and packaging on the farm, and even, given the focus on the farm family, the development of off-farm work for farm family labour.

The types of diversification considered in the empirical analysis below are only a subset of these types, and comprise:

(i) Diversification of agricultural enterprises (crops and livestock);
(ii) Diversification of marketing channels and the addition of non-agricultural enterprises (horse boarding, recreation, storage space); and
(iii) Development of off-farm work by the farm operator.

The measures of diversification used in the analysis below are crude (e.g. a net increase of 1, 2 or more enterprises). It is worth noting however that additional methods of measuring diversification must be investigated in subsequent research in order to measure reduction in dependency more directly (this would have to take account of the relative importance of each enterprise, not just the number of enterprises [Bryant (1970)].

Diversification of income sources is, of course, not the only possible outcome. The farmer and his/her family must evaluate their resources and opportunities in arriving at a decision to change. While diversification carries advantages in terms of income dependency, it may also have certain disadvantages. Indeed, specialisation on the farm has been one of the dominant trends within farming over the last 40 years; this specialisation was a product of farm consolidation, increasing size and specialisation of farm technology and the pressure to achieve scale economies in the face of the cost-price squeeze.

Our discussion of farm diversification and specialisation has focused at the individual farm level. In areas where specialised markets for certain farm products and services are very large, the possibility exists for the farm unit to become specialised without much risk of income loss even though dependency upon one or a limited number of enterprises or marketing channels is very high. In these circumstances, diversification within the rural-urban fringe may not be observable at the individual farm level but rather at some higher level of geographic aggregation, so that an area may be diversified by having specialised, yet different, farms.

The Study Area and Methodology

In the Montreal region, one of the three largest metropolitan regions in Canada, an area on the south bank of the St. Lawrence was initially selected for analysis. Prior analysis (Bryant, 1976; INRS. 1973) has established that this broad area contained a wide variety of situations with respect to urbanisation pressures as well as different agricultural environments. The whole of the area shown in Fig. 4.1 was analysed using population change indicators and agricultural structure indicators from the Canadian Census of Agriculture and Population for 1981 and 1986 (Marois et al., 1989).

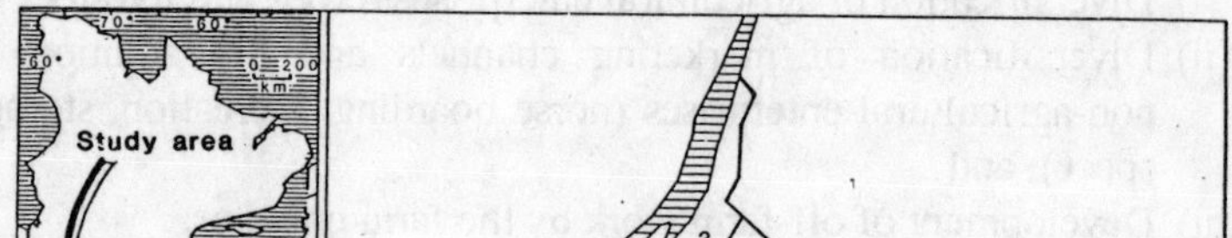

1- St-Dominique
2- St-Pie
3- Contrecoeur
4- St-Antoine-sur-Richelieu
5- St-Marc-sur-Richelieu
6- Verchères
7- La Présentation
8- St-Charles
9- St-Damase
10- St-Denis
11- St-Hyacinthe
12- Ange-Gardien
13- St-Césaire
14- St-Paul d'Abbotsford
15- St-Jean-Baptiste
16- St-Michel-Rougemont
17- St-Angèle-de-Monnoir
18- Ste-Marie-de-Monnoir
19- Granby
20- St-Alphonse
21- Ste-Cécile-de-Milton
22- St-Athanase
23- St-Grégoire
24- Ste-Brigide-d'Iberville
25- Rainville (Farnham)
26- L'Acadie
27- St-Jean-sur-Richelieu
28- St-Luc
29- Laprairie
30- St-Philippe
31- Boucherville
32- St-Bruno/Chambly
33- St-Hubert
34- Mont St-Hilaire
35- N.D.-de-Bon-Secours
36- St-Mathias
37- St-Amable
38- St-Mathieu-de-Beloeil
39- Varennes

I - Peri-urban II - Semi-rural III - Rural

FIG. 4.1: The Montreal study zones.

Then, three areas were chosen in which to undertake farm interviews (Fig. 4.1); the three areas includes eight smaller areas or Unified Census Sub-divisions. Zone I (Carignan, St. Bruno, St. Amable — Ste. Julie, Beloeil and Varennes) is the most urbanised and is most definitely "urban fringe" The combined units of St. Bruno — Chambly — Carignan — Ste. Basile had a 1986 population of almost 50,000 and had a 5 per cent population increase, 1981 to 1986. Population increases in the other units ranged from 20 per cent for Varennes (1986 population of about 10,000), 8 per cent for Ste. Julie — St. Amable (1986 population of 20,000) to 3 per cent for Beloeil (1986 population of 23,400). Population densities are high in Zone I and the role of the agricultural population is very low.

Zone II (St. Marc and Vercheres) is semi-rural with low population densities. Vercheres saw a 1.7 per cent population growth, 1981 to 1986 (1986 population of almost 5,000) and St. Marc 6.2 per cent (1986 population of 1,640). The agricultural population is much more important here (16 and 7 per cent in 1986 for St. Marc and Vercheres respectively. Zone III represents a rural zone or, in other words, a sort of control zone, lying to the east of the Richelieu River. Total population actually decreased in the two units here (St. Damase and St. Jean Baptiste) from 1981 to 1986 (–0.4 and –0.5 per cent in the two units respectively) and the agricultural population remains very significance (21 and 11 per cent in 1986 respectively).

A list of farmers was compiled for each Zone and, in all, a survey of 149 farmers was undertaken in the fall of 1988 and the winter 1989. Questions were posed concerning the changes in farm structure on each farm over the preceding eight years, changes in family circumstances and farmer's explanations of the various changes. But here only selected indicators of farm structure are presented and discussed.

Characteristics of Farm Structure and Change, 1981 to 1988

Three indicators of farm structure are briefly reviewed here: farm size in hectares and change in farm size; the extent of farmland renting; and farm type in terms of enterprise structure. Farm size and change are choosen because it has usually been assumed that larger farms and increases in farm size are associated with agricultural progress. To the extent that increases are made through purchase of additional farmland and to the extent that intense urban fringe areas are associated with higher land prices, one could anticipate that increases in farm size would be slower in Zone I and greatest in Zone III. Farmland renting has also been shown to increase in a number of urban fringe areas (Fielding, 1979; Johnston, 1989), representing an attempt by farmers to compact higher land prices by renting the additional land they need; evidently, this phenomenon can nullify the initial relationship discussed above between farm size, farm size increase and urban fringe pressures. Farm type is reviewed because one might expect relationships between farm type and farm size; in addition, some variations in farm type relating to intensity of production and marketing channels might be expected between Zones.

Farm Size

Zone I had the largest average farm size in 1988 and the gap in average farm size compared to the other two Zones widened significantly between 1981 and 1988 (Table 4.1). However, in terms of the proportion of farms increasing, remaining stable or decreasing their farm size over the period, Zones I and II are very similar in terms of the proportion of farms increasing in size, It is interesting to note that the proportion of "increasers" is much greater in zones I and II than in Zone III, showing that the "rural" control Zone (III) exhibits less indication of "progress" than the more intense urban fringe Zones. *This contrary to what might be expected from stereotyped interpretation of farm decline in the urban fringe, but it is consistent with the notions of adaptation under stress introduced earlier in this chapter.*

TABLE 4.1: Farm size (ha) in the Montreal study areas

Zone	# farms in sample	Average size		% change, 1981–88
		1981	1988	
I	65	87.24	119.5	+ 36.99
II	46	75.64	95.0	+ 26.41
III	38	88.32	95.1	+ 7.76

Zone	Change in farm size (# and (%))		
	Increase	Stable	Decrease
I	26 (41)	33 (52)	4 (6)
II	20 (43)	21 (46)	5 (11)
III	10 (26)	26 (68)	2 (5)

Note: The total number of farms varies from one table to another as each is based on the number of farms with no missing values for the relevant questions.

Farmland Rental

Given the pattern of farm size and change noted above, it can now be expected that farmland rental should be much higher in the urban fringe Zone (Table 4.2). On the one hand, the proportion of farmers renting land and the average level of renting per farm reporting is much higher in Zones I and II than in Zone III. Furthermore, of those farmers who increased their cultivated area, 52 per cent in Zone I did so wholly or partly by renting additional land compared to 47.4 per cent in Zone II and 43.8 per cent in Zone III (Deslauriers et al., 1990). On the other hand, the relative increase in the importance of rented land was much greater in Zone III compared to Zone I. The expansion of renting in Zone III is associated with the growing importance of cash cropping in this relatively favourable farming environment.

TABLE 4.2: Farmland rental in the Montreal study areas

Zones	# (and %) of farms reporting rented land		% change in # of renters	Average % land rented per farm reporting	
	1981	1988		1981	1988
I	29 (45)	35 (54)	20.7	42.5	47.9
II	23 (50)	25 (54)	8.7	36.3	42.0
III	11 (29)	13 (34)	18.2	20.1	26.9

Farm Type

Zones I and III have a higher proportion of field crop type farms (cash cropping as well as some of the farms in the "other" category) (Table 4.3). These are undoubtedly associated with the general movement towards more farmland rental in both these zones. The relative importance of dairying in Zone II reflects the presence of a "dairy belt" around Montreal (Marois et al., 1989). The urban fringe Zone (I) has the greatest concentration of "other" farms, which include specialised crop farms (fruit and vegetables, horticulturists, . . . and horse farms). Mixed farms are not of great importance in any of the zones. On a zone by zone basis, Zone I stands out as having the greatest diversity of farm types, while Zone II has the greatest degree of specialisation. As seen in Table 4.4, there have been no major changes in farm type during the 1980s.

TABLE 4.3: **Farm type in the Montreal study areas, 1988**

Zones	Farm type (# and (%))				
	Dairy	Livestock	Cash crop	Mixed	Other
I	12 (19)	0 (0)	24 (37)	5 (8)	24 (38)
II	23 (50)	7 (15)	8 (17)	2 (4)	6 (13)
III	8 (21)	2 (5)	16 (42)	4 (11)	8 (21)

TABLE 4.4: **Change in farm type in the Montreal study areas, 1981-88 (number of farms)**

Farm Type, 1981	Farm Type, 1988				
	Dairy	Livestock	Cash crop	Mixed	Other
Zone I					
Dairying	12		3	2	
Livestock					
Cash crop			20	1	
Mixed			1	2	4
Other					22
Zone II					
Dairying	23			1	
Livestock		7			
Cash crop			7		
Mixed			1	1	
Other					6
Zone III					
Dairying	8		2		
Livestock		2			
Cash crop			10		
Mixed			3	4	
Other			1		8

While the three indicator dimensions selected above so exhibit some differences between zones, notably the greater extent of farmland renting in zone one and a larger proportion of farms experiencing farm size increase in Zones I and II, it is also clear that there is a variety of farm responses in each of the zones to the complex of forces influencing agriculture. Furthermore, the three zones were not delimited to reflect the schematic of three landscapes discussed earlier but were based primarily upon an analysis of population densities and changes. Zone I is certainly not to be equated with a landscape of agricultural degeneration, even though some of the farms there are in decline. Even the cursory analysis presented above hints that the active landscapes of agricultural degeneration may be much more restricted in geographic extent that much of the literature of the 1960s and early 1970s suggested.

Diversification Strategies, 1981 to 1988

Crops and Livestock Enterprises

Based on net increases and decreases in the number of crop and livestock enterprises on each farm, the urban fringe (Zone I) and transition (Zone II) zones have over twice the relative number of farms diversifying than in Zone III (Table 4.5). Conversely, a much higher proportion of farms which experienced a reduction in the number of crops and livestock enterprises was observed in the rural zone (Zone III). Specialisation, one of the indicators frequently associated with traditional farms modernisation and progress certainly appears stronger, therefore, in the most rural areas; we must be quick to point out, however, that the diversification of farms in the other zones can also represent progress in terms of development of a more profitable enterprise structure. The differences between zones are consistent with the ideas of adaptation and opportunity differences discussed earlier.

TABLE 4.5: Farm diversification and specialisation in the Montreal study areas, 1981-88

Zone	# farms	Diversification # (and %)	Stable # (and %)	Specialisation # (and %)
I	65	30 (46)	26 (40)	9 (14)
II	46	22 (48)	19 (41)	5 (11)
III	38	8 (21)	18 (47)	12 (32)
Total		60	63	26

Diversification of crop and livestock enterprises has been mostly attained through a net addition of 1 or 2 enterprises (Table 4.6). Similarly, specialisation has occurred mainly through the net loss of 1 or 2 enterprises. The highest degree of diversification is found in the most intense urban fringe zone.

TABLE 4.6: **Farm level diversification and specialisation of crops and livestock enterprises, Montreal study areas, 1981 to 1988 (number of farms)**

Zone	Number of enterprises				
	Net loss		Stable	Net gain	
	less than −2	−2, −1		+1,+2	More
I	2	7	26	21	9
II	1	4	19	19	3
III	1	11	18	8	0

Detailed analysis reveals that this diversification is overwhelmingly the result of a net addition of crop enterprises rather than livestock ones, so the emphasis is now placed on crops. Wheat (for human consumption) and other extensive field crops account for the highest number of crops enterprises adoptions (Table 4.7). The proportion of adoptions of animal feed crops was particularly high in Zone II where dairy farms are most numerous. The significant increases in the number of farms reporting wheat production is explained by a number of factors present at the beginning of the 1980s: introduction of more adaptable varieties favourable prices and Quebec policy aimed at greater agricultural self-sufficiency mainly though subsidies and easy access to credit for storage facilities and farm machinery. Although the total is not high, it is noteworthy that the highest proportion of adoptions of specialised crops with a dominantly urban market was found in Zone I, the most intense urban fringe zone.

Marketing Strategies

To analyse marketing strategies, the emphasis has been placed on investigating differences between the most intense urban fringe Zone (I) and the two other Zones (II plus III) (Table 4.8). A larger proportion of the crop area in Zone I was devoted to enterprises normally associated with the urban market (vegetables, potatoes, tender fruits, nurseries and green houses), both in 1981 and 1988. The proportion actually increased in Zone I, while it decreased in the other two zones combined.

TABLE 4.7: Adoptions of crops enterprises, 1981 to 1988, in the Montreal study areas

Zone	# farms	# farms adopting	Hay	Grain corn	Wheat
I	65	43	4	7	21
II	46	27	4	10	14
III	38	19	1	2	13

Zone	Grains for animal feed*	Fruit Veg'bles	Nurseries Greenhouses	Other crops**
I	10	8	2	16
II	15	0	0	7
III	1	1	0	4

* includes corn, barley, oats, wheat, mixed grains.

** includes soya, buckwheat, potatoes, sugar beet.

TABLE 4.8: Urban market oriented enterprises and farm marketing strategies, Montreal study areas, 1981 to 1988

	ZONE I			
	1981		1988	
Enterprises	Area (ha)	%	Area (ha)	%
Vegetables	134.9	3.02	315.5	5.64
Potatoes	165.4	3.70	171.1	3.06
Tender fruit	27.9	0.62	31.8	0.57
Nurseries and greenhouses	18.2	0.41	14.9	0.27
	Total :	7.75	Total :	9.54
Strategies	# of farms	% farms* in sample	# of farms	% farms* in sample
Direct sales/ PYO	10		13	
Horse boarding	5	23.1	8	30.8
Sells directly in Montreal	3		3	

(Contd)

TABLE 4.8 (*Contd.*)

	ZONE II and III			
	1981		1988	
Enterprises	Area (ha)	%	Area (ha)	%
Vegetables	210.3	4.15	257.1	4.05
Potatoes	0.0	0.0	0.0	0.0
Tender fruit	18.0	0.35	13.9	0.22
Nurseries and greenhouses	66.7	1.32	94.3	1.49
	Total :	5.92	Total :	5.76
Strategies	# of farms	% farms* in sample	# of farms	% farms* in sample
Direct sales/ PYO	13		12	
Horse boarding	0	17.9	2	21.4
Sells directly in Montreal	3		5	

* Double-counting has been eliminated.

The proportion of farms with direct selling activities ("across-the-counter" or "pick-your-own") is considerably higher amongst Zone I farms than Zones (II plus III) farms. If two "non-agricultural" activities are added (boarding horses/riding school and selling produce direct in Montreal), the gap between Zone I and Zones (II plus III) is even greater and has actually increased, 1981 to 1988 [the 1981 gap being (23.1 — 17.9) per cent 5.2 per cent, while the 1988 gap was 9.4 per cent].

As one would expect based on the discussion earlier, a greater proportion of the sample farms has turned to different modes of marketing farm produce in the urban fringe compared to zones further away from the urban area. However, some of the possibilities for farm diversification observed elsewhere (not necessarily in urban fringe environments) were not observed in the sample, e.g. recreational vehicle storage and tourist oriented activities (Bryant, 1981; Gilg, 1990; Ilberry, 1990). However, horse boarding and riding have become significant activities on several farms in the St. Basile — Carignan area (Zone I) and local authorities there have seen them as part of the local strategy of economic development.

Off-farm Work

Both in 1981 and 1988, a much higher proportion of the farmers interviewed had some form of employment off the farm in Zone I than in the combined II and III zones (Table 4.9). However, this proportion decreased slightly in Zone I, while it increased slightly in the other two zones. Farmers working off the farm full-time are twice as significant relatively in Zone I. Most of the off-farm work is located locally or in the surrounding region, and only a relatively small proportion of farmers have work in Montreal. Off-farm work in the local and surrounding area has been indeed increased in importance in all zones, reflecting a certain decentralisation of employment opportunities in the greater Montreal region.

TABLE 4.9: Off-farm work (farmer) in the Montreal study areas, 1981 to 1988

	ZONE I				ZONES II and III			
	1981		1988		1981		1988	
	#*	%	#	%	#	%	#	%
Works on farm only	32	49.2	34	52.3	62	73.8	58	69.0
Works off the farm								
Full time	21	32.3	16	24.6	14	16.7	10	11.9
Part time	11	16.9	15	23.1	8	9.5	16	19.0
Location of off-farm work								
Local	11	34.4	15	48.4	5	22.7	9	34.6
In area	15	46.9	11	35.5	12	54.5	14	53.8
Montreal	6	18.8	5	16.1	4	18.2	3	11.5
Elsewhere	0	0.0	0	0.0	1	4.5	0	0.0

*information missing on one case.

Conclusion

The conceptual framework discussed in the first part of this paper emphasises progress made in how agriculture in the rural-urban fringe has been viewed during the 1970s and 1980s. Attention has been drawn increasingly to the complexity of forces of change influencing farming in the rural-urban fringe, as elsewhere. Urban development pressures are not the only forces that can destabilise agriculture in the rural-urban fringe. Furthermore, some forces related to the metropolitan presence, such as market opportunities, can have a positive impact on farming, as can other forces. The individual farmer, and farm family, represent the key to synthesising the complex of forces. Not only thus the framework suggests broad zonal types of differentiation in farm structure and change — and we should hasten to underline that this does not mean concentric zones. It also stresses variation between farms within any broader zone. What emerges is a mosaic of patterns of farm structure and change, with the relative mix of different types of farm change varying over broader zones.

By way of example, and without any attempts to link the patterns to sets of casual circumstances, some observations based on a sample survey of farms in the Montreal region were made. First, some comments were made on some selected indicators of farm structure and change, viz. farm size and change, farmland rental and farm type. Patterns of farm size and farmland rental change exhibit more of the positive relationship with urban pressures found in other recent work (cf. Bryant and Fielding, 1980), and a greater variety of farm types was present in the urban fringe zone than elsewhere. Deferences between study zones, therefore, existed, but we must also emphasise the variety of types of changes and structures present in all zones. Second, some examples of adaptation through diversification were examined. In the Montreal study area, farm land diversification has become an important phenomenon, especially in the urban fringe, although it is far from absent elsewhere. Unlike a similar study in the Toronto area (Bryant et al., 1989), the diversification has occurred at the farm level rather than through the development of a variety of specialised farms; the difference may well be related to a greater volume of market demand in the rural-urban fringe zones of agricultural produce direct from the farm around Toronto than around Montreal. While enterprises diversification might be considered a weakness from the perspective of scale economies in equipment utilisation (depending, of course, on the degree of technical substitution possible between enterprises), from a more ecological perspective it can impart more resilience to the production structure, both individually and at a regional level. Similarly, the adoption of new marketing channels based on tapping into the different types of market associated with major metropolitan areas is again a factor of strength and resilience, and off-farm work is certainly a long-standing way of diversifying (and augmenting) farm family income. Both these forms of adaptation and diversification were found in greater concentrations in the urban fringe zone.

The evidence presented above is consistent with the conceptual framework outlined earlier. However, the evidence has dealt solely with farm structure and change. The next stage of this research involves linking types of structural change to their formative circumstances, both on and off the farm, in more of a behavioural mode of investigation. This research direction, ultimately embedded in a longitudinal research programme, will provide an enhanced opportunity to advance our understanding of how the mosaic of farming structures in the rural-urban fringe evolves, particularly by permitting us to trace changes in the interaction patterns that link farmers, their families and farms into different "systems of exchange" at different scales of geographic analysis, bringing them into contact with different sets of forces — hence the dynamic mosaic.

REFERENCES

Bryant, C. R., 1970 : *Urbanisation and Agricultural Change since 1945: a Case Study from the Paris Region*, London : University of London.

———, 1976 : *Farm Generated Determinants Landuse Change in the Rural-Urban Fringe*, Ottawa : Lands Directorate, Environment Canada, Technical Report.

———, 1981 : "Agricultural change in an urbanising environment : A Case Study from the Paris Region, 1968 to 1975", *The Canadian Geographer*, 25 (1) : 27-45.

———, 1984 : "The recent evolution of farming landscapes in urban-centred regions," *Landscape Planning*, 11 : 307-26.

———, 1986 : "Agriculture and Urban development, in Pacione, M. (ed), *Progress in Agricultural Geography*, Beckenham, U.K. : Croom Helm.

———, 1989 : "Entrepreneurs in the rural environment, " *Journal of Rural Studies*, 5 (4) : 337-48.

Bryant, C. R. and Fielding, J. A., 1980 : Agricultural change and farmland rental in an urbanisation environment : Waterloo Region in Southern Ontario", *Cashiers de Geographie du Quebec*, 24 (62) : 277-98.

Bryant, C. R., et al., 1989 : "Agricultural intensification and change near Toronto and Montreal", Baltimore : Presentation at the annual meeting of the Association of American Geographers.

Bryant, C. R., Russwurm, L. H. and McLellan, A. G., 1982 : *The City's Countryside : Land and its Management in he Rural-Urban Fringe*, London : Longman.

Bryant, C. R., Russwurm, L. H. and Wong, S. Y., 1984 : "Agriculture in the Canadian urban field : an appreciation", in Bunce, M. F. and Troughton, M. J (eds.), *The Pressures of Change in Rural Canada*, Toronto : Department of Geography, Atkinson College, York University, Geographical Monograph 14.

Fielding, J. A., 1979 : *Farmland Rental in an Urbanising Environment : A Case Study of the Fringes of Kitchener, Waterloo and Cambridge, 1971 - 1975*, Waterloo : Department of a Geography, University of Waterloo, unpublished M.A. thesis.

Gilg, A. and Griffiths, A., 1990 : "Farm level responses to changes in the Common Agricultural Policy : A Report on three studies in South West England", Toronto : Presentation to the annual meeting of the Association of American Geographers.

Ilberry, B. W., 1988 : "Agricultural change on the West Midlands urban fringe", *Tijdschrift voor Economische en Sociale Geografie*, 79 (2) : 108-21.

———, 1990 : "Supply control measures and farm diversification in Britain", Toronto : Presentation to the annual meeting of the Association of American Geographers.

I. N. R. S. (Institut National de la Recherche Scientifique) 1973, *Region sud : agriculture*, Montreal : Office de planification et de development du Quebec and Universite du Quebec, Montreal.

Johnston, T. R. R., 1989 : *Farmers' Adaptive Behaviour in an Urbanising Environment : Guelph to Toronto Area*, Waterloo : Department of Geography, University of Waterloo, unpublished Ph.D. thesis.

Marois, C., Deslauriers, P. and Bryant, C. R., 1989 : "Changements structureis dans 1, agricultural de la rive-sub de Montreal, 1981 - 1986", Montreal presentation to the 57th, annual meeting of A.C.F.A.S.

Munton, R. J. C., 1974 : "Farming on the urban fringe", in Johnson, J. H. (ed.), *Suburban Growth : Geographical Processes at the Edge of the Western City*, London : Wiley.

Munton, R. J. C., Whatmore, S. and Marsden, T., 1988 : "Reconsidering urban fringe agriculture : a longitudinal analysis of capital restructuring on farms in the Metropolitan Green Belt", *Transactions of the Institute of British Geographers*, 13: 324-36.

Wong, S. Y., 1983 : *Agricultural change in Canada : 1941 - 1976*, Waterloo : Department of Geography, University of Waterloo, unpublished Ph.D. thesis.

5

DYNAMICS OF CROPPING PATTERN IN NORTHERN INDIA

SURENDER SINGH

THE LANDUSE pattern, of which the cropping pattern forms a part, has always been a dynamic phenomenon. The cropping pattern is, in fact, a reflection of the interplay of the complex social, economic and physical factors. All these factors themselves keep on changing, except the physical ones, which are comparatively static. Thus, under the influence of dynamic socio-economic factors, the cropping pattern also goes on getting modified, so much so, that, sometimes, it is wholly replaced after a sufficiently long span of time. Even minor social changes, e.g., a change in the dietary habits of the populace or the acceptance or rejection of a particular crop due to social and economic changes, may render any crop out of date and out of place, and *vice versa*.

However, ever since the times of Recardo, in the beginning of the 19th century, studies in the agricultural landuse have gained considerable importance and scholars from different disciplines, from time to time, have directed their efforts towards the establishment of some probable relationships between the landuse pattern and the prevailing socio-economic and physical environmental conditions.

The studies in the field of agricultural landuse emanated out of the economic problems, such as the problem of increasing prices of the agricultural products. Hence, economists, rather than geographers, were pioneers of the agricultural landuse studies. Among scholars other than economists, Von Thunen (1826) was, perhaps, the first who made an attempt towards a theory of agricultural landuse. He was a practical farm manager and hence, his postulates were based upon his experiences over a long period of time. Von Thunen's model, thus, was a major breakthrough towards a theory of agricultural landuse, and it gained a lot of support and was accepted by and large.

The basic theme of Von Thunen was that the market price of agricultural produce was largely dependent upon the distance between the market place and the centre of production. It was also accepted that the land rent also has a decisive bearing upon the types of crops grown in a particular area, and the amount of rent per unit of area is directly proportionate to the distance from a central market place, invariably a central city.

Von Thunen based his postulations on the assumption of "An Isolated State", an area of uniform fertility, surrounding a central city, and the entire produce of this region was marketed in the central city. (The only means of transportation available to the farmers was the horse-driven cart.) He assumed that the cost of transport was directly proportionate to the distance travelled. On the assumption of such an isolated state, he arrived at a theory of agricultural landuse and devised some circular zones around the central city, intensitywise. His first zone surrounding the central city was the zone of most intensive cultivation, followed by zones of lesser intensity of farming. Thus, there emerged a system of concentric circles.*

By postulating the system of concentric rings for farming activities, Von Thunen deduced that the fields nearer to the central cities or the places of consumption are cultivated more intensively and *vice versa.*

The ideas of Von Thunen seem to be true in today's world. Jonnason (1925) tried the concept of Von Thunen in his study and found almost identical rings in the area of Edward Plateau. There have been found similar situations in many historical studies as well, especially in those parts of the world where the means of transport are not very well developed and the animal-driven transport is more commonly employed. In such cases the cropping pattern has responded to the distance factor. According to a study of the North Indian Villages by Blakie (1971) some adjustments in landuse have taken place on the basis of distance from the settlement site (*Abadi*). Blakie has established that the intensity of cropping decreases away from an *Abadi*.

---

* Von Thunen devised the following Zones:
(1) Market gardening; (2) Firewood and timber production; (3) Crop farming without fallow; (4) Crop farming with fallow and pasture; (5) Three-field system of farming; and (6) I ivestock ranching.

Similar attempts have been made on a global scale also. Peet (1969) established that if Western European countries are considered as a core (an area of high density of population and hence an area of mass consumption), the intensity of cropping is found to be decreasing outwards, i.e. towards American and Asian countries. There have been numerous other studies as well which have confirmed the idea of decreasing intensity of cropping away from central places especially in commercial farming regions. To name only a few, Dalton (1971), Horwath (1969) and Harvey (1963 and 1966), who have proved almost the same at local scale.

As regards the situation in India, Chisholm (1962), apart from Blakie, accepts, to a great extent, the ideas put forth by Von Thunen, and he admits that there is certainly a tendency of intensive landuse nearer to the settlements. Thus, what Von Thunen hypothesized in the early 19th century seems to be true, to a large extent, and holds good even today at all levels. However, the author has made an attempt to probe further in depth this problem by taking the Union Territory of Delhi as a case study.

The present study is an empirical one based upon the evidence available in the revenue records of six sample villages of the northern part of Delhi. The study takes into account the situation as depicted by the landuse data for a period of 13 years, from 1965 to 1978, and seeks to establish the relationship between the changes in the cropping pattern and the socio-economic changes, if any. Thus, the aim of this study is to see whether the dynamics in the cropping pattern which seems to be a spontaneous phenomenon has a random or a particular trend.

Selection of the Union Territory of Delhi for this study is based on the idea of Von Thunen's "Isolated State". Although it is not an isolated State in the true sense of the term, it can be a rough representative of the same and the metropolitan core of Delhi serving as the centre of mass consumption. The villages which have been included in the study have been selected through purposive sampling, and are at different distance limits from Delhi and enjoy varied levels of connectivity with the urban core. The time period of 13 years seems to be enough to establish the dynamism in the cropping pattern and the data pertaining to the sample villages are available for this period and they are comparable. Table 5.1 gives an insight into the distance of the villages from the city and their connectivity with the urban core.

TABLE 5.1: Connectivity index of sample villages

Sl. No.	Name of village	Distance from Delhi (from Red Fort) in kms.	Frequency of bus trips per day	Frequency of train trips per day
1.	Khera Kalan	17.2	36	7
2.	Bakhtawarpur	20.4	32	—
3.	Tigipur	20.4	32	—
4.	Palla	22.8	27	—
5.	Bankner	26.0	25	—
6.	Ochandi	32.4	20	—

Source: (1) DTC Time-table—1979.
(2) Time-table, Northern Railway—1979.

Table 5.1 shows that the distances at which the villages involved in the study are located vary from 17.2 kms to 32.4 kms and the frequency of bus transport decreased with increasing distance from the city. The rail serves only one village.

The landuse data obtained and analysed for the purpose of this study, for the six selected villages, show definite differences not only in the cropping pattern in these villages but also portray very clearly the dynamics of the cropping pattern under changing urban influence.

The Dynamics of the Cropping Pattern

For the sake of simplicity, all the crops have been classified into certain groups. The basis of this classification is the same as adopted by the revenue departments and the entire classification is almost identical to that accepted for the revenue purposes. The classification is two-fold : Firstly, all the crops have been grouped into two categories, Kharif and Rabi, on the basis of their season of growing. Then, all the crops have been categorised into five groups on the basis of their use and economic value, irrespective of the season of their growing. These groups are : (1) Cereals—wheat and rice; (2) Millets and Pulses—jowar, bajra, maize, gram and other pulses; (3) Fruits and vegetables—fruits, vegetables, flowers, spices, etc.; (4) Cash crops—sugarcane, cotton, tobacco, oilseeds, etc.; and (5) Fodder crops.

The total cultivated area of the total cropped area comprises the total of the area under different crops which is then broken into Kharif cropped area and Rabi cropped area. Percentagewise the area under

Rabi crops accounts for a larger part of the total cropped area. In some cases this percentage is as high as 77, the lowest ever percentage being 47.5. During most of the years the area under Rabi crops is something between 50 and 70 per cent of the total cultivated area, the remaining 30 to 50 per cent being under the Kharif crops. It reflects the dominance of the Rabi crops over the Kharif ones. If further analysed, the landuse data reveal that in some cases up to 50 per cent or even more of the total cropped area is covered by a single crop of Rabi season, namely, wheat. Thus, wheat is the first-ranking crop in this area; and this is true in all the villages. However, there is a close competition for the next rank between the millets, vegetables and fodder crops. During the dry years the millets take the lead, while during the rainy years they are surpassed by vegetables, closely followed by fodder crops. This is more so in the case of four villages in closer vicinity of the urban area, because of easy access to the city market.

The figures for the total cropped area have undergone dramatic changes during 1966—78 in the caśe of four villages, the total cropped area has registered an increase of about 50 per cent. The other two villages have experienced a series of ups and downs, the net result being almost no change in the total cultivated area during the last 13 years.

If the total cropped area figures are analysed seasonwise, some striking facts emerge. Such an analysis at the outset reveals that almost all the variations in the figures of total cropped area have been due to the variations in the Kharif cropped area. The area under Rabi crops has changed to a considerably lesser extent comparatively, and these figures certainly do not show any significant change.

The upward trend of the figures of total cultivated area in majority of the cases reveals the mounting pressure on land or the increasing intensity of cropping. Although it cannot be supported statistically due to non-availability of all the relevant data, it is true, again in majority of the instances, that the double-cropped area has increased considerably and the close observations in the field reveal that a lot of marginal areas have been brought under plough, as a result of increased needs for farmland. The practice of rotation of crops with fallows in between has broken down and nowadays, the short-period vegetable crops are obtained during the period in between the Kharif and the Rabi seasons. This has all been caused due to the increasing urban influence and increased demands for agricultural products to satisfy the needs of a larger urban populace. That is why the encroachment upon marginal lands is, to a greater degree, in the villages nearer to the city.

The changing situation of connectivity and urban influences affecting the market situation seem to be the most potent factor having a bearing upon the cropping pattern of the villages under study. The changes being observed and the existing cropping pattern seem to be strengthening the views forwarded by Von Thunen and later by other scientists like Blakie, regarding the agricultural landuse. For the following analyses the figures of the percentage of the area under particular crop groups to the total cropped area have been computed which reflect some definite trends, explained below.

Cereal Crops

Generally, occupying more than 50 per cent of the total cropped area, cereals rank the highest among the crop groups. Figure 5.1 shows the pattern of dynamics in the area under cereals (percentage to total cropped area) in all the six villages. The seasonwise analysis of the same data reveals that during the Rabi season the cereals (wheat) cover well above 85 per cent and even up to 90 per cent of the total cropped area. On the other hand the same crop group during the Kharif season (rice) has gained importance only during the recent years. During the period covered by this study, only a small fraction of the total cultivated area has been assigned to this crop. However, the area under rice crop has been going on increasing slowly but steadily.

The graphs for all the six villages (Fig. 5.1) display one common characteristic — an upward trend in each case. In 1966, in none of the villages was the area under cereals more than 40 per cent of the total cultivated area, while in 1978, in none of the cases it was less than 50 per cent. Another striking fact is that during the last 13 years, the highest percentage of area under cereals has been during 1977 in all the villages.

These changes have been caused by factors like mechanisation of agriculture and increased demand for foodgrains. Modern technology can, perhaps, be best suited to wheat farming. Also there has been an effect of the Government policies in the form of fixation of minimum support prices of wheat and rice. The role of a lot of research work which has brought revolutionary changes through the development of higher yielding varieties, during the past one decade or so, cannot be

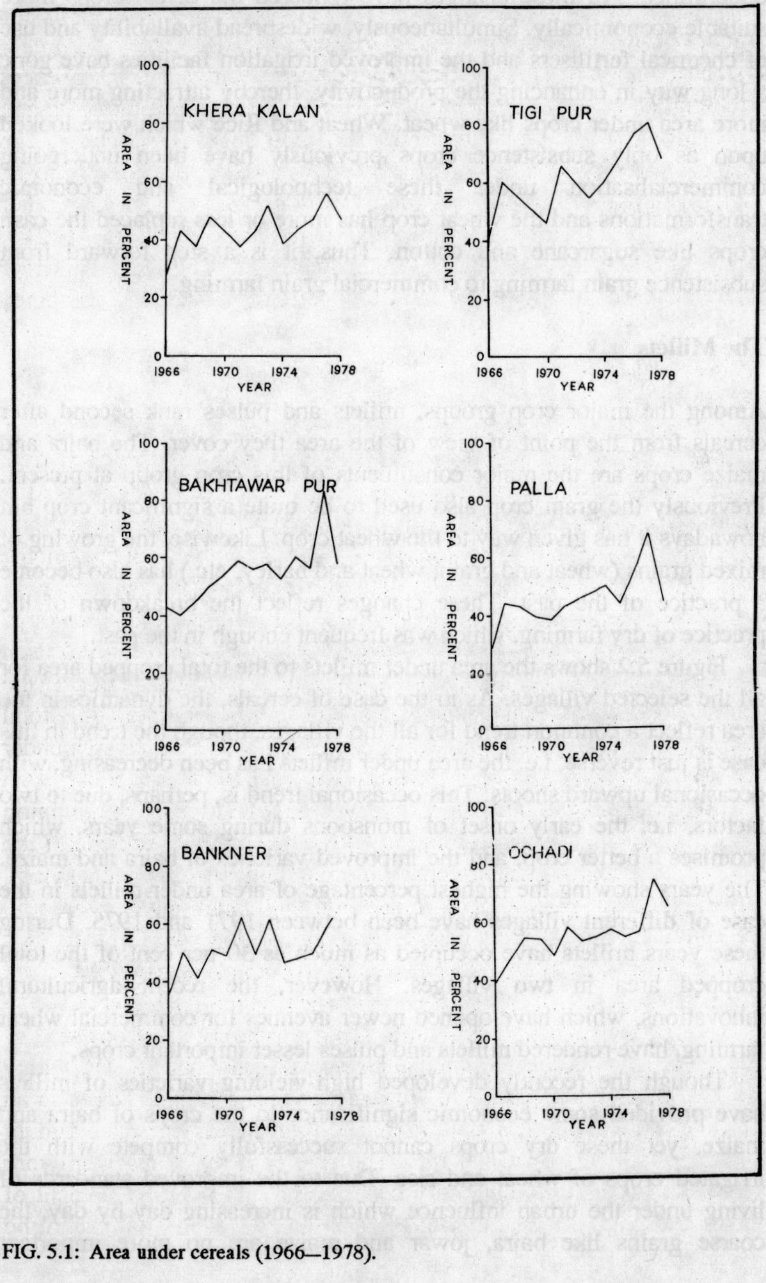

FIG. 5.1: Area under cereals (1966—1978).

undermined. All these changes have rendered the cereal crops more suitable economically. Simultaneously, widespread availability and use of chemical fertilisers and the improved irrigation facilities have gone a long way in enhancing the productivity, thereby attracting more and more area under crops like wheat. Wheat and Rice which were looked upon as only subsistence crops previously have been undergoing commercialisation under these technological and economic transformations and the wheat crop has more or less replaced the cash crops like sugarcane and cotton. Thus, it is a step forward from subsistence grain farming to commercial grain farming.

The Millets

Among the major crop groups, millets and pulses rank second after cereals from the point of view of the area they cover. The bajra and maize crops are the major constituents of this crop group at present. Previously the gram crop also used to be quite a significant crop but nowadays it has given way to the wheat crop. Likewise, the growing of mixed grains (wheat and grain; wheat and barley, etc.) has also become a practice of the past. These changes reflect the breakdown of the practice of dry farming, which was frequent enough in the past.

Figure 5.2 shows the area under millets to the total cropped area for all the selected villages. As in the case of cereals, the dynamics in the area reflect a common trend for all the villages, though the trend in this case is just reverse, i.e. the area under millets has been decreasing, with occasional upward shoots. This occasional trend is, perhaps, due to two factors, i.e. the early onset of monsoons during some years, which promises a better crop, and the improved varieties of bajra and maize. The years showing the highest percentage of area under millets in the case of different villages have been between 1971 and 1975. During these years millets have occupied as much as 30 per cent of the total cropped area in two villages. However, the recent agricultural innovations, which have opened newer avenues for commercial wheat farming, have rendered millets and pulses lesser important crops.

Though the recently developed high-yielding varieties of millets have provided some economic significance to the crops of bajra and maize, yet these dry crops cannot successfully compete with the irrigated crops of wheat and rice. Due to the improved standards of living under the urban influence which is increasing day by day, the coarse grains like bajra, jowar and maize are no more important

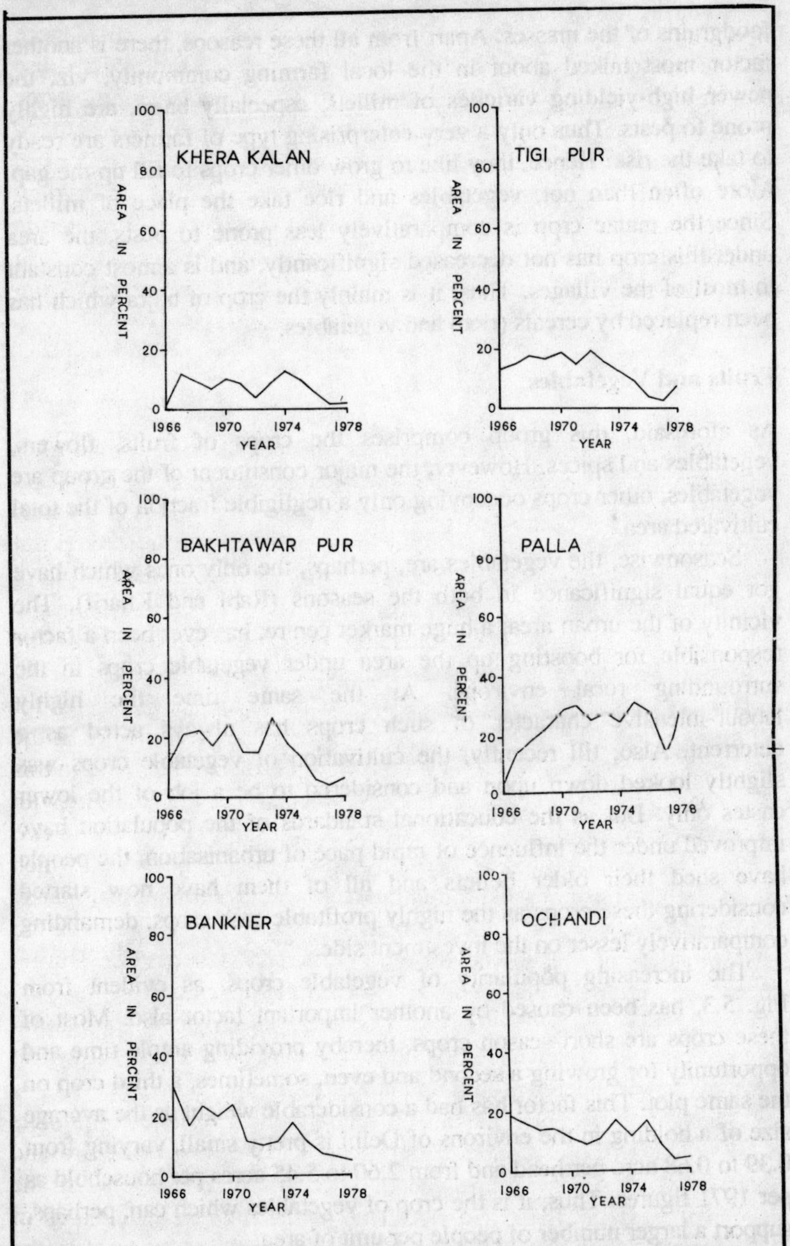

FIG. 5.2: Area under millets (1966—1978).

foodgrains of the masses. Apart from all these reasons, there is another factor most talked about in the local farming community, viz. the newer high-yielding varieties of millets, especially bajra, are highly prone to pests. Thus only a very enterprising type of farmers are ready to take the risk. Hence, they like to grow other crops to fill up the gap. More often than not, vegetables and rice take the place of millets. Since the maize crop is comparatively less prone to pests, the area under this crop has not decreased significantly, and is almost constant in most of the villages. Thus, it is mainly the crop of bajra which has been replaced by cereals (rice) and vegetables.

Fruits and Vegetables

As aforesaid, this group comprises the crops of fruits, flowers, vegetables and spices. However, the major constituent of the group are vegetables, other crops occupying only a negligible fraction of the total cultivated area.

Seasonwise, the vegetables are, perhaps, the only ones which have got equal significance in both the seasons (Rabi and Kharif). The vicinity of the urban area, a huge market centre, has ever been a factor responsible for boosting up the area under vegetable crops in the surrounding rural environs. At the same time the highly labour-intensive character of such crops has always acted as a deterrent. Also, till recently, the cultivation of vegetable crops was slightly looked down upon and considered to be a job of the lower castes only. But as the educational standards of the population have improved under the influence of rapid pace of urbanisation, the people have shed their older beliefs and all of them have now started considering these crops as the highly profitable cash crops, demanding comparatively lesser on the investment side.

The increasing popularity of vegetable crops, as evident from Fig. 5.3, has been caused by another important factor also. Most of these crops are short-season crops, thereby providing ample time and opportunity for growing a second and even, sometimes, a third crop on the same plot. This factor has had a considerable weight as the average size of a holding in the environs of Delhi is pretty small, varying from 0.39 to 0.68 acre per head and from 2.60 to 5.45 acres per household as per 1971 figures. Thus, it is the crop of vegetables which can, perhaps, support a larger number of people per unit of area.

Figure 5.3 provides an insight into the dynamics of the area under this group of crops in all the villages. The first indication in this regard is that the area under vegetables fluctuates a great deal year after year and season after season. The most important among the causes is the variability in the outset of the monsoons, determining, to a great extent, the success of crops during the Kharif season. Also, among rural folks, even today, there is a tendency of collective decision-making. Thus during some of the years people en masse decide to grow vegetable crops in one particular area, thereby causing an upsurge in the area figures. This increased production has a negative effect on the prices of crops which acts as a deterrent to growing these crops during the coming season. The supply restricted in this way again pushes up the prices, thereby again attracting large areas under these crops and this trend continues.

Another trend revealed by the analyses is that the area under vegetable crops has increased in all the villages. However, this increase is lesser in the case of far-off villages. Also the trend has started comparatively recently in such villages. Again, the fluctuation in the absolute acreage is more in the nearer villages as they are better aware of the market position. This increasing trend reflects almost the commercialisation of vegetable crops in the vicinity of the city and these crops seem to be competing well with cereal crops as an increasing trend is evident in most cases, decreasing in none of the villages.

Cash Crops

Figure 5.4 shows the trends in the cropping of cash crops—sugarcane, cotton and oilseeds. The area under these crops during the past 13 years, as shown by the graphical analysis, is getting reduced so drastically that it would not be an overstatement to say that these crops have today lost their importance altogether. There are certain factors which have carried these crops to such a low rank in these villages. Due to reasons unknown to the farmers, the sugarcane and cotton have become perhaps the most unsuccessful crops. The older well-adapted varieties of sugarcane have become highly prone to pests of several kinds and the newer varieties have not been able to adapt that successfully. The cotton crop has also become equally prone to pests. Thus these crops have been rendered rather uneconomical as they have a long maturing season thereby ruling out the possibility of multi-cropping. Also these crops are comparatively more labour- and capital-intensive and the yields are very low. Thus the returns are, more often than not, comparatively lower.

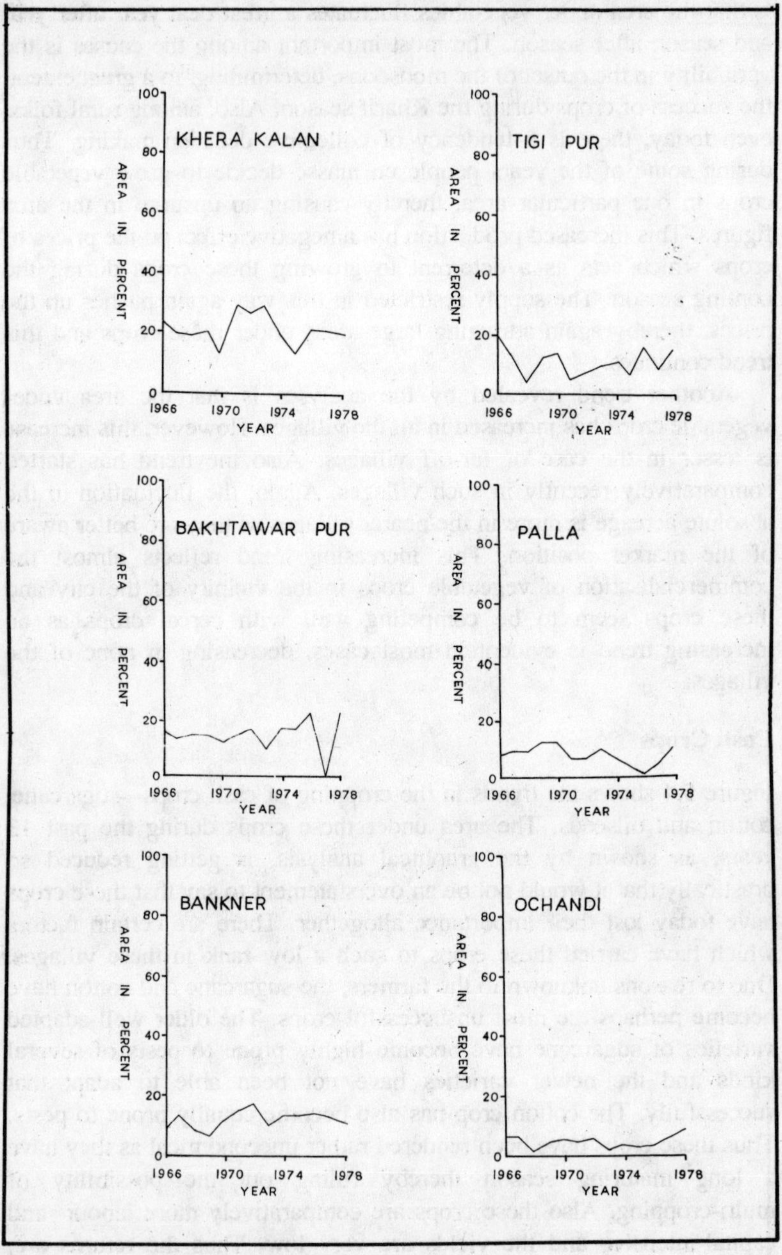

FIG. 5.3: Area under fruits and vegetables (1966—1978).

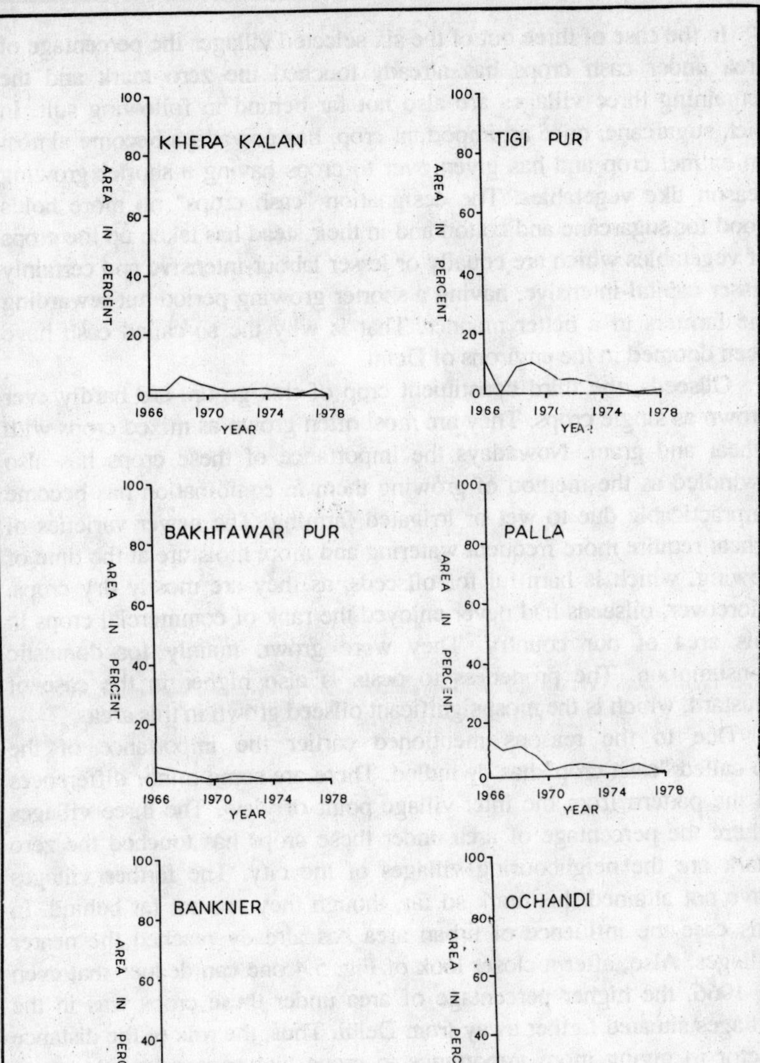

FIG. 5.4: Area under cash-crops (1966—1978).

In the case of three out of the six selected villages the percentage of area under cash crops has already touched the zero mark and the remaining three villages are also not far behind in following suit. In fact, sugarcane, once an important crop, has nowadays become almost an extinct crop and has given over to crops having a shorter growing season like vegetables. The designation "cash crops" no more holds good for sugarcane and cotton and in their stead has taken up the crops of vegetables which are equally or lesser labour-intensive and certainly lesser capital-intensive, having a shorter growing period but rewarding the farmers in a better manner. That is why the so-called cash have been doomed in the environs of Delhi.

Oilseeds, the third constituent crop of this group, are hardly ever grown as single crops. They are most often grown as mixed crops with wheat and gram. Nowadays the importance of these crops has also dwindled as the method of growing them in combination has become impracticable due to wet or irrigated farming. The newer varieties of wheat require more frequent watering and more moisture at the time of sowing, which is harmful for oilseeds, as they are mostly dry crops. Moreover, oilseeds had never enjoyed the rank of commercial crops in this area of our country. They were grown mainly for domestic consumption. The proneness to pests is also higher in the case of mustard, which is the most significant oilseed grown in this area.

Due to the reasons mentioned earlier the importance of the so-called "cash crop" has dwindled. There are some minor differences in the pattern from the inter-village point of view. The three villages where the percentage of area under these crops has touched the zero mark are the neighbouring villages of the city. The farther villages have not attained this mark so far, though they are not far behind. In this case the influence of urban area has already reached the nearer villages. Also, after a closer look of Fig. 5.4 one can deduce that even in 1966, the higher percentage of area under these crops was in the villages situated farther away from Delhi. Thus, the role of the distance factor in giving more importance to crops like vegetables, etc., over cash crops like sugarcane is evident.

Fodder Crops

An analysis of the data, as depicted in Fig. 5.5, shows that the area under fodder crops has been undergoing rapid changes. However, the pattern of change differs from that in the case of cereal crops or

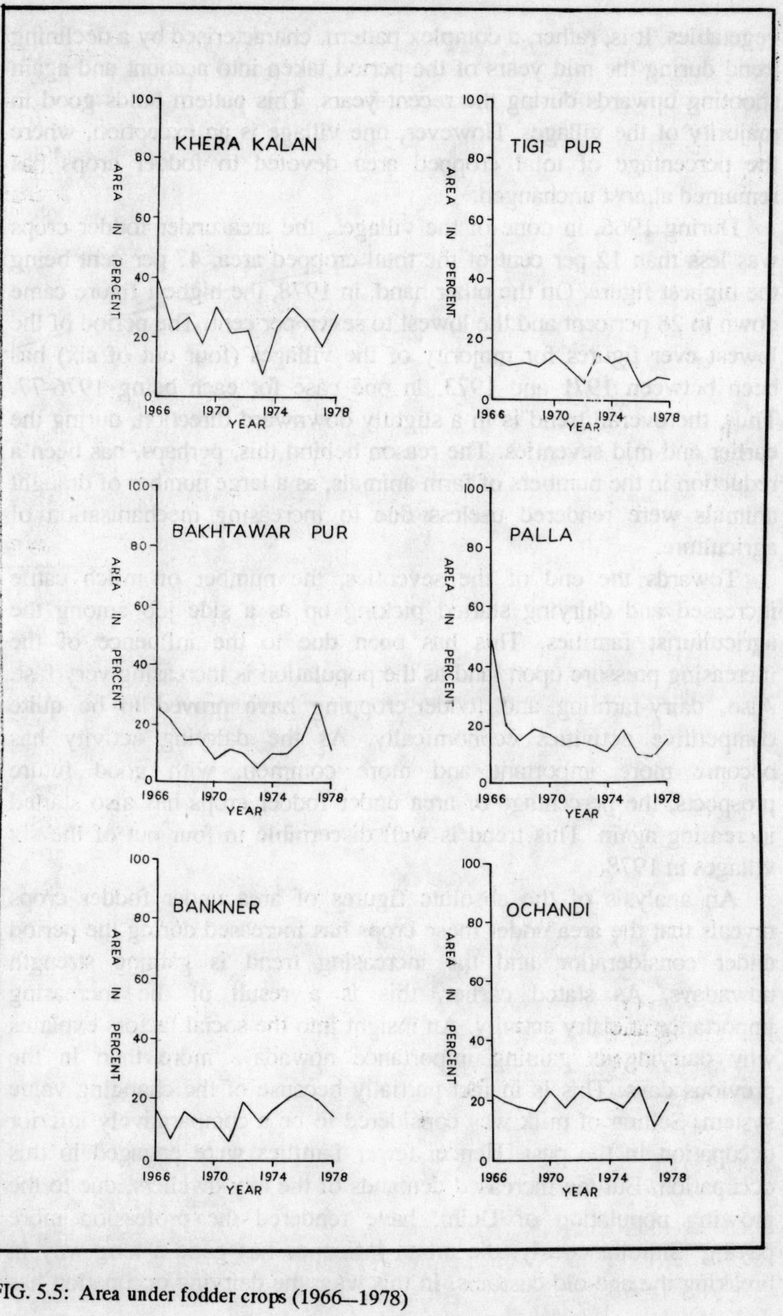

FIG. 5.5: Area under fodder crops (1966—1978)

vegetables. It is, rather, a complex pattern, characterised by a declining trend during the mid years of the period taken into account and again shooting upwards during the recent years. This pattern holds good in majority of the villages. However, one village is an exception, where the percentage of total cropped area devoted to fodder crops has remained almost unchanged.

During 1966, in none of the villages, the area under fodder crops was less than 12 per cent of the total cropped area, 47 per cent being the highest figure. On the other hand, in 1978, the highest figure came down to 28 per cent and the lowest to seven per cent. The period of the lowest ever figures for majority of the villages (four out of six) has been between 1971 and 1973, in one case for each being 1976-77. Thus, the overall trend is in a slightly downward direction, during the earlier and mid seventies. The reason behind this, perhaps, has been a reduction in the numbers of farm animals, as a large number of draught animals were rendered useless due to increasing mechanisation of agriculture.

Towards the end of the seventies, the number of milch cattle increased and dairying started picking up as a side job among the agriculturist families. This has been due to the influence of the increasing pressure upon land as the population is increasing very fast. Also, dairy-farming and fodder-cropping have proved to be quite competitive activities economically. As the dairying activity has become more important and more common, with good future prospects, the percentage of area under fodder crops has also started increasing again. This trend is well discernible in four out of the six villages in 1978.

An analysis of the absolute figures of area under fodder crops reveals that the area under these crops has increased during the period under consideration and the increasing trend is gaining strength nowadays. As stated earlier, this is a result of the increasing importance of dairy activity. An insight into the social factors explains why dairying is gaining importance nowadays more than in the previous days. This is in fact partially because of the changing value system. Selling of milk was considered to be a comparatively inferior occupation in the past. Hence, fewer families were engaged in this occupation. But the increased demands of the city dwellers, due to the growing population of Delhi, have rendered the profession more paying. Simultaneously, the urban influence has gone a long way in breaking the age-old customs. In this way, the dairying occupation has now been accepted as any other occupation by the rural folks, thereby bringing more and more people in the trade. All this has had a direct bearing upon the area under fodder crops. As the villages nearer to the city are comparatively more prone to urban influences, this also explains the higher percentage of area under these crops in the neighbouring villages and *vice versa.*

Summary and Conclusions

In the foregoing discussion the major changes in the cropping pattern in the northern part of Delhi (Union Territory of Delhi) and the factors which have wrought these changes have been observed. From this discussion it can be deduced that the cropping pattern in this area has witnessed great changes since 1966. Also, the changes seem to be conforming to the age-old postulates of the agricultural scientists, although not fully.

The first thing to emerge out of this discussion is that the total cropped area has increased due to the increasing pressure on land. The population of the Union Territory which was only 4,05,819 in 1911 has gone up to 4,065,698 in 1971. Table 5.2 gives an insight into the pattern of growth of population in the Union Territory which has tremendously increased the pressure on land.

The second major change has been in the area under cereal and vegetable crops, i.e. an increase in both. Among the cereals the incidence of irrigated crops has increased and crops of wheat and rice have rendered the dry crops like gram and mixture crops out of place and out of date. This is due to the commercialisation of cereal crops. In fact there have come revolutionary changes through mechanisation of agriculture and introduction of new varieties of wheat and rice. Similarly, the area under vegetable crops, though not changed much percentagewise, has increased in absolute terms, reflecting a clear spatial pattern. Villages better connected with the metropolitan city of Delhi have a higher percentage of the total cultivated area under these crops. The improving connectivity during the recent past has increased the rate of growth of the area under such crops. The shifting of the main vegetable market from a central to a peripheral location towards the north has increased the area under vegetable crops in neighbouring villages. Also, the factor of increased demand due to an increase in the population of the urban area has had its bearing upon the area figures.

TABLE 5.2: Decinal percentage variation in population

Decade	Percentage of variation of population		
	Total	Rural	Urban
1901-11	+ 1.98	− 8.24	+ 11.13
1911-21	+ 18.03	+ 4.62	+ 27.94
1921-31	+ 30.26	+ 2.59	+ 46.98
1931-41	+ 44.27	+ 17.72	+ 55.48
1941-51	+ 90.00	+ 38.10	+106.58
1951-61	+ 52.44	− 2.52	+ 64.17
1961-71	+ 52.93	+ 39.93	+ 54.57

Source: Census of India—1971, Delhi. Pt. II-A. General population Tables.

As the area under cereals and vegetables has gone up, the area under millets and cash crops has registered a decrease, so much so that cash crops have become almost extinct. The decrease in the area under millets has been phenomenal in villages nearer to Delhi, distancewise as well as connectivitywise. This change has carried the landuse pattern closer to that expected by Von Thunen and his followers. While dry millets have been replaced by wet crops grown with the help of irrigation, the fate of cash crops has been slightly different. In this case the value of crops of sugarcane, cotton, etc., has decreased comparatively, rendering these crops almost extinct and more so in the neighbouring villages of Delhi. As explained in the earlier discussion the erstwhile cash crops no more stand competition with vegetable crops.

The percentage of area under fodder crops has also undergone changes, decreasing in the beginning but again picking up during the later years. Area under these crops which had decreased due to decreasing numbers of farm animals under the influence of technological changes has again started picking up due to increasing importance of dairy activity. This increasing trend is also stronger in villages in close vicinity of the city.

From what has preceded, certain conclusions can be drawn. Of the most important one is that the intensity of cropping in the villages surrounding the urban area is increasing. Although, intensitywise clearcut zones for different types of crops, as envisaged by Von Thunen and others, cannot be demarcated, the general pattern does conform to the same. The inter-relationship between connectivity to the central city and the place of production and the intensity of cropping is quite clear. Thus the concept of distance from the market-place needs revision and the idea of connectivity involving physical distance and time and effort in completing the journey will perhaps stand a better test in the situations prevailing in the modern times. The total replacement of long-season crops like sugarcane and cotton in the villages nearer to city and their persistence in the far-off villages, although only marginally, confirms the concept of decreasing intensity of landuse away from the central market centre. However, the role of increasing pressure of population on land cannot be neglected. In fact this factor has gone a long way in changing the cropping pattern from lesser intensive to more intensive in the environs of Delhi. It is the increasing population of Delhi urban area which has increased the requirements for milk and vegetables, thereby attracting more and more people to grow vegetable crops and take up dairying as an occupation.

In a nut-shell, it can be safely confirmed that the cropping pattern in the villages in the northern part of the Union Territory of Delhi, though undergoing rapid changes under the influence of urbanisation and technological changes, still conforms to that, generally envisaged by agricultural scientists, and the age-old ideas of Von Thunen and his followers.

REFERENCES

Amani, K. Z., 1968 : "Harnarayanpur : A Study in the Changes in Agricultural Landuse". *Geographical Outlook*, 5, p. 71.

Anderson, J. R., 1973 : *A Geography of Agriculture in the United States Southeast.*

Best, R. H. & Coppock, J. T., 1962 : *The Changing use of Land in Great Britain.*

Blakie, P. M., 1971 : "Organisation of Indian Villages", *Transactions of the Institute of British Geographers*, No. 52, London, pp. 15.

Bradford, M. G. & Kent, W. A., 1977 : *Human Geography : Theories and their Applications.*

Census of India, 1971 : Delhi, Pts. X A & B. *Town and Village Directory.*

———, 1971 : Delhi Pt. II-A—*General Population Tables.*

Chisholm, M., 1962 : *Rural Settlements and Landuse.*

Dalton, R. T., 1971 : "Peas for Freezing — a recent development in Lincolnshire agriculture", *East Midland Geographer*, 35, p. 133-41.

Day, R. R. & Tinney, E. H., 1969 : "A Dynamic Von Thunen Model", *Geographical Analysis*, 1, pp. 137-51.

Dunn, E. S., 1954 : *The Location of Agricultural Production.*

Fielding, G. J., 1965 : "The Role of Government of New Zealand Wheat Growing", *Annals of the Association of American Geographers*, Vol. 55, pp. 87-97.

Found, W. L., 1971 : *A Theoretical Approach to Landuse Patterns.*

Garrison, W. and Marble, D. F., 1957 : "The Spatial Structure of Agricultural Activities", *Annals of the Association of American Geographers*, 47, pp. 137-43.

Grotewold, A., 1959 : "Von Thunen in Retrospect," *Economic Geography,* 35, pp. 346-55.

Haggett, P., 1965 : *Locational Analysis in Human Geography.*

Harvey, D. W., 1963 : "Locational Change in the Kentish Hop Industry and the Analysis of Landuse Patterns", *Transactions of the Institute of British Geographers,* pp. 123-44.

———, 1966 : "Theoretical Concepts and the Analysis of Agricultural Landuse Patterns in Geography", *Annals of the Association of American Geographers,* 56, pp. 361-74.

Hoover, E. M., 1963 : *The Location of Economic Activity.*

Horwath, R. J., 1969 : "Von Thunen's Isolated State and the Area Around Addis Ababa", *Annals of the Association of American Geographers,* 59, pp. 308-23

Jasbir Singh, 1974 : *Agricultural Atlas of India : A Geographical Analysis.*

———, 1976 : *Agricultural Geography of Haryana.*

Johnson, H. B., 1962 : "A Note on Thunen's Circles", *Annals of the Association of American Geographers*, 52, pp. 213-20.

Jonnason, O., 1925 : "Agricultural Regions of Europe", *Economic Geography*, pp. 277-314.

Mead, W. R., 1958 : *An Economic Geography of the Scandinavian States and Finland.*

Mohammad, Noor, 1978 : *Agricultural Landuse in India*, Delhi.

Morgan, W. B., 1969 : "The Zoning of Landuse around Rural Settlements in Tropical Africa", Thoman, M. F. and Whittington G. W. (eds.), *Environment and Landuse in Africa.*

———, 1973 : "The Doctrine of Rings", *Geography*, 58, pp. 301-12.

Morgan, W. B., and Munton, R. J., 1971 : *Agricultural Geography.*

Peet, J. R., 1969 : "The Spatial Expansion of Commercial Agriculture in the Nineteenth Century—a Von Thunen Interpretation"—*Economic Geography*, 45, pp. 283-301.

Rutherford, J., Logan, I. and Misseon, G.J., 1969 : *New Viewpoints in Economic Geography.*

Sauer, O., 1919 : "Mapping the Utilisation of the Land", *Geographical Review,* 8.

Symons, L., 1968 : *Agricultural Geography.*

Tidswell, Vincent, 1976 : *Pattern and Process in Human Geography.*

Von Thunen, 1826 : *Thunnen Isolated State* (Translated by Hall, P.) (ed.), 1966.

Walker, Michael, 1977 : *Agricultural Location.*

6

CHANGING PATTERNS OF CROPS AND LANDUSE IN HARYANA[1]

J. P. SINGH

I

CHANGES IN pattern of crops overtime are continuous and dynamic processes. The pace and pattern of these changes depend on various factors, such as technological economic, institutional and physical. From pure economic angle, a farmer's decision to go for a particular farm enterprise is primarily based on its relative profitability vis-a-vis other competing enterprises. The relative profitability of a crop depends upon technology, prices of inputs and output, quantity of inputs used, yield per hectare and facilities for its market clearance vis-a-vis competing enterprises. Apart from relative profitability, irrigation, rainfall, soil type, topography, temperature and related weather factors, closeness to market centres, subsistence requirements of farm families, etc. also exert considerable pressure on changes in cropping pattern. As there is no instrument available with the State to force farmers to adopt a particular type of cropping pattern, the change in cropping pattern would be effected mainly through irrigation, technological change, and price policy. These aspects would be looked into while examining cropping pattern changes in Haryana State.

The extent of changing in cropping pattern varies considerably among different regions of Haryana depending upon variations in their agro-climatic factors and infrastructural facilities. In view of these considerations, this paper examines the extent of change in cropping pattern in different agro-climatic regions of Haryana, economic rationale of changing cropping pattern in these regions and scope for its diversification in future. Attempt has also been made to understand the reasons for existence of a particular type of cropping pattern at a particular point of time and for its change over the years.

II

METHODOLOGY AND SOURCE OF DATA

The State of Haryana has been divided into four regions on the basis of broad similarity in agro-climatic conditions and cropping patterns of districts within each region. These are:

1. Karnal, Kurukshetra and Ambala region,
2. Rohtak, Sonepat and Jind region,
3. Hissar, Bhiwani and Sirsa region,
4. Gurgaon, Faridabad and Mahendargarh region.

For the sake of convenience, these have been nicknamed as Karnal region, Rohtak region, Hissar region and Gurgaon region. The districts coming in each region are contiguous.

Changes in cropping patterns have been examined in terms of changes in percentage share of individual crops in gross cropped area at four points of time. At each point of time, three yearly averages of data on area under individual crops have been taken and used for working out their share in gross cropped area. The minor crops, for which separate analysis could not be done, have been grouped in the category of residual crops. The points at which the analysis of cropping pattern and related aspects have been done are 1969-72, 1974-77, 1979-82, and 1983-86. The growth rates of area, production and yield of different crops at the State level have been worked out between the first and the last point of time. Irrigation, high yielding varieties, price and price policy, yield and relative profitability of crops have been taken into consideration while examining changes in cropping pattern.

The possibility of area shift from one crop to another has been examined on the basis of economic return. This has been done by comparing the value productivity and profitability of competing crops. The value productivity per hectare has been obtained by multiplying yield by farm harvest price of individual crops. This will help in comparing productivity of different crops, which is not feasible in terms of physical yields. Some idea about relative profitability of different crops has been obtained by deducting cost of production from value productivity per hectare. The amount of cost of production of individual crops has been obtained on the basis of the ratio of cash and kind expenses (cost A_2) to the value of output per hectare at whatever points of time cost of production figures are available.[2] In view of non-availability of data on cost of production of all the crops for recent years, this method has to be adopted to work out a rough measure of profitability of individual crops at the last point of time. The examination of prospect of shift of area on the basis of relative productivity and profitability of crops has some good reasons, as both these measures take into account the impact of price and yield of individual crops and hence, incorporate the influence of price policy and level of production technology. In addition, the measure of relative profitability accounts for the cost of production of different crops. At the State level, coefficients of variation of yield of individual crops during 1969-70 to 1985-86 have been worked out in order to have an idea about the extent of uncertainty in their yield and its relation with changes in cropping pattern.

III

SPATIAL CROP AND LANDUSE PRACTICES

Landuse and Irrigation

In this section, the change in extent or area under cultivation and irrigation and that in intensity of irrigation and intensity of cropping between 1969-72 and 1983-86 have been examined (Table 6.1). During this period, the net sown area as percentage of geographical area increased from 80.9 to 82.3, net irrigated area as percentage of net sown area increased from 42.2 to 61.9, gross irrigated area as percentage of gross cropped area increased from 44.9 to 65.0, irrigated area by tubewells as percentage of net irrigated area increased from 26.0 to 44.6, intensity of irrigation increased from 149 to 163 per cent and intensity of cropping increased from 140 to 155 per cent in the State of Haryana. These improvements would certainly have a salutary effect on production performance of agriculture in Haryana.

TABLE 6.1: Percentage of net sown area, irrigated area, intensity of irrigation, and intensity of cropping in regions of Haryana in 1969-72 and 1983-86.

Regions	Net Sown area as % of village paper area		Net Irrigated area as % of Net Sown area		Gross Irrigated area as % of gross cropped area		Irrigated area by tubewells as % of NIA		Intensity of Irrigation		Intensity of Cropping	
	1969-72	1983-86	1969-72	1983-86	1969-72	1983-86	1969-72	1983-86	1969-72	1983-86	1969-72	1983-86
1. Karnal region	75.53	81.33	54.02	78.67	58.73	86.60	53.29	74.76	156.5	170.8	143.9	160.8
2. Rohtak region	83.27	81.26	45.15	71.84	47.95	69.04	6.75	29.31	154.5	152.5	145.5	158.70
3. Hissar region	84.35	86.04	43.70	55.97	49.61	61.80	3.66	14.45	150.0	167.3	132.2	151.5
4. Gurgaon region	80.15	77.28	23.28	38.32	18.76	38.03	52.33	83.41	114.5	149.8	142.0	151.0
Haryana	80.86	82.33	42.18	61.92	44.86	65.01	25.99	44.63	148.8	163.0	139.9	155.2

So far as the regional profile is concerned, the net sown area as percentage of village paper area increased from 75.5 to 81.3 in Karnal region and from 84.4 to 86.0 in Hissar region but declined from 83.3 to 81.3 in Rohtak region and from 82.2 to 77.3 in Gurgaon region. The declining share of net sown area in the latter two regions are because of fast increase in cultivable waste and current fallows in Jind district of Rohtak region and in area under forest, non-agricultural uses and current fallows in Gurgaon district and under forest, non-agricultural uses and barren and unculturable land in Mahendargarh district of Gurgaon region. The net irrigated area as percentage of net sown area increased very fast in all the regions and stood at 78.7 in Karnal region, 71.8 in Rohtak region, 56.0 in Hissar region and 38.3 in Gurgaon region in the mid-Eighties. The percentage of gross cropped area irrigated also follows similar pattern. The irrigated area by tubewells as percentage of net irrigated area increased from 53.3 to 74.8 in Karnal region, 6.8 to 29.3 in Rohtak region, 3.7 to 14.5 in Hissar region and 52.3 to 83.4 in Gurgaon region. Thus, not only that the extent of irrigation is very high in Karnal region but also tubewells are the major source of irrigation. On the other hand, the extent of irrigation is quite low in Gurgaon region, but it is largely met through tubewells. On the contrary, canal irrigation is more important in Rohtak and Hissar region. It is thus clear that irrigation infrastructure is quite high in Karnal and Rohtak region as compared to Hissar and Gurgaon region. This would, of course, have considerable impact on change in cropping pattern and agricultural growth of these regions.

The intensity of irrigation increased considerably in all the regions except Rohtak region. It stands at around 170 in Karnal and Hissar regions and around 150 in Rohtak and Gurgaon regions. The increase in intensity of cropping has been quite substantial in all the regions and stands around 160 in Karnal and Rohtak regions, but around 151 in Hissar and Gurgaon regions. As expected, the intensity of irrigation seems to bear fairly good relation with the extent of net sown area irrigated and area irrigated by wells and tubewells. The intensity of cropping is very closely related with the intensity of irrigation. This clearly implies that unless intensity of irrigation is raised, the intensity of cropping would not increase. Since wells and tubewells, which are largely privately owned, provide greater control over the source of irrigation, the large availability of irrigation through these sources will shift cropping pattern in favour of those crops that need frequent, timely and controlled irrigation, such as rice and wheat. The region with low extent of irrigation will naturally go for dry land crops, to a large extent.

Crop Pattern Changes

The changes in cropping pattern in regions of Haryana since the early Seventies are examined in the following:

Karnal Region

The percentage shares of individual crops in the gross cropped area in Karnal region at four points of time are presented in Table 6.2, which indicates that rice-wheat cropping system was prominent in the region in 1969-72, as rice accounted for 17.3 per cent and wheat accounted for 35.4 per cent of the gross cropped area. Maize in Kharif season and gram in rabi season were the next two important crops and accounted for about 7.4 per cent and 9.7 per cent of the gross cropped area, respectively. Sugarcane and *Bajra* were also quite important as they, respectively, accounted for about 4.6 per cent and 3.5 per cent of the gross cropped area. Pulses, excluding gram and all the oilseeds accounted for a very small proportion of gross cropped area in 1969-72. The share of area under *Masur* was 1.64 per cent, *Urad* 0.62 per cent and other pulses 0.62 per cent. So far as oilseeds are concerned, share of area under groundnut was 0.8 per cent, rapeseed-mustard 1.6 per cent, sesamum 0.08 per cent and linseed 0.04 per cent. The area share of residual crops was 12.4 per cent, *jowar* 1.5 per cent and remaining crops less than 0.5 per cent. It is thus clear from the above that the cropping pattern was dominated by rice, maize, sugarcane and *bajra* in kharif season and wheat and gram in rabi season in 1969-72.

An examination of change in cropping pattern since 1969-72 indicates that the change was very slow between 1969-72 and 1974-77, but very sharp after 1974-77. Among the important crops, share of area under rice and wheat increased very fast, while that under maize and gram declined very sharply. The share of gross cropped area under rice increased to 29.8 per cent and wheat to 42.6 per cent, while that under maize declined to 3.3 per cent and gram to 0.86 per cent by 1983-86. The area shares of most of the remaining crops declined substantially, while these remained at around the same level in the case of a few crops. So far as the total pulses and oilseeds are concerned, their area shares declined, respectively, from 12.54 per cent and 2.52 per cent in 1969-72 to 2.60 per cent and 1.73 per cent in 1983-86.

TABLE 6.2: **Percentage share of individual crops in gross cropped area in Karnal region.**

Crops	1969-72	1974-77	1979-82	1983-86
1. Rice	17.28	18.75	27.89	29.77
2. Jowar	1.51	0.80	0.48	0.39
3. Bajra	3.52	2.61	1.41	1.39
4. Maize	7.35	8.14	4.40	3.28
5. Wheat	35.43	35.44	39.73	42.56
6. Barley	1.34	2.17	0.69	0.32
7. Gram	9.66	7.22	3.17	0.86
8. Urad	0.62	0.58	0.63	0.35
9. Moong	0.02	0.02	0.02	0.02
10. Masur	1.64	1.24	1.13	0.74
11. Other Pulses	0.62	0.44	0.56	0.63
12. Groundnut	0.78	0.73	0.41	0.28
13. Rapeseed-Mustard	1.62	1.52	1.11	1.41
14. Sesamum	0.08	0.14	0.07	0.07
15. Linseed	0.04	0.03	0.03	0.01
16. Sugarcane	4.56	4.89	4.39	4.30
17. Tobacco	0.01	0.01	0.01	0.02
18. Potato	0.29	0.77	0.76	0.58
19. Cotton American	0.42	0.15	0.31	0.08
20. Cotton (*Desi*)	0.83	0.77	0.70	0.66
Total (above crops)	87.62	86.42	87.83	87.66
Residual crops	12.38	13.58	12.17	12.34
Total (GCA)	100.00	100.00	100.00	100.00
Gross Cropped area (in thousand hectare)	1279.34	1342.00	1401.00	1457.00

Rohtak Region

The cropping pattern of Rohtak region at four points of time is presented in Table 6.3. It is clear from the table that the cropping pattern of this region was dominated by *bajra* and *jowar* in Kharif season and wheat and gram in Rabi season in 1969-72. The share of gross cropped area under wheat was 25.66 per cent, gram 22.55 per cent, bajra 16.91 per cent, jowar 10.48 per cent and sugarcane 5.29 per cent in 1969-72. The area under rice and cotton each was less than 2.50 per cent. The share of area under oilseed crops was less than 2.0 per cent and that under pulses excluding gram was less than one per cent. None of the remaining crops except barley occupied even one per cent of the gross cropped area in 1969-72.

The cropping pattern prevailing in 1969-72 got substantially changed during the following years. The change was the most spectacular in the case of gram whose area share declined to 8.90 per cent and wheat whose area share increased to 37.00 per cent in 1983-86. Changes in area shares of some other crops are also quite important. The percentage share of area under rice increased from 2.40 to 6.02, other pulses from 0.44 to 1.33, rapeseed-mustard from 1.63 to 4.31 and cotton (American) from 0.44 to 1.52 between 1969-72 and 1983-86, while that under maize declined from 0.92 to 0.28, barley from 1.74 to 1.28 and cotton (*Desi*) from 2.02 to 1.63 during the same period. On the other hand, the percentage share of area under jowar declined to 7.49, bajra to 14.19 and sugarcane to 3.70 in 1983-86. The area share of remaining minor crops either remained constant or declined to a certain extent. The share of residual crops increased from about 9.0 per cent to 11.8 per cent between 1969-72 and 1983-86.

Hissar Region

The cropping pattern of this region at four points of time is presented in Table 6.4. It is important to note from the table that gram is the most important crop, followed by wheat and bajra. But in 1969-72, bajra was much more important than wheat. Although, gram still remains the most important crop in this region, its area share has consistently declined over the years. The area share of bajra also declined substantially in the region. Between 1969-72 and 1983-86, the percentage share of area under gram declined from 29.64 to 21.94, bajra from 21.95 to 16.69, jowar from 1.87 to 0.40, barley from 0.88 to 0.58, moong from 0.82 to 0.26, other pulses from 0.66 to 0.34, sugarcane from 1.10 to 0.29 and cotton (*Desi*) from 5.12 to 3.14. On the other hand, percentage share of area under wheat increased from 13.33 to 19.93, rapeseed-mustard from 3.56 to 6.73, cotton (American) from 5.54 to 11.69 and rice from 1.18 to 2.67. So far as the total pulses are concerned, their area share declined from 31.22 per cent to 22.64 per cent. On the other hand, share of area under oilseeds increased from 3.59 per cent to 6.86 per cent. This region of Haryana appears to have the best potential for increasing area in favour of oilseeds and pulses. In this region, area under cotton (American) and rapeseed-mustard has shown the sharpest increase.

TABLE 6.3: **Percentage share of individual crops in gross cropped area in Rohtak region**

Crops	1969-72	1974-77	1979-82	1983-86
1. Rice	2.39	2.21	5.59	6.02
2. Jowar	10.48	8.42	6.29	7.49
3. Bajra	16.91	18.87	16.63	14.19
4. Maize	0.92	0.91	0.49	0.28
5. Wheat	25.66	27.88	33.58	37.00
6. Barley	1.74	2.66	1.68	1.28
7. Gram	22.55	17.72	12.89	8.90
8. Urad	0.02	0.03	0.05	0.02
9. Moong	0.03	0.04	0.04	0.04
10. Masur	0.40	0.42	0.41	0.33
11. Other Pulses	0.44	0.62	0.41	1.33
12. Groundnut	—	—	0.01	0.01
13. Rapeseed-Mustard	1.63	2.37	2.31	4.31
14. Sesamum	0.06	0.08	0.07	0.07
15. Linseed	—	—	—	—
16. Sugarcane	5.29	6.11	4.44	3.70
17. Tobacco	0.01	0.004	0.02	—
18. Potato	0.04	0.06	0.06	0.06
19. Cotton (American)	0.44	0.37	0.87	1.52
20. Cotton (*Desi*)	2.02	2.12	2.07	1.62
Total (above crops)	91.04	90.90	87.91	88.16
Residual crops	8.96	9.10	12.09	11.84
Total (GCA)	100.00	100.00	100.00	100.00
Gross Cropped area (in thousand hectare)	1058.33	1150.00	1139.67	1190.67

Gurgaon Region

The data on percentage share of individual crops in gross cropped area at four points of time are presented in Table 6.5 for this region. An examination of data presented in the table reveals that the cropping pattern of this region has shifted substantially in favour of wheat and rapeseed-mustard and against gram, bajra and barley crops. Most of other crops also lost their share in gross cropped area. Among the important crops, the percentage share of area under wheat increased from 16.11 to 28.89 and rapeseed-mustard from 4.06 to 9.04, while it declined under bajra from 30.49 to 23.49, gram from 24.37 to 12.38, jowar from 4.72 to 4.01, barley from 6.05 to 4.81 and sugarcane from 1.38 to 0.54 between 1969-72 and 1983-86. The share of area under total pulses declined from 26.62 per cent to 14.24 per cent and that under total oilseeds increased from 4.12 per cent to 9.44 per cent during the period. The share of area under residual crops increased from 9.5 per cent to 13.2 per cent.

TABLE 6.4: **Percentage share of individual crops in gross cropped area in Hissar region**

Crops	1969-72	1974-77	1979-82	1983-86
1. Rice	1.18	1.27	2.00	2.67
2. Jowar	1.87	0.92	0.47	0.40
3. Bajra	21.95	26.01	18.54	16.69
4. Maize	0.26	0.17	0.11	0.10
5. Wheat	13.33	12.83	15.18	19.93
6. Barley	0.88	1.58	0.91	0.58
7. Gram	29.64	26.72	25.78	21.94
8. Urad	0.01	0.02	0.02	0.01
9. Moong	0.82	0.76	0.14	0.26
10. Masur	0.09	0.10	0.12	0.09
11. Other Pulses	0.66	0.50	0.20	0.34
12. Groundnut	0.01	0.01	0.02	0.18
13. Rapeseed-Mustard	3.56	2.95	6.16	6.73
14. Sesamum	0.02	0.03	0.03	0.03
15. Linseed	—	—	0.19	—
16. Sugarcane	1.10	0.81	0.53	0.29
17. Tobacco	0.01	0.02	0.00	0.00
18. Potato	0.01	0.02	0.04	0.02
19. Cotton (American)	5.54	6.13	9.94	11.69
20. Cotton (*Desi*)	5.12	5.27	3.64	3.14
Total (above crops)	86.09	86.1´	83.99	85.10
Residual crops	13.91	13.89	16.01	14.90
Total (GCA)	100.00	100.00	100.00	100.00
Gross Cropped area (in thousand hectare)	1556.00	1812.00	1999.00	2016.13

TABLE 6.5: Percentage share of individual crops in gross cropped area in Gurgaon region

Crops	1969-72	1974-77	1979-82	1983-86
1. Rice	0.20	0.33	0.55	0.81
2. Jowar	4.72	5.06	4.86	4.01
3. Bajra	30.49	26.34	30.01	23.49
4. Maize	0.49	0.64	0.37	0.40
5. Wheat	16.11	22.74	30.00	28.89
6. Barley	6.05	9.06	7.07	4.81
7. Gram	24.37	18.23	7.46	12.38
8. Urad	0.07	0.09	0.19	0.11
9. Moong	0.58	0.16	0.08	0.10
10. Masur	0.13	0.23	0.31	0.42
11. Other Pulses	1.47	1.21	0.89	1.23
12. Groundnut	0.02	0.10	0.10	0.10
13. Rapeseed-Mustard	4.06	5.08	5.22	9.04
14. Sesamum	0.04	0.12	0.15	0.30
15. Linseed	—	—	—	—
16. Sugarcane	1.38	1.39	0.95	0.54
17. Tobacco	0.02	0.01	0.00	0.01
18. Potato	0.03	0.04	0.06	0.07
19. Cotton (American)	0.14	0.07	0.10	0.07
20. Cotton (*Desi*)	0.10	0.04	0.10	0.03
Total (above crops)	90.49	90.94	88.51	86.82
Residual crops	9.51	9.06	11.49	13.18
Total (GCA)	100.00	100.00	100.00	100.00
Gross Cropped area (in thousand hectare)	1088.33	888.00	877.33	936.67

Haryana State

The cropping patterns of Haryana State at four points of time are present in Table 6.6. It is clear from the table that rice in kharif season and wheat in Rabi season have gained substantial proportion of gross cropped area. Among the main losers are bajra in Kharif season and gram in Rabi season. The area share of wheat increased from 22.24 per cent to 30.94 per cent and rice from 5.36 per cent to 10.13 per cent between 1969-72 and 1983-86, while the area share of bajra declined from 18.0 per cent to 13.3 per cent and gram from 21.9 per cent to 12 per cent. Most of other crops also lost their relative share in gross cropped area. The area shares of jowar, maize and barley declined substantially. The area share of six important cereal crops increased from 54.42 per cent in 1969-72 to 59.26 per cent in 1983-86. On the other hand, the area share of pulses, taken together, declined consistently from 23.77 per cent in 1969-72 to 13.43 per cent in 1983-86, the decline being the sharpest in the case of gram. Although the share of area of individual oilseeds is very small, except rapeseed-mustard, it is encouraging to note that area shares of rapeseed-mustard and sesamum showed considerable increase over the period of time. The percentage share of area of rapeseed-mustard increased from 2.76 to 4.77 during 1969-72 and 1983-86. As a result, the area share of all the oilseeds crops, taken together, increased from 3.02 per cent in 1969-72 to 5.01 per cent in 1983-86. The area share of American cotton increased substantially, while that of Desi cotton declined. The overall share of cotton crops increased from 4.21 per cent to 6.21 per cent between 1969-72 and 1983-86. Among the other important cash crops, area share of sugarcane declined, while that of potato increased substantially. The area share of residual crops increased consistently over the years and stood at about 14 per cent in 1983-86.

An examination of data on the magnitude of area gained and lost by individual crops between 1969-72 and 1983-86 indicates that the actual gain in cropped area in Kharif season was 120.4 thousand hectares, in Rabi season 301.3 thousand hectares, and by residual crops 196.8 thousand hectares, thus amounting to total increase of about 619 thousand hectares. But, out of this total increase, only 50 thousand hectares have come through the increase in net sown area and the remaining 569 thousand hectares through increase in intensity of cropping. So far as the individual crops are concerned, they have gained area on account of shift in area from other crops, intensity of cropping and increase in net sown area. Among the major crops gaining large amount area in Kharif season during the period covered in the study are rice and cotton (American), each gaining, respectively, about 300 thousand hectares and 158 thousand hectares. Among the crops losing substantial amount of cropped area are bajra, jowar, maize, sugarcane, cotton (*Desi*) and moong. Sesamum and other pulses gained about 3.0 thousand hectares each. So far as the Rabi crops are concerned, wheat gained as much as 625.20 thousand hectares, rapeseed-mustard 129.6 thousand hectares, and potato 5.3 thousand hectares. On the other hand, gram lost 412.1 thousand hectares, barley 38.3 thousand hectares, and masur 7.6 thousand hectares. The shifts in area were much more important during 1974-77 and 1979-82 than the sub-periods preceding and succeeding it. When the area increase of crops that have gained area between 1969-72 and 1983-86 is added up, it works out to be about 1402 thousand hectares. Since the increase in gross cropped area was only about 619 thousand hectares, the area gained by crops through shifts of area from other crops works out to be 783 thousand hectares. Thus, shift in area has been relatively more important source of change in cropping pattern than the increase in gross cropped area in Haryana State.

TABLE 6.6: **Percentage of gross cropped area under different crops at four points of time-Haryana state**

Crops	1969-72	1974-77	1979-82	1983-86
1. Rice	5.36	5.83	9.27	10.13
2. Jowar	4.23	3.26	2.42	2.50
3. Bajra	18.01	18.44	15.71	13.31
4. Maize	2.27	2.47	1.35	1.01
5. Wheat	22.24	23.71	27.98	30.94
6. Barley	2.31	3.25	2.01	1.37
7. Gram	21.86	18.23	14.34	12.08
8. Urad	0.18	0.18	0.20	0.14
9. Moong	0.39	0.30	0.07	0.10
10. Masur	0.56	0.49	0.48	0.36
11. Other Pulses	0.78	0.59	0.45	0.75
12. Groundnut	0.21	0.21	0.13	0.15
13. Rapeseed-Mustard	2.76	2.82	3.91	4.77
14. Sesamum	0.04	0.08	0.07	0.09
15. Linseed	0.01	0.01	0.01	0.00
16. Sugarcane	2.94	3.13	2.41	2.10
17. Tobacco	0.01	0.01	0.00	0.00
18. Potato	0.09	0.23	0.23	0.18
19. Cotton (American)	1.95	2.27	3.95	4.56
20. Cotton (*Desi*)	2.26	2.51	1.99	1.65
Total (above crops)	88.48	88.02	87.01	86.23
Residual crops	11.52	11.98	12.99	13.77
All Crops	100.00	100.00	100.00	100.00

A summary view of the performance of crops in Haryana could be had from compound growth rate per annum of area, production and yield of individual crops between 1969-72 and 1983-86. The output, area and yield of all crops, taken together, have been growing at the rate of 3.2 per cent, 0.65 per cent and 2.5 per cent per annum, respectively. The majority of the crops show negative growth of their output and area and about two-fifths of them show negative growth of their yield also. In fact, the crops showing negative growth of yield invariably show negative growth of area and hence, negative growth of their output. It is important to note that some of the important crops losing area at a substantially high rate have also shown considerable decline in their yield. This is particularly true for gram, bajra, groundnut and cotton (*Desi*) crops. This could have resulted due to shift of better quality land from these crops to other crops. Out of twenty crops for which growth rates have been examined, only rice, wheat, other pulses, rapeseed-mustard, sesamum, potato and cotton (American) show positive growth of their area. All these crops invariably show positive rate of growth of yield, which is relatively higher than their competing crops. The growth of yield thus seems to be important for attracting area under a crop, provided it is relatively more prominent than the competing crops and its (Crop's) level of productivity is relatively high. Since major crops like rice, wheat and cotton (American), despite having fairly high level of productivity at the base, showed fairly good growth of their yield during the period of analysis, their gaining of substantial amount of area is quite understandable.

As regards the magnitude of growth of important crops, the growth per annum of output of wheat was 5.3 per cent, rice 8.6 per cent, and cotton (American) 7.9 per cent. Among the relatively less important crops, the rate of growth of output per annum was 7.7 per cent for rapeseed-mustard, 5.2 per cent for other pulses, 6.7 per cent for potato and 7.5 per cent for sesamum. The rate of growth of area was the major source of output growth for all these crops. Among the important crops, the rate of decline of output per annum was 5.1 per cent for gram, 2.7 per cent for bajra, 4.6 per cent for maize, 1.4 per cent for groundnut, 1.8 per cent for sugarcane and 3.0 per cent for cotton (*Desi*). In the case of all these crops, the main source of decline in output was the decline in their area, the decline in yield being relatively much less important.

IV

FACTORS INFLUENCING CROPPING PATTERN CHANGES

The causes for changes in cropping pattern have been examined at the State level in terms of prices, productivity, profitability, technology and irrigation.

An examination of rate of growth of farm harvest prices of individual crops between 1969-72 and 1982-85 indicates that the crops gaining substantial amount of area have shown, in general, a lower growth of prices than their competing crops losing substantial amount of area. But the coefficient of variation of prices of the former group of crops is definitely lower than the latter group. The crops gaining area have shown relatively much higher rates of growth of their yield. Hence, stability of their prices appear to be more important from the point of view of gaining area than the increase in prices with large fluctuations. The higher increase in prices of crops losing substantial amount of area appears to be due to decline in their supply. The price policy of the Government particularly in respect of rice and wheat, for which procurement operations are carried out effectively and for which farmers are assured to receive procurement prices fixed in advance by the Government, has definitely had a salutary effect on the shift in area in their favour.

The value productivity per hectare at 1969-72 prices for all crops, taken together, increased from Rs. 1144 in 1962-72 to Rs. 1623 in 1983-86. The value productivity of coarse cereals in Kharif season, i.e. jowar, bajra and maize, is not only far too lower than rice but also has been declining or stagnating. The level of productivity of Kharif pulses and oilseeds have also been much lower than rice. It is thus natural that rice has been gaining area at the cost of these crops. The productivity of cotton (American) has been higher than rice and far too higher than its competing crops including cotton (*Desi*) throughout the period and hence, substantial shift in area in its favour is quite understandable. In Rabi season, the value productivity of wheat has been substantially higher than gram, barley, other Rabi pulses and oilseeds at both the points of time. As compared to increasing productivity of wheat, the productivity of gram, the crop losing largest amount of area to wheat, has been consistantly declining and stood at one fourth of the productivity of wheat in 1983-86. Thus, the shift of cropping pattern substantially in favour of wheat is very much justified. In general, the productivity levels of pulse crops have been low, although moong and other pulses showed substantial improvement over the years. Their later year's productivity compares well with coarse Kharif cereals. So far as oilseeds are concerned, rapeseed-mustard seems to have done well. Although the level of productivity of sugarcane and potato have been substantially higher than all the other crops, they have been losing area during recent years. This could be attributed to their high cost of production and poor marketing facilities. The relative profitability of crops gaining area has substantially been higher than those losing area. It is thus clear that relative levels of value productivity and profitability of crops have mainly been responsible for the shift of cropping pattern in Haryana.

There has been very fast increase in the percentage of cropped area under high yielding varieties of rice, wheat, maize and bajra crops, but the increase was much faster in the case of first two crops than the latter two crops. It stood at 82 and 94, respectively, in the case of rice and wheat as compared to 41 and 62 in the case of maize and bajra in 1983-86. The increase in area under high yielding varieties had led to substantial increase in both area and yield of rice and wheat, but there was considerable decline in both area and yield of bajra, while area under maize declined substantially despite a good increase in its yield. The decline in area under bajra and maize was mainly because the relative productivity of these two crops has been much lower than their competing crops. Thus, the availability of high yielding varieties of a crop and their adoption may not guarantee the shift in area in its favour unless its cultivation is relatively more profitable than its competing crops. These varieties would, however, be adopted so far as their yield remains higher than the traditional varieties. This is what has happened in the case of bajra and maize. The increase in area under rapeseed-mustard and Tur (*Arhar*) was mainly because of the availability of their improved varieties, which led to a fast increase in their yield.

The percentage of irrigated area under all crops increased from 45 in 1969-72 to 65 in 1982-85. In view of increased irrigation facilities and fast adoption of high yielding varieties of rice and wheat, oilseeds and pulses, as groups of crops, which respond poorly to irrigation, were major losers of their area to rice and wheat crops, whose response to irrigation is quite good. The decline in unirrigated area under all crops, taken together, between 1969-72 and 1982-85, comes to 788 thousand hectares, whereas the increase in total irrigated area during the period equals to 1173 thousand hectares. In fact, the loss of unirrigated area by crops was mainly on account of its being brought under irrigation. It is important to mention that unirrigated gram lost maximum amount of area i.e. 404 thousand hectares. Irrigated gram also lost about 93 thousand hectares of area. The increase in irrigated area under wheat during the above mentioned period comes to 683 thousand hectares. Thus, virtually all the area lost by gram seems to have gone to wheat with the increase in irrigation facilities. The second important crop having gained irrigated area is rice. The irrigated area gained by rice has mainly come from the reduction in unirrigated area of important Kharif crops, i.e. Kharif cereals, pulses, oilseeds and sugarcane, as a result of bringing it under irrigation and also shifting irrigated area from some of these crops. Thus, the increase in irrigation facilities has mainly been responsible for switching area from low value crops to high value crops. Some important crops largely grown under unirrigated condition do not provide sufficient returns to irrigation and hence, unirrigated area under these crops after receiving irrigation shifts to other crops having higher response to irrigation. Thus, the increase in irrigation facilities has been one of the most important factors effecting drastic changes in cropping pattern in Haryana. Whereas the percentage of irrigated area stood at 96 for rice, 95 for wheat, 92 for sugarcane and 99 for potato and cotton (American), it remained less than 30 in the case of bajra, jowar, maize, gram, urad, groundnut and sesamum in 1982-85.

Conclusions

The main objectives of this paper have been to examine changes in cropping pattern in different agro-climatic regions of Haryana State since the early seventies and to find out factors responsible for and rationale behind such changes. The analysis have been done by categorising the State of Haryana into four agro-climatic regions. The period covered in the study, is from 1960-70 to 1985-86. The analysis have been done at four points of time, i.e. 1969-72, 1974-77, 1979-82, and 1983-86, by taking three yearly average of data on various aspects.

The production performance of crops in Haryana State has been quite good. The output, area and yield of all crops, taken together, have been growing at an annual compound rate of 3.2 per cent, 0.65 per cent and 2.5 per cent, respectively, between 1969-72 and 1983-86. The majority of the crops show negative growth of their output and area and about two-thirds of the crops show negative growth of their yield. In fact, the crops showing negative growth of yield invariably show negative growth of area and hence, negative growth of their output. Out of twenty crops for which growth rates have been examined, only rice, wheat, other pulses, rapeseed-mustard, sesamum, potato and cotton (American) have shown positive growth rate of their area. The growth of yield is observed to be important for attracting area under a crop, provided it is relatively more prominent and its level of productivity is relatively higher than competing crops.

The land utilisation statistics of Haryana State indicate that more than 82.0 per cent of the geographical area is under cultivation, about 65.0 per cent of the gross cropped area is irrigated, the intensity of irrigation is 163 per cent and intensity of cropping is 155 per cent. As regards changes in cropping pattern, the share of area under wheat increased from 22.24 per cent to 30.94 per cent and rice from 5.36 per cent to 10.13 per cent between 1969-72 and 1983-86, while the share of area under bajra declined from 18.0 per cent to 13.3 per cent and gram from 21.9 per cent to 12.0 per cent. Most of the other crops, except a few, such as cotton (American), rapeseed-mustard, etc. also lost their share in gross cropped area. The area share of all pulses, taken together, declined from 23.77 per cent in 1969-72 to 13.43 per cent in 1983-86.

As regards regional profile, the net area sown as percentage of the village paper area stood at 81.3 in Karnal and Rohtak regions, 86.0 in Hissar region and 77.3 in Gurgaon regions in 1983-86, thus indicating that the expansion of the cultivation has almost reached to a maximum limit in all the regions. The net irrigated area as a percentage of net sown area increased very fast in all the regions and stood at 78.7 in Karnal region, 71.8 in Rohtak region, 56.0 in Hissar region, and 38.3 in Gurgaon region in the mid-eighties. Tubewells are the major source of irrigation in Karnal and Gurgaon regions, while canals are more important in Rohtak and Hissar regions. The increase in intensity of cropping has been quite substantial in all the regions and stood at around 160 in Karnal and Rohtak regions, and around 151 in Hissar and Gurgaon regions in 1983-86. As expected, the intensity of cropping is very closely related with the intensity of irrigation.

As regards the cropping pattern of individual regions, the rice-wheat system of cropping was prominent in Karnal region in 1969-72 and became much more prominent by 1983-86. The share in gross cropped area of rice increased from 17.3 per cent to 29.8 per cent and wheat from 35.4 per cent to 42.6 per cent between 1969-72 and 1983-86. Maize in Kharif season and gram in Rabi season were the next important crops whose share in gross cropped area declined from 7.4 per cent to 3.3 per cent and 9.7 per cent to 0.86 per cent, respectively, during the period. The share of area under most of the remaining crops also declined substantially over the years.

The cropping pattern of Rohtak region, which was dominated by *bajra* and *jowar* in Kharif season and wheat and gram in Rabi season in 1969-72, got substantially changed during the subsequent one and half decades. The change was the most spectacular in the case of gram whose share in gross cropped area declined from 22.50 per cent to 8.90 per cent and wheat whose share in gross cropped area increased from 22.7 per cent to 37.0 per cent during 1969-72 and 1983-86. The percentage share of area under bajra declined from 16.91 per cent to 14.19 per cent, jowar from 10.48 per cent to 7.49 per cent and sugarcane from 5.29 per cent to 3.70 per cent during the period, but the area share of rice increased from 2.40 per cent to 6.02 per cent and rapeseed-mustard from 1.63 per cent to 4.31 per cent. Thus, wheat, rice, and rapeseed-mustard were the main gainers of area in the region, as the relative productivity and profitability of these crops were higher than their competing crops.

In the cropping pattern of Hissar region, gram was the most important crop followed by bajra, wheat and cotton in 1969-72, their respective share in gross cropped area being 29.64 per cent, 21.95 per cent, 13.33 per cent, and 10.66 per cent. But, by 1983-86, the percentage share of these crops changed to 21.94, 16.69, 19.93 and 14.83, respectively. The share of rapeseed-mustard increased from 3.56 per cent to 6.73 per cent, and rice from 1.18 per cent to 2.67 per cent, between 1969-72 and 1983-86, while it showed a decline in the case of the most of the remaining crops. This region seems to be showing a good prospect for increasing area under oilseeds, in particular under rapeseed-mustard.

The cropping pattern of Gurgaon region has shifted substantially in favour of wheat and rapeseed-mustard and against gram, bajra and barley crops. Most of the other crops also lost their share in the gross cropped area. Among the important crops, the percentage share of area under wheat increased from 16.11 to 28.89 and rapeseed-mustard from 4.06 to 9.04, while it declined under bajra from 30.49 to 23.49, gram from 24.37 to 12.38, jowar from 4.72 to 4.01, barley from 6.05 to 4.81 and sugarcane from 1.38 to 0.54 between 1969-72 and 1983-86. The dry land crops still remain quite important in the region. Rapeseed-mustard seems to be having very good prospect of gaining area in the region.

Among the factors influencing change in cropping pattern, value productivity and profitability of crops have been observed to be the most important. These are, of course, influenced by irrigation, technology and prices. The stability in prices appear to be more important from the point of gaining area than the increase in them with large fluctuations. The increase in irrigation facilities had led to a shift in cropping pattern in favour of those crops that have higher response to irrigation, such as rice, wheat and cotton (American). Similarly, availability of high yielding/improved varieties has also been responsible in bringing change in cropping pattern.

NOTES

1. This paper is extracted from the authors report entitled "Prospects of Changing Cropping Pattern in favour of Oilseeds and Pulses in Punjab and Haryana", Research Study No. 89/1, Agricultural Economics Research Centre, University of Delhi, Delhi-110007 (Mimeo).
2. For details of methodology and source of data, see J.P. Singh, op. cit.

7

PATTERNS OF AGRICULTURAL LANDUSE IN GHAGHARA-RAPTI DOAB*

NOOR MOHAMMAD

GHAGHARA-RAPTI DOAB is essentially an agricultural region where nearly 90 per cent of the population directly or indirectly depends on agriculture. Thus the agriculture in this region is not only a source of livelihood but also a tradition and way of life.

The agricultural year is divided into two main seasons, i.e. Kharif — the season of summer crops, and Rabi — the season of winter crops (Shafi, 1960). The sowing of Kharif crops starts with the advent of the monsoon, generally in the second week of June. The principal Kharif crops are rice (broadcast and transplanted), small millets, maize, sugarcane[1] and arhar[2] (*Cajanus indicus*). Almost all the Kharif crops are produced without irrigation with the exception of sugarcane which is sown in the months of May and June when the amount of rainfall is negligible. The Rabi crops are usually sown in October and are harvested in March and April. The major crops of this season are wheat, gram, peas, wheat mixed with barley, and lentil. As during this season, the rainfall is meagre and variable, all the Rabi crops rely either on the moisture available in the soil after the monsoon rains or on irrigation from various sources. With the exception of lentil and gram almost all the Rabi crops are irrigated at least once during the growing period.

General Landuse

An attempt has been made here to work out the area and percentages of various uses of land in 1964-65 and the resulting figures have been

---

* Reprinted from Mohammad, N., *Agricultural Landuse in India*, Delhi, 1971.

given in Table 7.1. The percentages of the various uses of the land to the total area of the region have been shown in Fig. 7.1.[3]

The Table 7.1 gives an overall picture of landuse in Ghaghara-Rapti Doab. The total area of the region is 1,907,724 hectares, out of which 1,450,477 hectares or 76 per cent is the net sown area. About 11 per cent of the total area is classified as area not available for cultivation, and the other uncultivated land excluding fallow land covers about nine per cent of the total area. The remaining four per cent of the total area is left fallow.

Out of this 205,418 hectares classified as land not available for cultivation, 78,876 hectares or about 38 per cent is put to non-agricultural use, while the remaining 126,542 hectares or about 62 per cent is barren and uncultivable land. The former category includes settlements, roads, railways, cemeteries, cremation grounds, cart-tracks, embankments, *khuds* etc., while the latter includes all the area covered with water. It will be seen from Fig. 7.1 that the area under barren and uncultivable land generally increases from lower Doab towards upper Doab, while the cultivated land increases inversely. The area of the land not available for cultivation is comparatively higher in those parganas which are along the bank of the river Ghaghara. This is due to many reasons: firstly, most of the sandy deposits of the river is infertile and unprofitable for agricultural operations; secondly, the water of the Ghaghara spreads over a considerable distance on both sides during the rainy season; thirdly, there are innumerable rivulets, ox-bow lakes and shallow depressions along the sides of the Ghaghara; and lastly, the belt of the *usar* soil which passes along the southern edge of the *bhangar* land, runs through those parganas. The maximum expanse of such land occurs in parganas Chillupar, Digsir, Mahuli West, Dharmapur, Fakharpur and Hissampur.

The other uncultivated land excluding fallow land comprises groves, pastures, grasslands and other lands which have not been included in fallow as well as in lands not available for cultivation. Out of 170,234 hectares under this group, 47,474 hectares or about 28 per cent is under groves and 122,760 hectares or about 72 per cent is under culturable waste.[4] The possibilities of agricultural extension, whatsoever they are, should be directed towards such lands. Groves and pastures should be scientifically maintained in a reasonable area and the extra and unmanaged areas should be given to the cultivation of grain crops. Special attention should be focussed in parganas Dharmapur, Chardra, Sadullahnagar, Babhnipar and Nawabganj.

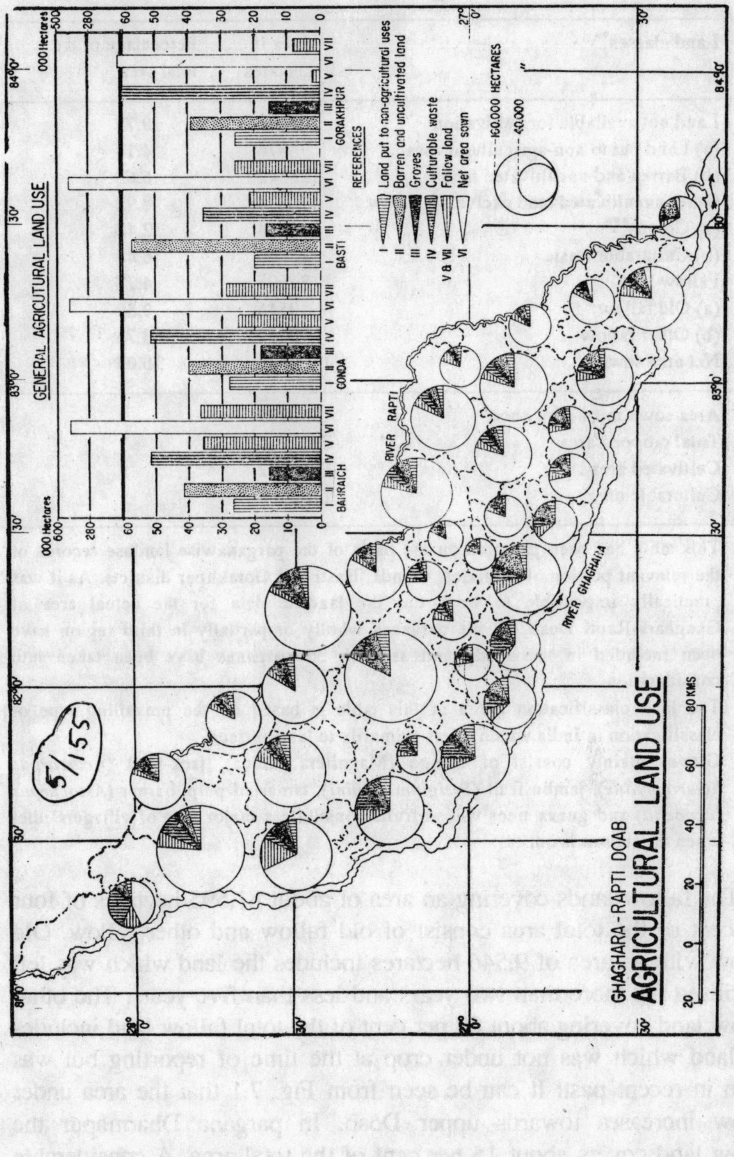

FIG. 7.1: Ghaghara-Rapti doab: agricultural landuse.

TABLE 7.1: General landuse in the region (total area 1,907,727 hectares)*

Land classes**	Area in hectares	Percentage of the total area
1. Land not available for cultivation	205418	10.77
(a) Land put to non-agricultural uses	78876	4.14
(b) Barren and uncultivated land	126542	6.63
2. Other uncultivated land excluding fallow	170234	8.92
(a) Groves***	47474	2.48
(b) Culturable waste	122760	6.44
3. Fallow land	81595	4.28
(a) Old fallow	9546	0.50
(b) Other fallow	72049	3.78
4. Net area sown	1450477	76.03
Area sown more than once	552239	
Total cropped area	2002716	
Cultivated area	1532072	
Culturable area	1702306	

* This table has been prepared on the basis of the parganawise landuse records of the relevant portion of Bahraich, Gonda, Basti and Gorakhpur districts. As it was practically impossible to work out the landuse data for the actual area of Ghaghara-Rapti Doab, all the parganas wholly or partially in third region have been included in this study and as such 36 parganas have been taken into consideration.

** The land classification given in this table is based on the prevailing type of classification in India which refers primarily to land surface.

*** Groves mainly consist of mango (Magnifera indica), jack-fruit (*Artocarpus heterophyllus*), jambu fruit (*Syzigium cumini*), tamarind pulp, barhar (*Artocarpus lakoocha*) and guava trees whose fruits constitute a major part of villagers' diet when the season is on.

The fallow lands covering an area of about 81,595 hectares of four per cent of the total area consist of old fallow and other fallow. Old fallow with an area of 9,546 hectares includes the land which was left unutilised for more than two years and less than five years. The other fallow land covering about 88 per cent of the total fallow land includes the land which was not under crop at the time of reporting but was sown in recent past. It can be seen from Fig. 7.1 that the area under fallow increases towards upper Doab. In pargana Dharmapur the fallow land covers about 15 per cent of the total area. A considerable area is also covered in parganas, Charda, Nanpara, Hissampur, Guwarich, Paharpur and Mahdeva. These fallow lands can easily be brought under cultivation with the help of manures, fertilisers and irrigation facilities.

The per capita cultivated land in this region is only 0.347 hectares (0.86 acre). In view of low production per hectare, it provides very low level of sustenance. The cultivated area can be increased to 2,754,832 hectares by bringing 122,760 hectares of culturable wasteland and 81,595 hectares of fallow land under cultivation. However, if the total cropped area (during the Kharif and Rabi seasons) is taken into consideration it amounts to 2,002,916 hectares and as such 0.477 hectare of 1.1 acres support one person in the region.

In Table 7.1 an attempt has been made to estimate the total cropped area, the cultivated area and the culturable area by adjustment in the given classes and sub-classes. The area sown more than once in an agricultural year is termed as double-cropped land which is 552,239 hectares or about 38 per cent of the net sown area or a little less than one-fourth of the total area. The total cropped area, therefore, includes net sown area and double cropped area (1,450,477 + 522,239 = 1,972,716 hectares) and as such it is 105 per cent of the region. These lands are also known as gross cultivated lands. Thus in all, the cultivated land includes the net sown area and fallow lands, i.e. 1,450,477 + 81,595 = 1,532,072 hectares which is about 80 per cent of the total area of the region. The estimate of culturable area is made by adding the last three broad landuse classes or by subtracting the lands not available for cultivation from the total area of the region. Thus it is 1,702,306 hectares (1,907,724–205,418 = 1,702,306 hectares) or about nine-tenths of the total area.

Cropping Patterns in Kharif Season

Kharif is an important agricultural season of the region when nearly three-fourths of the cultivated land is sown. A statistical statement of the area together with the percentage of land occupied by each crop is given in Table 7.2. The Fig. 7.2 shows the percentage of the area occupied by each crop in each pargana.

Table 7.2 shows many interesting features regarding the use of agricultural land in the *kharif* season. The overall cropping pattern in this season shows that out of the net sown area nearly 91 per cent is devoted to cereals and only 9 per cent is given to non-grain crops which include sugarcane, fodder, vegetables, oilseeds and other crops. Broadcast and transplanted rice, maize, small millets and pulses are the principal cereals of this region. The principal Kharif crops are:

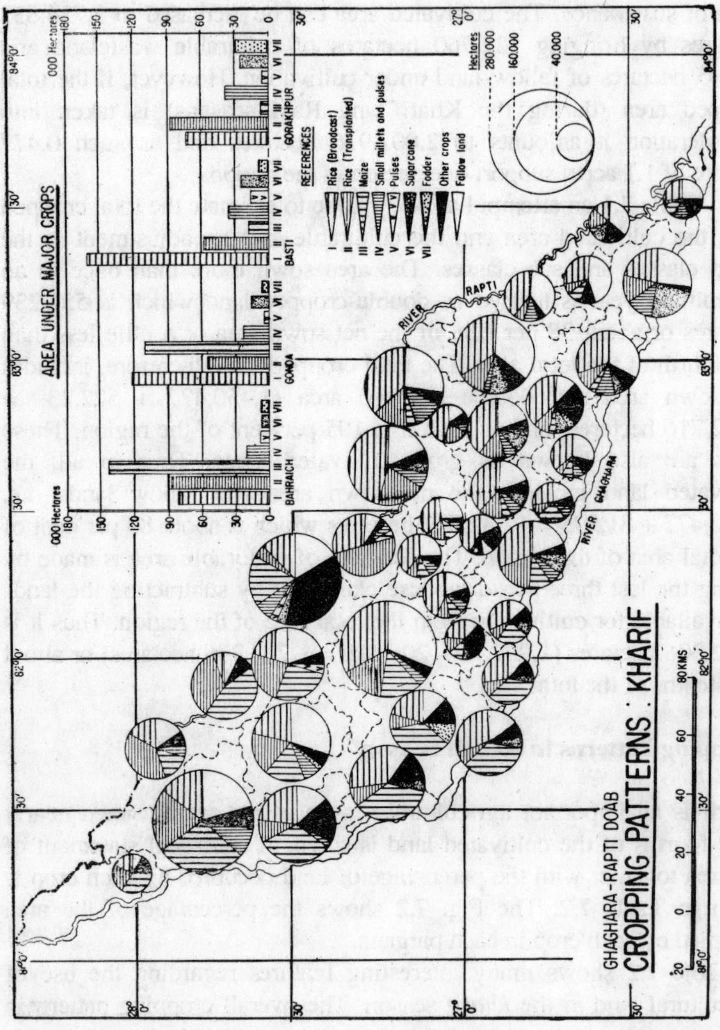

FIG. 7.2: Ghaghara-Rapti doab: cropping patterns — kharif.

TABLE 7.2*: Agricultural landuse in the Kharif season: Cultivated land . . . . 1,532,072 hectares net cropped land in the Kharif season 1,117,541 hectares

Crops	Area in hectares	Percentage of the cultivated land	Percentage of the net cropped land	Total percentage of the cultivated land	Total percentage of the net cropped land
Grain Crops				64.88	88.87
Rice (Broadcast)	400867	26.16	35.85		
Rice (Transplanted)	212894	13.89	19.05		
Maize	257619	16.84	23.04		
Small millet & pulses	96082	6.29	8.59		
Other cereals	26218	1.70	2.34		
Pulses				0.96	1.38
Arhar	4641	0.30	0.41		
Other pulses	10056	0.66	0.97		
Other Crops				7.10	9.75
Sugarcane	50625	3.30	4.53		
Fodder	24743	1.61	2.21		
Vegetables	5818	0.37	0.52		
Oilseeds	7322	0.47	0.65		
Others	20656	1.35	1.84		
Fallow	414531	27.06		27.06	
Total	1532072	100.00	100.00	100.00	100.00

*** The Table has been computed by the author on the basis of the parganawise unpublished record of the Kharif crops of 1964, obtained from the Board of Revenue, Uttar Pradesh (Lucknow).**

Rice (Oyza Sativa)

Rice has the distinction of being the most extensively cultivated crop of the region and of being the staple food of a major portion of the population. Although the fodder quality of the straw from this plant is poor, it is used as the main source of cattle feed in this region.

There are mainly two varieties of rice grown in this region viz. the early maturing variety which is sown from the middle of June to mid-July and is harvested by the end of October, and the late maturing variety which is sown in July and is harvested in December. A third kind of rice known as boro is also grown but on a very limited area around ponds or along the river banks. It is sown in March and is reaped in May.

Early and late maturing varieties of rice taken together cover a little more than half of the net sown area in the Kharif season. The broadcast rice matures in about three months and is locally called *bhadai* or *kunwari* as it is harvested in the months of *Bhadon* and *Kunwar*.[5] There are two prevailing methods of sowing the early maturing variety of rice in the region, i.e. the puddle cultivation and the dry cultivation. The cultivation in the puddle system is carried on in three different ways. In the first method, the seed is sown after being sprouted slightly; in the second, the seed is sown broadcast; in the third, the seedlings are separately raised in a nursery and then are transplanted in the puddled field at an interval of about 12 × 20 cm. However, in all the methods preparation of field is carried out in the same manner. Ploughing and cross-ploughing are repeated with short intervals of a few days. The water is retained in the field with the help of *mendh* (small earthen embankment) and the land is ploughed three or four times while it is still under water, and finally after harrowing, sprouted seeds or seedlings are broadcast or transplanted respectively.

In the dry cultivation the field is ploughed soon after the harvest of the previous crop, if possible, or with the first rain. The ploughings are repeated and at least four ploughings are given before the onset of the monsoon. The weeds are removed and burnt and manure is spread. After manuring the fields are once again ploughed after which they are considered to have been brought under proper tilth. With the first heavy shower of rain in the middle or third week of June, the seeds are sown or harrowed in. Sometimes the seeds are sown after an early shower or rain which makes the soil moist. In each case the seeds lie in the ground and do not germinate until after more rainfall. This practice locally known as *Dhuriana* saves the time needed for germination as the crop comes up after the first regular rain.

Broadcast rice covers 400,867 hectares or about one-third of the net sown area. It is generally sown in the best fields and is usually followed by peas and other staples in the Rabi season. The quality of rice is, however, poor and the yield is low. The important species are *sathi, babsam, mutmuri* and *anjuna*. Frequently *anjana* is mixed with *kodon* (Paspalm scrobiculatam) and this mixed crop is called *Dhankodwa*. There is a general practice in the region to cultivate broadcast rice in combination with pulses mostly *arhar*.

The per-hectare yield of broadcast rice varies form 800 kg to 1300 kg from the upper Doab to the lower Doab. This discrepancy in per-hectare yield is owing to the variation in the amount of rainfall and the difference in the quality of soil. There is a close relationship between the quality of soil and rice cultivation. It dominates the scene everywhere in the lower Doab owing to the clayey loam soil, but in the upper Doab, specially in the north-western portion, its area diminishes and is replaced by maize or such other light crops owing to the sandy soil, low water-table, and uncertain rainfall. The parganas with high intensity of rice cultivation are Maghar East, Bansi East, Rasoolpur, Babhnipar and Budhapar.

The late maturing variety of rice, covering 212,894 hectares or 19 per cent of the net sown area, is the next crop in importance. There are two methods of its cultivation, viz. broadcasting and transplanting. The former method of its cultivation is locally known as *Bowari* in which the fields are ploughed twice or thrice after the first shower of rain. When sufficient rain falls, the fields are once again ploughed and the seeds are spread at the rate of 120 kg per hectare. In the latter method, the seedlings are raised in small plots of well-manured good quality land. The seed-bed is ploughed twice or thrice under dry conditions and then puddled by giving two or three more ploughings. Finally, after harrowing germinated seeds are sown. The sowing is done very closely. After three or four weeks the seedlings are uprooted and transplanted on the same or next day as it is not desirable to keep uprooted seedlings longer. The seedlings are planted in regular lines with a wide space of 10 × 20 cm with three to four seedlings per hill. The early planting, by the end of July, is often conducive to higher yield (Kamath, 1954).

Late maturing variety of rice is usually grown in clayey soil which is found in shallow depressions and around the lakes and ponds which are locally known as *Dabar*. The position of land as regards the drainage is just as important as the character of the soil. The plant needs adequate supply of water until the end of October. Fields have, therefore, to be levelled and surrounded by firm, and fairly watertight bunds.

The per hectare yield of late maturing variety of rice varies from 850 kg in the upper Doab where the soil and the amount of rainfall are not very much conducive for its cultivation, to 1300 kg in the lower Doab where both the climatic and edaphic factors are favourable for its cultivation.

It will be seen from Fig. 7.2 that in 10 out of 36 parganas of the region, transplanted rice is the leading crop. There is a close relationship between the quality of soil and the cultivation of transplanted rice. It covers more than 15 per cent of the gross cultivated land in six parganas, viz. Charda, Dharmapur, Maghar East, Bansi East, Rasoolpur and Basti East, where the soil is clayey loam. There are 10 parganas where it covers more than 10 per cent of the gross cultivated land. These are Nanpara, Gonda, Mankapur, Budhapar, Babhnipar, Sadullahnagar, Maghar East, Amroha, Basti West and Maghar West. The area under transplanted rice in north-western Doab is negligible.

During the last five years considerable emphasis is being given to the Japanese method of rice cultivation. It is a method which includes improved agricultural practices with judicious manuring and proper plant protection measures. This method has given encouraging results by increasing per-hectare yield to 900 to 2000 kg from the upper to the lower Doab and is gradually gaining popularity. During the year of survey the area sown by the Japanese method was 44,707 hectares while in 1954, a decade earlier, it was only 12,773 hectares.

Maize

Maize which is also known as Indian corn is one of the staple foods of the people of the region. It can be reaped in a short time and is, therefore, of special value as one of the first crops available for food. It provides the cultivator and his family with food at a time when the reserves of the Rabi grains are running low. Sometimes the villagers pick up the immature cobs to be roasted for food.

Maize is sown generally after the first shower of the monsoon in early or middle of June. The crop matures in September. A long break in the rainfall or excessive fall and floods damage the crop. On the other hand, waterlogging is equally harmful and thus the crop is never sown in low-lying fields. Well-drained and generally light soils are best suited for maize in preference to heavy clay or clayey loam. Experiments on manuring maize have shown that it responds vigorously to nitrogenous manures. At the Agricultural Research Institute, Pusa, it was found that 40 lbs of nitrogen per acre in the form of rapeseed cakes was the most suitable manure and was superior to 8,000 lbs of farm-yard manure per acre (Burns, 1944).

Interculture is its chief feature. Several vegetables, like pumpkins, cucumber, various kinds of pulses like urd and arhar and grown with it, but all except arhar (an annual crop) are harvested before the main crop. The interculture has two distinct advantages : firstly, pulses like urd and arhar are deep-rooted crops which develop bacteria that enrich the soil; and secondly, vegetables and cereals mature quickly and at a time when the farmer's meagre stocks are getting more meagre.

Maize cobs are generally picked up for being used partly as green cobs and partly as ripe cobs. In the upper Doab where it is cultivated on a large scale, the cobs are gathered dead ripe to be hulled and stored or sold. Ordinarily only one cob can be reckoned on each plant, but on a rich soil many plants may carry more than one cob.

The cultivation of maize is a somewhat laborious business. In order to get a good yield of grain, the distance should be about 30 cm and, therefore, the seeds have to be sown carefully. Weeding the earthing are other necessary operations. After the cobs have formed, they have to be protected from birds, stray animals and even from human beings, and thus the crop has to be watched regularly till the harvest is ready. The stalks of the maize provide a nutritious fodder, if they are cut while green, but as the crop is grown for its grain the stalks are used for fuel.

Maize, being an important crop of the upper Doab, covers 257,619 hectares or 23 per cent of the net sown area. The cultivation of maize is mainly governed by the quality of the soil as well as the amount of rainfall. In the upper and north-eastern portion of the Doab comprising the parganas of Dharmapur, Nanpara, Fakharpur, Hissampur, Paharpur and Mahdeva, maize is the dominating crop. Maize is the first ranking crop in 11 parganas, while it 2nd ranks in four and 3rd in six parganas.

As the climate and soil of the upper Doab are favourable for the successful cultivation of maize as compared to the lower Doab, the per-hectare yield in the upper Doab is also high. It varies from 500 kg per hectare in the lower Doab to 1100 kg per hectare in the upper Doab.

Small Millets

Small millets covering an area of about 96,082 hectares is the other important cereal of the region. It is generally sown in combination with pulses. The millets grown in the region are of small variety and are called *kodon* and *sawan*. The seed is broadcast at the rate of 20 kg per hectare as soon as the rains break and it matures in about eight weeks from the time of its sowing. It flourishes well when the rainfall is scanty but suffers severely from an excess of rainfall. It may be grown on a variety of soils but it flourishes best on the light sandy soil. It is generally grown on the *Palo* soils of the village. It covers more than 10 per cent of the gross cultivated land in Fakharpur, Sadullahnagar and Dhuriapar parganas where the soil is generally light.

The yield per-hectare varies from 500 kg in the lower Doab to 700 kg in the upper Doab owing to the difference in edaphic and climatic conditions. It responds well to the liberal doses of nitrogen. At the Maize Research Institute, Chindwara (M.P.), it is found that about 18 kg of nitrogen per hectare gives an average yield of about 950 kg per hectare (Mandloi and Tiwari, 1966).

Pulses

The major pulses of the region in Kharif are arhar, urd and moong which collectively occupy about 10 per cent of the net sown area. *Arhar* is a deep-rooted and soil-building crop and is scarcely grown alone. It is generally grown in combination with small millets, broadcast rice, or maize. In the early parts of the growing season, it grows slowly, specially when it is mixed with a cereal crop but after the harvest of the cereal crop, it grows rapidly and covers the entire space. It is sown in the middle of June, flowers in cold weather and ripens in March. It may be grown on almost all types of soil but it flourishes well on light moist soil where its long roots may develop freely. Excessive rainfall and even more waterlogging harm the crop but it survives best in deficient rainfall. Frost is the main danger to it and single ground frost may destroy the entire crop. The plant flourishes well if there is sufficient humidity in the atmosphere during its growing period, but it needs a cool, dry and light sunny weather during its flowering and ripening period. Cloudy weather at the time of flowering damages the flowers and encourages the attack of pod borers.

The green and dry pods of *arhar* are used for preparing *dal* and vegetable dishes. The outer husk of the seed is used as food for milch animals. The stalks are put to various uses such as fuel and thatching purposes.

In the parganas of Rasoolpur, Basti West, Nagar East and Budhapar, it covers more than 35 per cent of the net sown area either in combination with rice or any other Kharif crop. Its area is higher in those parganas which lie in a well-drained level plain of sandy loam to loam soil.

There is much variation in the per-hectare yield of *arhar*. Its yield is higher in *bhangar* land which are free from flood and waterlogging. However, per-hectare yield varies from 650 kg to 1,000 kg.

In addition to arhar, moong and urd (*Phaseolus mungo*), other Kharif pulses, collectively, occupy 10,056 hectares. All these pulses are sown in July and they ripen in September. There are two varieties of *urd* grown in this region. One is sown in July and harvested in September while the other is sown in September after the harvest of early maturing rice and is harvested in January.

Sugarcane

Sugarcane is the most important cash crop of the region. The juice is extracted from sugarcane by a press (locally known as *kolhoo*) and is used for preparing gur, sugar, jaggery and a thick syrup called *Rab*. The molasses obtained can either be used as poultry feed, livestock feed or may be fermented and converted into alcohol. The bagasses are used as fuel, poultry litter, or for preparing compost. The green leaves are used as fodder for animals.

Before the sowing of cane, land is repeatedly ploughed and the soil is well-manured and pulverised. Sowing begins in March when the temperature is sufficient for the growth of crop, but the low rainfall makes irrigation necessary. Sugarcane requires constant attention during the first four or six months of its growth. Soon after planting, weeding is done which is generally repeated four or five times. It requires frequent watering until the rains break. The growing period of sugarcane varies from nine months as a minimum in the early maturing varieties to as many as 13 months in late maturing varieties (Ayer, 1954).

Sugarcane is quite adaptable to varying soil conditions but loamy soil with good drainage and slightly acidic to alkaline soil are best suited for this crop. However, a growing period of 10 months with adequate moisture followed by dry winter are the best conditions for high yield, but severe cold checks the growth and frost kills the cane.

Sometimes sugarcane is ratooned, i.e. the stalk is allowed to grow, the land is hoed, manured and irrigated and a second crop is raised.

Sugarcane covers 50,625 hectares or about six per cent of the net sown area. Out of this 37,905 hectares or 75 per cent is irrigated and the remaining 12,720 hectares or 25 per cent is unirrigated. Sugarcane is produced without irrigation in those parganas which lie in the *khadar* of the rivers Ghaghara and Rapti. Here the water-table is high and the sub-soil remains moist all the year round. It will be seen from Fig. 7.2 that Nawabganj, Balrampur, Basti East and Mahuli West are the leading sugarcane producers of the region. The most compact region of sugarcane-producing parganas lies in the middle of the Doab where more than five per cent of the gross cultivated land is given to cane cultivation. Irrigation facilities and good quality loamy soil are the chief causes of this concentration. There are 12 parganas lying mostly in the upper Doab which do not produce sugarcane at all owing to relatively poor soils, and inadequate irrigation facilities.

Other Crops

Foddeı, vegetables and oil-seeds are the other non-grain crops which collectively cover 58,539 hectares or five per cent of the net sown area. Although more than half of the population is vegetarian, it is surprising that the area under vegetables is only 0.5 per cent of the total cropped area.

Fallow

It will also be observed from Table 7.2 that more than one-fourth of the cultivated land is left fallow owing to its incapability to produce any crop in both the seasons. Of the total fallow land about 9,564 hectares have been left fallow for more than two years. The 18 per cent of the fallow lands have been left fallow only for one year, while the remaining 82 per cent have been left fallow only during the Kharif season, and it is utilised in the following Rabi season, for the cultivation of wheat and wheat mixed with barley. The area and percentages under fallow lands increase towards the upper Doab and north-western Doab. Hissampur, Fakharpur, Nanpara, Dharmapur, Guwarich and Bahraich are the parganas where there are higher percentage of the fallow land owing to the infertile soil and inadequate irrigation facilities. On the contrary in the lower and central Doab the percentages of fallow land decreases due to its fertile soil and adequate irrigation facilities.

Cropping Patterns in Rabi Season

Rabi is the other important agricultural season of the region and 966,968 hectares or about 63 per cent of the cultivated area is devoted to the cultivation of Rabi crops. The actual areas and percentage under Rabi crops are given in Table 7.3. The Fig. 7.3 shows the use of cultivated land in the year 1964-65 on the pargana basis.

It will be seen from Table 7.3 that about 63 per cent of the total area is given to the cultivation of Rabi crops and about one-fourth is covered with the continual Kharif crops.[6] The remaining 13 per cent is left fallow owing to infertility of the soil. Out of the total sown area in the Rabi season, about 57 per cent is given to the cultivation of grain crops, more than one-fourth is devoted to pulses and the remaining 15 per cent is covered with crops like fodder, oilseeds and vegetables. The major cereals of the region are wheat, barley *bejhar* and wheat mixed with barley, whereas peas, gram and lentil are important pulses. The principal Rabi crops are:

Wheat (Triticum Sativum)

Wheat covers an area of 339,020 hectares or about 35 per cent of the net cropped land in the Rabi season. It is an important staple crop of this region and stands next to rice in acreage. It is usually sown at the end of October. The crop takes about five months to mature and requires at least one watering during the growing period. Wheat, which is grown in the *Khadar* areas of the region, hardly needs irrigation because of the stored moisture in the soil. However, 113,006 hectares — about one-third of the area under wheat, is without irrigation, while the remaining two-thirds area is irrigated through various sources. It ripens in March and April and is generally very dry when harvested. Threshing is done by trampling the ears of the grains under the feet of cattle and finally the produce is screened and winnowed thoroughly before it is stored.

Wheat is generally grown in the fields which have remained fallow in the previous Kharif season but the practice of sowing it after maize or fodder appears to be spreading. As it requires a firm, fine seed-bed, the fields are thoroughly prepared before sowing the seed. Frequent ploughings and harrowings are given during the monsoon season for getting a firmer seedbed. There are two common methods of wheat-sowing in the region: broadcast method and line-sowing method

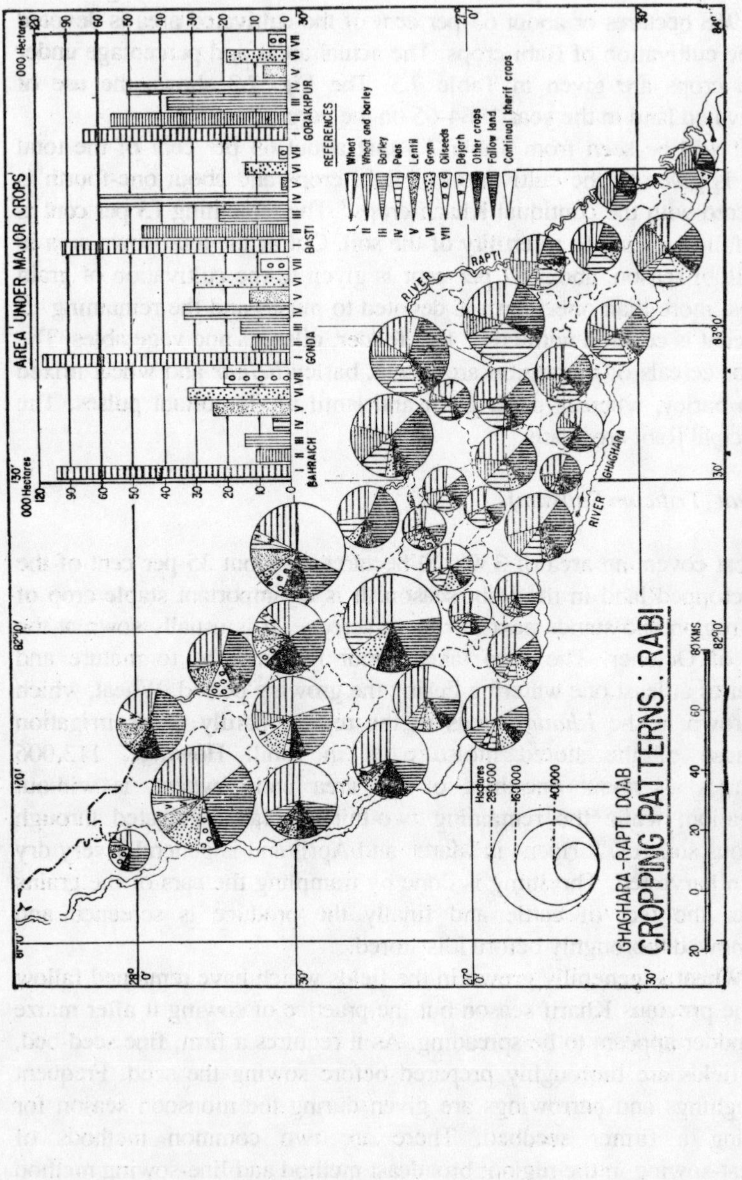

FIG. 7.3: Ghaghara-Rapti doab: cropping patterns: Rabi.
Note: This has been prepared on the basis of land percentages occupied by each crop in the pargana.

TABLE 7.3: **Agricultural landuse in the Rabi season: Cultivated land . . . 1,532,072 hectares net cropped land in the Rabi season 966,968 hectares**

Crops	Area in hectares	Percentage of the cultivated land	Percentage of the net cropped land	Total percentage of the cultivated land	Total percentage of the net cropped land
Grain Crops				35.96	56.99
Wheat	339,020	22.13	35.06		
Wheat & barley	102,654	6.70	10.62		
Barley	65,270	4.26	6.75		
Bejhar	36,045	2.35	3.75		
Other cereals	7,993	0.52	0.83		
Pulses				17.77	28.14
Peas	136,700	8.92	14.13		
Gram	87,154	2.68	4.26		
Lentil	41,020	5.70	9.01		
Other pulses	7,195	0.47	0.74		
Other Crops				9.37	14.87
Oil Seeds	34,815	2.28	3.60		
Vegetables	8,542	0.56	0.88		
Fodder	571	—	0.05		
Others	99,992	6.53	10.34		
Fallow	190,861	13.12	—	12.13	
Continual Kharif Crops	374,243	23.78	—	23.78	
Total	1,532,072	100.00	100.00	100.00	100.00

* **This table has been computed by the writer on the basis of parganawise unpublished records of Rabi crops of 1964-65 obtained from the Board of Revenue, Uttar Pradesh (Lucknow).**

(in this the seed is sown through a seed tube attached to the plough or is dropped by hand in the furrows). A third method which is becoming popular is the dibbling method in which two or three seeds are dropped in the holes made by a dibbler.[7]

The average yield of wheat in the region ranges from 600 kg to 1,000 kg per hectare which is certainly lower than that of the western districts of Uttar Pradesh and is mainly because of poor agricultural practices. The yield, however, varies from 600 kg in the upper Doab to 1,000 kg in the lower Doab. It is because of the difference in the quality of the soil, means of irrigation facilities and the agricultural practices.

It can be seen in Fig. 7.3 that wheat is mostly grown on the *bhangar* land as compared to the lower parts of the *khadar*. Fakharpur, Paharpur, Mankapur, Sadullahnagar, Budhapar and Guwarich are the main wheat producing parganas where it covers more than 20 per cent of the gross cultivated land. In Chillupar, Dhuriapar and Maghar East parganas where rice covers a considerable area wheat occupies a very small area amounting to one to five per cent of the net sown area.

Wheat Mixed with Barley[8] *(Gojai)*

It covers 102,654 hectares or about 11 per cent of the net sown area and is next important Rabi cereals of the region. The climatic requirements are more or less the same as in wheat. It requires a temperature of about 15°C during the growing season and 26°C when it matures. A slight rainfall amounting to 30 mm is beneficial during its period of growth. Heavy rainfall or frost is highly injurious for the crop. Hot winds during the period of maturity are harmful as they bruise the flowers and the grains are shrivelled. The crop is irrigated at least once during the growing season with the exception of those areas which lie in the *khadar*.

Gojai is grown in deep fertile soil of more loamy sand rather than clayey loam. It grows well on light soil as compared to wheat. About 50 per cent of *gojai* is sown on lands which are let fallow in the previous Kharif season and the remaining half is grown on fields cleared from the Kharif crops. The fields for *gojai* are ploughed several times, clods are broken and weeds and stubbles are removed. The fields are harrowed in order to find fine tilth and good seed-bed. Fields are heavily manured with cattle manures at the rate of 15 cartloads per hectare along with chemical fertilisers and manures.

The sowing of *gojai* usually takes place during October and is done by casting the seeds in the wake of ploughing operations. The seed is sown either directly from the hand in the furrows or through a bamboo tube fastened to the plough stilt. The other method is to broadcast the seeds with hand at the rate of 120 kg per hectare. It is harvested in the months of March and April when the crop is dead ripe as at this stage the grains are hard and the skin is wrinkled. The yield of *gojai* varies from 650 kg in the north-eastern Doab to 850 kg in the lower Doab depending on the quality of soil and irrigation facilities.

It may be observed from 7.3 that in the parganas of Hasanpur Maghar, Chillupar and Dhuriapar, it occupies a place of importance covering about 17 to 20 per cent of the gross cultivated land. In these parganas which lie at the confluence of the rivers Ghaghara and Rapti, the soil is of silty deposits of light character. Here owing to the stagnation of flood water, the sowing is delayed. The farmers, therefore, prefer to grow wheat in combination with barley which is a king of assurance against the hot winds of March which bruise the flowers of wheat.

Barley (Hordeum Vulgare)

Barley grown on an area of 65,270 hectares or about 7 per cent of the net sown area is the staple diet of the poor people of the villages. It is grown mostly in the same climatic conditions as wheat, but it is more adaptable and has a wider range of cultivation than wheat. Moderate temperature with an average well-distributed rainfall and plenty of sunshine during the ripening period are the suitable climatic conditions.

Barley being a shallow rooted crop grows well on light loam. About 90 per cent of barley is grown on fields which are occupied by either broadcast rice, maize, or millets in the previous Kharif season. Fields are ploughed several times after the harvest of the Kharif crops, and manures are supplied at the rate of about 8 cartloads per hectare to obtain a fine tilth and firm seed-bed.

Barley is sown by the middle of October. Seed is sown either in rows or is broadcast at the rate of 120 kg per hectare. After three or four weeks of its sowing, the fields are irrigated in the ***bhangar*** areas while in the ***khadar*** land, irrigation is not needed. Barley is an early maturing crop and it becomes ready for harvest within 110 days and as such provides grain to the farmer when his foodgrain stocks are running low.

It will be seen from Fig. 7.3 that barley is generally grown all over the area but the zone of its high concentration lies in the extreme east of the region consisting of the parganas of Bansi East, Maghar West, Maghar East, Hasanpur Maghar, Unual, Bhauapar and Chillupar. The other zone of concentration is in the south-east consisting of the parganas of Guwarich, Digsir and Paharpur where it covers about 5 to 10 per cent of gross cultivated land.

Bejhar (Wheat, Barley, Peas and Gram)[9]

Bejhar is the other Rabi crop of considerable importance because it acts as an insurance against the vagaries of the climate. It covers 36,042 hectares or about four per cent of the net sown area. The climatic requirements are the same as in wheat mixed with barley. It grows well in *khadar* areas as compared to *bhangar* land. Clayey loam to loamy soil are most suitable for its cultivation.

It is sown usually in the month of October when the monsoon retreats. The seed is sown either by broadcast or in lines. The crop is grown without irrigation in *khadar* land while it is necessarily irrigated in *bhangar* land. It matures within four months and is ready for harvest in the last week of March or in early April.

It will be seen from Fig. 7.3 that about two-thirds area of *bejhar* lies in the *khadar* as it flourishes well in the *khadar* soil as compared to *bhangar* soil. Chillupar, Balrampur, Nanpara, Dharmapur and Hissampur are important *bejhar* growing parganas where it occupies nearly five per cent of the cultivated land. The crop totally disappears from the scene in the parganas of Maghar East, Amorha, Basti West, Nagar West, Basti East, Mahuli West and Dhuriapar. Most of these parganas lie in the *bhangar* areas.

Peas

Peas is an important Rabi pulse of the region and covers 1,36,700 hectares or 14 per cent of the net sown area. It is grown on heavy soil such as loam or clayey loam, but open well drained soils adequately supplied with lime are ideal for peas. The crop is usually grown on the fields which were occupied by broadcast rice, millets or maize. After the harvest of the Kharif crop, the field is ploughed and cross-ploughed and moderate doses of manures and fertilisers are provided. A heavy shower is beneficial prior to the sowing of crop as it helps in preparing seed-bed and tilth.

The crop is sown in the third week of October and is ready for harvest in February or March. When the crop attains a height of about eight inches, the end of the stem is nipped off so that the plants may produce offshoots vigorously. The crop is grown on the *bhangar* land and is irrigated after three or four weeks of its sowing but it does not require as much water as is needed by wheat and barley, because excessive moisture promotes abundant vegetative growth at the cost of seed-buds and this leads to a poor yield of the crop. The crop is very much susceptible to weather conditions and the frost during its flowering period is highly injurious to it (Moree, 1944), but luckily frost in this region is rare and the crop does not suffer.

Peas are grown with several purposes in view. The immature pods are often picked and are used either as fresh vegetables or are roasted. The ripe crop is harvested in the middle of February, and the grains are obtained by threshing it out by bullocks. Dry grains are used either for stock-feeding or are turned into flour locally known as *sattu.*

The cultivation of peas is concentrated in the east of Utraula, Gonda, Mankapur and Nawabganj parganas. In all the eight parganas of lower *bhangar* land excluding Bhauapar, peas is the important Rabi pulse occupying more than 10 per cent of the net sown area. The main reason for such concentration is the loamy clay soil and adequate irrigation facilities. The area under peas decreases towards the upper Doab. In the district of Bahraich it almost disappears because of the inferior quality of soil and an inadequate amount of rainfall.

The per-hectare yield of peas varies from 600 kg in the upper Doab to 1000 kg in the lower Doab depending on the quality of soil, irrigation facilities and agricultural practices.

Gram (Cicer Hrietinum)

Gram, covering an area of 87,145 hectares or about nine per cent of the net sown area, is one of the principal Rabi pulses grown in the region. It is used for human consumption both in the green form and in the dry state, and its tender leaves are used as leafy vegetables. Gram is grown on sandy to sandy loam soil. It is fairly sensitive to variations of rainfall during October. During this month, if the rainfall is heavy, the gram which is sown earlier is destroyed and the ground is so much chilled by the rain that the seeds sown later do not germinate properly. A light shower is helpful in January but it cannot stand frost at any stage of its growth. Frost during the flowering time retards the development of flowers and destroys the seed inside the pod. Moreover, cloudy weather with rain at flowering time damages the crop in two ways: firstly, by preventing the setting of the seeds; and secondly, by interfering with the aeration of the roots and nodules, and causes the plants to wilt.

Gram is always the first crop to be sown in the Rabi season. The sowing usually begins in October after the heavy rains are over. It is rarely grown as a sole crop. Generally it is grown on the fields occupied by any Kharif crop. The fields are ploughed twice or thrice after the harvest of the Kharif crop and the seeds of gram are spread at the rate of 20 kg per hectare. When the crop attains the height of three inches, the end of the stem is compulsorily nipped off so that the offshoots of the plants may multiply vigorously. Almost all the crop is produced without irrigation.

The distribution of gram is closely related to the natural aeration of the soil. Wherever the roots of gram can obtain an abundant air supply the acreage is large (Howard, Howard and Khan). The cultivators recognise that an open soil is more suited for gram and that the air in the soil is a more important factor than the presence of moisture. However, gram is a more significant crop in the upper Doab than in the lower and middle Doab. Its area of concentration extends almost over the whole of the central upland but it is generally concentrated in the parganas of Nanpara, Charda, Bahraich and Gonda, where it covers more than eight per cent of the gross cultivated land. However, the government should give incentives for the cultivation of gram in the fields which are left fallow in the Rabi season.

Lentil

Lentil covers 41,020 hectares or four per cent of the net area sown. It grows on a variety of soils such as light loam soil of the upper Doab and the clayey soil of the lower Doab. It is, however, usual to cultivate rice as the first crop of the year and to follow it with lentil in the late season. Very often when the soil is still most late in the *Kharif* season, lentil is sown in the standing crop of rice. It is grown as a dry and unirrigated crop with the help of available moisture and the heavy dews of the season. The crop receives no further attention by way of weeding and interculture. It becomes ready for harvest in about three to four months after its sowing. It will be seen from Fig. 7.3 that Bhinga, Nanpara, Charda and Dharmapur are the leading parganas for the production of lentil.

Oilseeds

Oilseeds occupying 34,815 hectares or about four per cent of the net sown area constitute one of the prime sources of oil and fat for the predominantly vegetarian diets of the region. Oilcakes from most oilseeds are valued highly as fertilisers and as feed for poultry and livestock. The major oilseeds of the Rabi season are mustard, rapeseed, *til* and linseed. Nearly 50 per cent of the oilseed is grown in combination with the Rabi crops. It requires a highly manured soil and is generally grown in the good quality land around the settlements. The crop is sown in October and ripens in February. At least one watering is necessary for its successful cultivation. The leading oilseed-producing parganas are Dharmapur, Ikauna, Nanpara and Balrampur. In the lower Doab the crop is generally grown in combination with Rabi crops, and therefore, the data are not available.

It will be seen from Table 7.3 that 190,861 hectares or about 13 per cent of the cultivated land is left fallow owing to their infertility to produce any Rabi crop after producing a crop in the Kharif season. It is clear that the area of fallow land is much more in the upper and north-western Doab where the soil is of poor quality than in the Lower Doab. It is however, advisable to bring as much land out of this as possible by cultivating leguminous crops like gram, peas, or lentil which, in addition to food and fodder, shall also increase the fertility of soil, with the result that a good crop in the following season may be produced.

Table 7.3 further shows that out of 1,532,072 hectares of cultivated land, only 966,968 hectares or a little more than half of it is actually cultivated during the Rabi season. Nearly 364,243 hectares or a little less than one-fourth of the cultivated land remained occupied under the continual crops such as sugarcane, *arhar* and transplanted rice. It will certainly help in increasing per-capita cultivated land, if the fields under such crops are also sown with early maturing leguminous Rabi crops which will also increase the fertility of the soil.

NOTES

1. Sugarcane is a continual Kharif crop which is sown in the months of May and June and is harvested from December to March.
2. *Arhar (Cajanus indicus)* is an annual crop which is sown in the Kharif season and is harvested with the Rabi crops.
3. Figure 7.1 has been prepared on the basis of the percentage occupied by each land type in each pargana during the year 1964-65. The parganawise statistics have been obtained from the Board of Revenue, Uttar Pradesh (Lucknow).
4. The Waste Land Survey and Reclamation Committee includes in the 'culturable waste' all lands available for cultivation but not taken up for cultivation or abandoned after a few years owing to some reasons. Such lands may be fallow for more than five years and may be covered with shrubs. It does not include the

lands which are reserved for pastures or grazing.

5. *Bhadai* is an adjective from *Bhadon* — a month in Hindi calendar which usually corresponds with the months of August and September. *Kunwari* is also an adjective from *Kunwar* — a month in Hindi calendar corresponding with the months of September and October. As the broadcast rice is harvested in two months, it is known as *bhadai* or *kunwari*.
6. Continual Kharif crops include transplanted rice, sugarcane and pulses in the Kharif season.
7. Dibbler is a wooden implement having 24 pegs at regular intervals in a rectangular shape.
8. Wheat sown in combination with barley or at places in combination with barley and peas is locally known as *Gojai*.
9. The combination of *bejhar* changes according to the localities. In the *khadar* land it includes black gram, green gram, barley and wheat, while in the *bhangar* land it generally comprises barley, wheat, peas and gram.

REFERENCES

Ayer, A. K. Y. N., 1954 : *Field Crops of India*, Bangalore, p. 209.

Burns, W., 1944 : *Technological Possibilities of Agricultural Development in India*, Lahore, p. 67.

Howard, A., Howard, G. L. C. and Khan, A. R. : "Some Varieties of Indian Gram", *Memoirs of the Department of Agriculture of India*, Botanical Series, Vol. VII, p. 217.

Kamath, M. G., 1954 : *Rice Cultivation in India*, Indian Council of Agricultural Research, New Delhi, p. 19.

Mandloi, K. K. and Tiwari, P. K., 1966 : "Kodon Responds to Nitrogen", *Indian Farming*, No. 1, Vol. XVI, New Delhi.

Moree, H. I., 1944 : *Crops and Cropping*, London, p. 90.

Shafi, M., 1960 : *Land Utilisation in Eastern Uttar Pradesh*, Aligarh, p. 48.

8

CHANGING AGRICULTURAL LANDUSE IN THE HILLY TRACTS OF RAJASTHAN

INDRA PAL AND LAKSHMI SHUKLA

NO EMPHASIS is too heavy to hammer home the fact that land is a resource of primary importance in the economy of any country and no one land unit can, by nature, be like another. But the physical attributes and the socio-economic structure of a land unit, no doubt, put a stamp upon its use pattern. However, for the primary physiological need of man — food — land has to be put to optimum use.

The regional imbalance in agriculture, the future prospects and potentialities and the problems that confront their development are some of the vital issues of the study of landuse in the district of Chittaurgarh. In fact quite a large portion of its land due to the hilly and rugged topography is lying as barren and cultivable waste. Infertile soils, paucity of rainfall, social backwardness and low accessibility are other factors which have arrested its agricultural development. But no one has yet touched this area for such studies and no work appears to exist presenting a synthesis of the various characteristics of its landuse.

Agriculture is providing livelihood to nearly 79 per cent of its total population (944,981 persons, 1971 Census), 20 per cent of whom are tribal. Of the total population 38.6 per cent are workers, out of whom 30.0 per cent are engaged in agriculture and allied activities. Twenty-two per cent of the total geographical area is lying as culturable waste. This has reduced the per capita cultivated land only to 0.41 hectares, and, for all these reasons, the district has been selected for an "Integrated Development Programme".

Geographical Background

Located between 23°32′ and 25°07′ N Latitudes and 74°13′ and 75°49′

E Longitudes, Chittaurgarh district is situated in the south-eastern quadrangle of Rajasthan. It is sandwiched between the semi-arid and semi-humid areas of Rajasthan and Madhya Pradesh and shares the landuse and cropping characteristics of both the regions. The southern part of the district has a larger tribal population than the northern. Thus the transitional nature of both climate and demography of the district shows some significant impact on its cropping pattern.

Geomorphic Divisions

Chittaurgarh district may be divided into the following geomorphic divisions (illustrated in Fig. 8.1):

A. Uplifted Plain : (Central-cum-north-western parts)

It is a part of Chhappan Plain

(i) Central High Plain (Chhoti Sadri, Bari Sadri, Dungla and Bhadesar Tehsils).
(ii) North-western High Plain (Rashmi and part of Kapasan).
(iii) Riverian Low Plain (Gangrar, part of Chittaurgarh, Nimbahera and Kapasan).

B. Vindhyan Hilly and Plateau Region : (North-eastern part)

(i) Vindhyan Hills (Eastern Chittaurgarh).
(ii) Vindhyan Plateau (Begun and part of Nimbahera).

C. Lava Plateau : (Southern Part) (Pratapgarh)

The Soils

The soils of the district are classified in four main types:

1. Hilly regosoles and lithosoles (310,423 hectares).
2. Light yellow-brown soils (312,171 hectares).
3. Black soils (245,374 hectares).
4. Red loams (72,172 hectares).

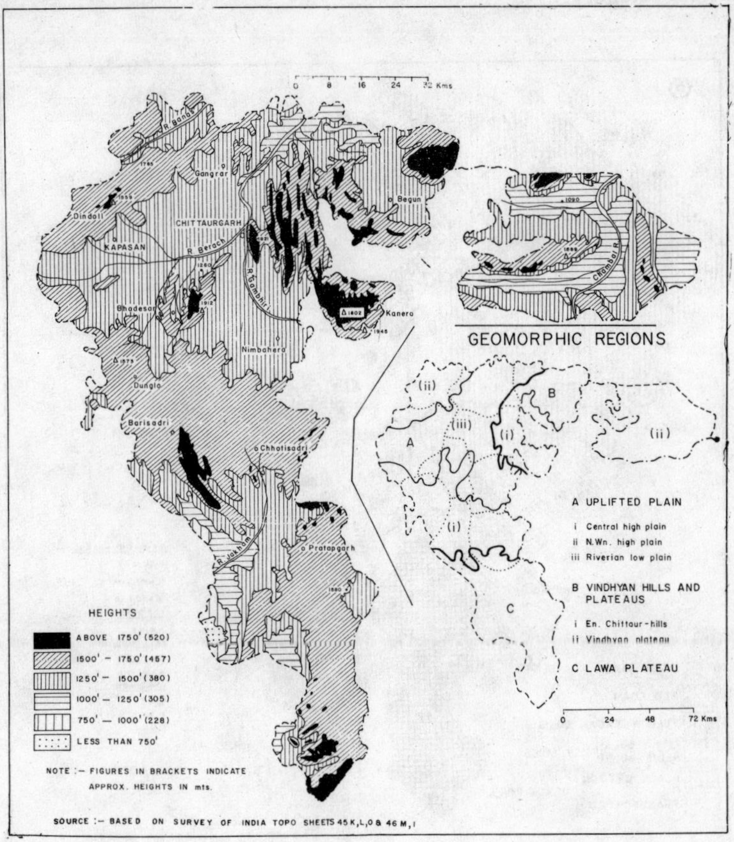

FIG. 8.1: Chittaurgarh district : relief.

Their distribution and characteristics are shown in Fig. 8.2.

Landuse Pattern

As customary the total geographical area of the district is divided into two major classes: (i) arable, and (ii) Non-arable. The first category includes the land actually cultivated (or net sown area), fallow, cultivable waste, permanent pastures and grazing grounds. All such lands which are declared as forests or barren and the other uncultivated land and land not available for agriculture, etc., are included in the second category.

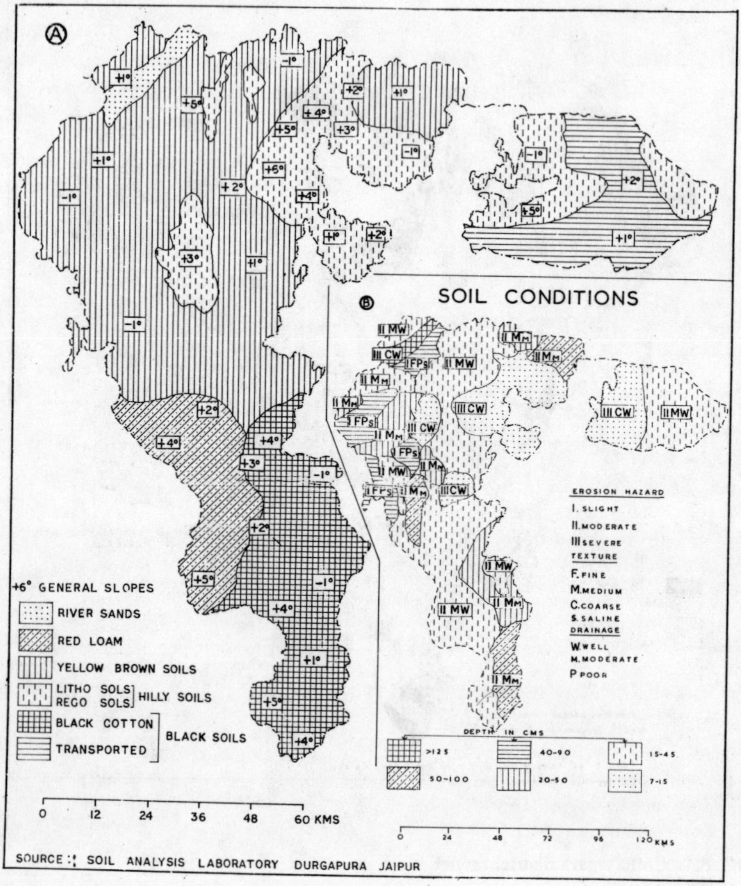

FIG. 8.2: **Chittaurgarh district : soil.**

The present study of landuse in Chittaurgarh district is based on the averages of the three-year data for 1972, 1973 and 1974 and the study of landuse changes is made for a 20 year period from 1954 to 1974 taking quinquennial data for the years 1954, 1959, 1964, 1969 and 1974.

On the basis of averages for the years 1972, 1973 and 1974 their percentages are 80.86, 81.02 and 81.98 respectively to the total geographical area of the district. The percentage of the arable land in the district is slightly higher than 78.47 of the State as a whole. It may be contrasted with the maximum percentage of 90.50 in Bikaner district and the minimum of 44.23 per cent in Udaipur district. Similarly, the non-arable percentage of the district may be compared with 21.53 of the State.

Table 8.1 shows the average percentage shares of the arable and non-arable lands in the district and their sub-divisions.

TABLE 8.1: General landuse of Chittaurgarh district (Average of the years 1972, 1973 and 1974)

Category	Percentage	
A. *Arable Land*	81.02	to the total geographical area of the district.
1. Net sown	50.00	to the total arable land.
2. Current fallow	3.06	-do-
3. Old fallow	4.00	-do-
4. Land under miscellaneous trees and groves, etc.	0.02	-do-
5. Cultivable waste (usar)	29.11	-do-
6. Pasture land and grazing grounds	13.81	-do-
B. *Non-arable Land*	18.98	to the total geographical area of the district.
1. Forest	20.77	to the total non-arable land.
2. Barren (gravely, stony, undulated, etc.) and other uncultivated land	67.69	-do-
3. Land not available for agriculture: (settlements, roads, railways, dams and water bodies, etc.)	11.54	-do-

Non-Arable Land

The area under the non-arable category in the district has decreased by 1.36 per cent within the study period of 20 years from 1954 to 1974. The greatest decrease of 2.86 per cent occurred during the last quinquennial period in 1969 to 1974. The area under forests and barren lands has been constantly decreasing especially in the Uplifted Plain, while it is increasing mainly on the Vindhyan Hills and Plateau and Lava Plateau which are covered with shallow soils, not suitable for cultivation or due to the utilisation of the cultivable wastes and old fallows for the construction of buildings, roads, railways and reservoirs. The average percentages of the components of non-arable land are represented in Fig. 8.3.

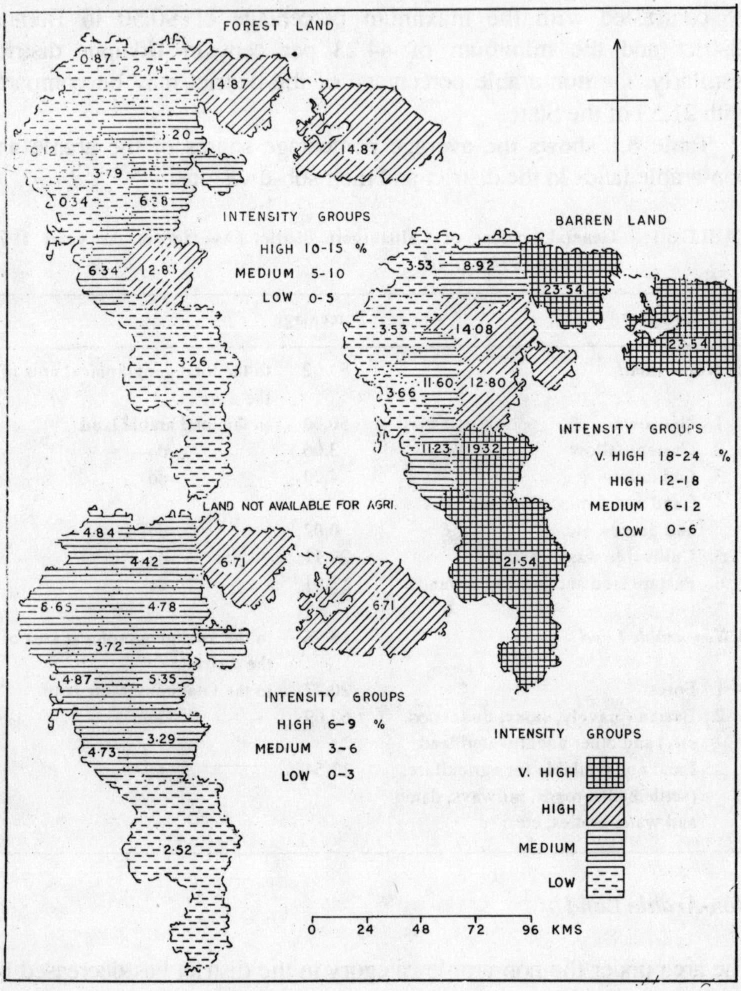

FIG. 8.3: Chittaurgarh district : elements of non-arable land (average percentages to total geographical area (1972-1974)

Forests

The forests cover only 4.26 per cent of the total geographical area or

20.77 per cent of the district's total non-arable area. They are of semi- or sub-humid edaphic types. Teak zone (Anogeissus latefolia) lies in the Lava Plateau in the south where the rainfall is higher (averaging 700 mm per annum) and the soils are of basaltic origin, while in the north where the rainfall is less than 700 mm. the "dhokra" (Anogeissus pendula) is found on the Vindhyan Hills and Plateau and mixed forests are found on the Central High Plain. The lowest 4.8 per cent forest area of the non-arable land and 0.12 per cent of the total geographical area is found in Kapasan tehsil (Riverian Low Plain) while the highest percentage under forest (52.0; 14.87) is found in Begun tehsil (Vindhyan Plateau).

The total change in forest area is +7.68 per cent during the study period. The greatest change of +5.94 per cent is brought about in the last quinquennial period, through the conservation programmes in the areas declared as "Protected and Reserved Forests". Contrarily, by unauthorised tillage of the common grazing grounds in villages and encouraging the herds on the hill slopes adjacent to the plain, they have enhanced the erosion as can be seen in the Uplifted Plain. In Rashmi (North-western High Plain) the ridges have been depleted so much that the villagers locally call them *Tantiya Magras* (hills without trees or grass cover).

In the Lava Plateau region the tribals have been burning trees and cutting lumber in the forests indiscriminately. Hence, they have incurred a great national loss in the teak zone. They wrecklessly cut even the tender trees for fuel and for making carts and implements in spite of strict vigilance by the forest guards. Though the tribals have been allowed by the Forest Act, 1955, to cut timber from the unreserved areas with licences to be issued by the guards, yet they cut *ballis* or long poles for unauthorised selling. The permissible limit for cultivation by the Forest Department is on a maximum slope of eight degrees, but they would stealthily cut trees on greater slopes and then till it. This has further increased the erosion hazard in these areas. Over the above this, the secondary growth of vegetation on such tracts, which grows after two or three years of cultivation, is not as good as the primary or original growth.

Barren Land

The district as a whole has an average of 67.69 per cent area of the total non-arable land under this class. This is too much and requires attention of the planners.

Generally the large barren areas are lying in the Vindhyan and Lava Plateau regions. They have increased in the tehsils of the Plain region as the denudation of vegetation has resulted in the rolling of boulders from the hill slopes and washing of the foothills below. But the barren land has decreased significantly in those hilly and plateau regions where slopes and less fertile tracts of shallow soils have been afforested. The greatest variation during the study period is seen in the Vindhyan Plateau (–14.92%) and Lava Plateau (–8.88%).

Land not Available for Agriculture

A general trend is that of a constant increase in all the tehsils under this head. The district has witnessed a total increase of 4.8 per cent of land in non-agricultural uses during the study period, with the greatest change of +7.9 per cent in Nimbahera, followed by Chittaurgarh (+5.9%), Rashmi (+3.6%) and Gangrar (+3.0%). This is mainly due to the establishment of factories, expansion of settlements, construction of many roads and irrigation projects, which have taken cultivable wastes but mostly from the barren land and thus have increased the area under this category.

Arable Land

The arable or cultivable land undergoes frequent changes as it is influenced by a number of physical and socio-economic factors. The percentage of the arable land to the total geographical area averages 81.02. Table 8.2 gives the quinquennial percentage changes of the arable land in the district which is only +1.36 per cent.

It indicates a very slow rate of development as compared with other parts of the State, especially in the Eastern Plain of Rajasthan. Not only that, three tehsils of the Uplifted Plain have shown a negative variation. Even the positive variation is less than four per cent in most of the tehsils. Pratapgarh (Lava Plateau) and Begun (Vindhyan Plateau) are the only units where arable land has increased significantly. Table 8.3 shows a balancesheet of the increase/decrease in the arable land and the names of donors/receivers of the non-arable land.

TABLE 8.2: Tehsilwise quinquennial percentage variation of arable land (total tehsil area — 100%)

Tehsils	1954	1959	1964	1969	1974	Average % for 3 years (1972-74)	Percentage change during 20 years (1954-74)
Chittaurgarh	79.05	76.75	75.03	76.79	77.25	77.68	– 1.80
Gangrar	82.93	85.93	86.86	85.57	85.40	85.48	+ 2.47
Kapasan	90.88	82.48	92.92	95.45	80.83	83.14	– 10.05
Rashmi	88.53	90.44	91.76	94.90	90.06	92.48	+ 1.53
Begun	70.45	65.23	66.05	60.33	75.05	67.94	+ 4.60
Pratapgarh	67.52	75.87	76.48	77.33	77.40	77.49	+ 9.88
Nimbahera	75.00	76.96	77.97	78.30	77.00	78.50	+ 2.00
Bhadesar	85.00	82.59	83.32	84.19	81.55	82.00	– 3.45
Dungla	87.88	91.56	92.12	91.67	91.91	91.68	+ 4.03
Bari Sadri	76.52	80.00	80.12	80.20	80.31	80.00	+ 3.79
Chhoti Sadri	72.00	75.00	75.20	75.80	75.90	75.85	+ 3.90
District	79.66	80.23	80.71	80.90	81.13	81.02	+ 1.47

TABLE 8.3: The increase/decrease in arable land over 20 years — 1954—1974

Tehsils	% increase/ decrease in arable land	Donors/Receivers of non-arable land		
		Forest	Barren land	Land not available for agriculture
Pratapgarh	+9.88	−2.0	−8.88	+1.00
Begun	+4.60	+6.0	−14.92	+4.32
Dungla	+4.03	−1.0	−5.03	+2.00
Chhoti Sadri	+3.90	+1.07	−5.38	+0.41
Bari Sadri	+3.79	+2.0	−6.12	+0.33
Gangrar	+2.47	−1.87	−2.69	+2.09
Nimbahera	+2.00	−1.37	−3.00	+2.37
Rashmi	+1.53	−0.53	−2.00	+1.06
Kapasan	−10.05	+1.37	+4.25	+4.38
Bhadesar	−3.45	+2.35	+2.35	+2.10
Chittaurgarh	−2.80	+2.88	−5.08	+5.00

It is encouraging that the barren land has decreased in 9 out of 11 tehsils. This has contributed to the heads of arable land in eight tehsils, forests in six tehsils and non-agricultural uses in all the tehsils.

On the basis of average figures in Table 8.4 the 11 tehsils are classified into four intensity groups.

TABLE 8.4: Intensity groups of arable land

Intensity Groups	Average % to the total geographical area (1972-74)	No. of Tehsils	Name of Tehsils
Very High	More than 90	2	Rashmi, Dungla (N.Wn. & Central High Plains)
High	80-90	3	Bhadesar (Central High Plain) Gangrar, Kapasan (Riverian Low Plain)
Moderate	70-80	5	Bari Sadri, Chhoti Sadri (Central High Plain) Chittaurgarh & Nimbahera (Riverian Low Plain) (Pratapgarh (Lava Plateau)
Low	Less than 70 (District Avegare 81.02%)	1	Begun (Vindhyan Plateau)

In all the cases, the occurrence of high percentage of arable land is not always an indication of a higher share of the net cultivated land. There may be large amounts of cultivable wasteland in some of the tehsils.

With regard to Table 8.4 it may be stated that the percentages of the tehsils within the intensity groups may have slightly fluctuated from time to time but they have retained their original intensity groups as is clear from Fig. 8.4. The fluctuations in the percentages of arable land are due to several reasons. It may be due to the tilling of the forest or the barren lands for raising crops or to develop pasture lands. But if the arable land has decreased it is because the cultivable wastelands, pasture lands and old fallow lands have gone to the non-arable category.

It is evident that, as a result of planned development since independence, there is general reduction in the barren land. This has added area to the arable land, to forest and to non-agricultural uses. In recent years, the pastures have been converted into forests and the cultivable wasteland has been put to non-agricultural uses.

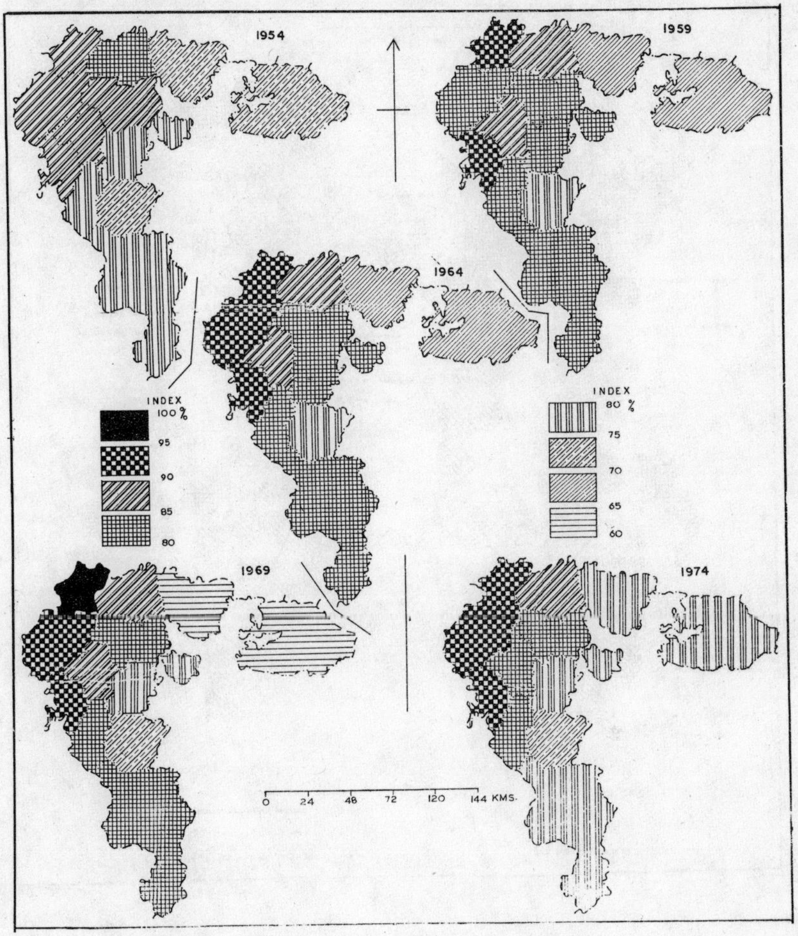

FIG. 8.4: Chittaurgarh district : arable land percentage to total area.

For a comprehensive appraisal of the arable land, it is important to discuss in detail its component parts individually. The intensity groups of these components are shown in Fig. 8.5. The quinquennial differentials of the component elements of the arable land in district are represented in Table 8.5.

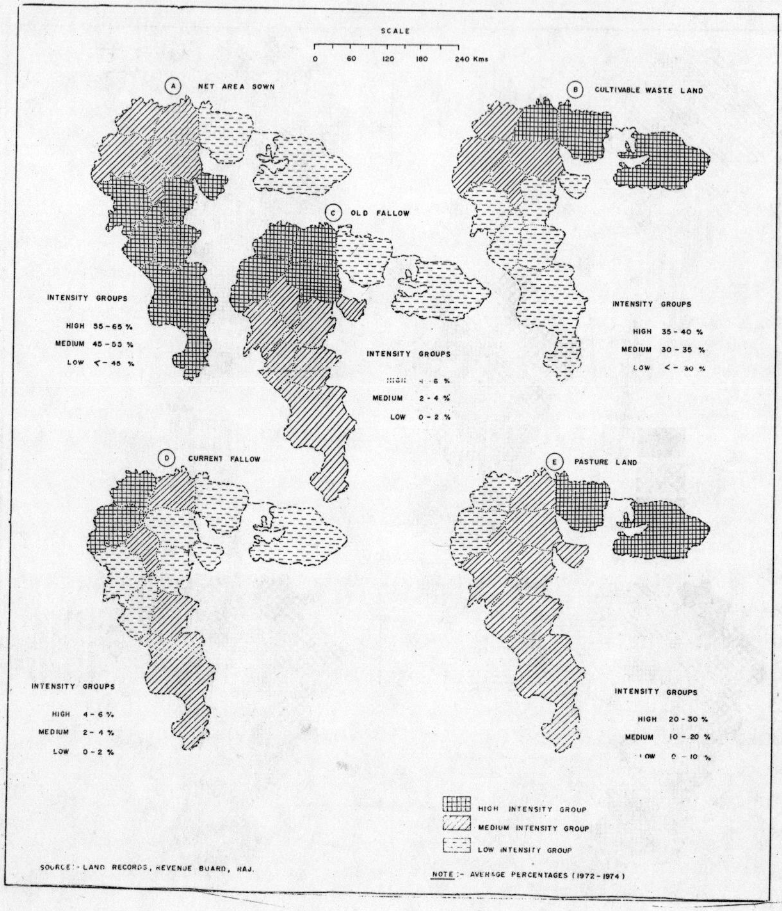

FIG. 8.5: Chittaurgarh district: elements of arable lands

Table 8.5 gives an encouraging picture of the cultivable land for a poor rural economy. For any agrarian development, the net cultivated area and pastures and grazing grounds should increase. All these characteristics are observed here. The miscellaneous trees and groves which should have also increased a little, have not decreased significantly.

TABLE 8.5: **Quinquennial differentials of the components of arable land**

Sl. No.	Component Elements	1954	1959	1964	1969	1974	Average % to arable land (1972-74)	Percentage change during years (1954-74)
1.	Net area sown	41.44	44.70	47.57	50.99	53.18	50.00	+11.74
2.	Current fallow	2.90	2.42	1.98	2.18	2.26	3.46	−0.64
3.	Old fallow	12.26	6.83	5.48	4.19	3.48	4.01	−8.42
4.	Misc. trees and groves	0.90	0.61	0.06	0.04	0.01	0.02	−0.07
5.	Cultivable wasteland	33.00	34.54	31.52	30.08	28.35	29.11	−4.65
6.	Pastures and grazing grounds	10.46	11.10	12.39	11.42	13.51	12.46	+3.05

Net Sown Area

The net sown area is the land which is being actually tilled for raising any type of crop like food, cash or fodder, etc. It covers, on an average, 50 per cent of the total arable land of the district, which is a little higher than the State's average of 46.58. The percentage ranges from a maximum of 64.17 in Dungla (Central High Plain) to a minimum of 25.91 in Begun (Vindhyan Plateau). A total of 11.74 per cent increase took place in the 20 years of the study period by a slow and continuous process, having the highest increase of 3.42 per cent in the quinquennial period of 1964-69. Kapasan tehsil recorded the highest percentage variation of +23.64 and Begun the lowest, +4.50.

When the cultivable waste and the old fallow lands decreased the net sown area increased. For example, in the Lava Plateau region out of a decrease of 26.97 per cent in the cultivable waste land, 14.69 per cent went to the net sown area.

A decrease of 16.07 per cent of the cultivable waste and fallow land was on account of increased the irrigation facilities. By 1969-74 majority of the irrigation projects were completed and major soil conservation programmes were undertaken. The expansion of the Bhopal Sagar Project not only increased the irrigation facilities but also helped in the utilisation of the waste and fallow land for cultivation. Consequently Kapasan tehsil recorded the highest plus variation of 23.64 in the net sown area.

The very slow rate of increase in the net sown area is due to several factors, such as high variability of rainfall, lack of extensive irrigation facilities, poor accessibility, and lack of incentives.

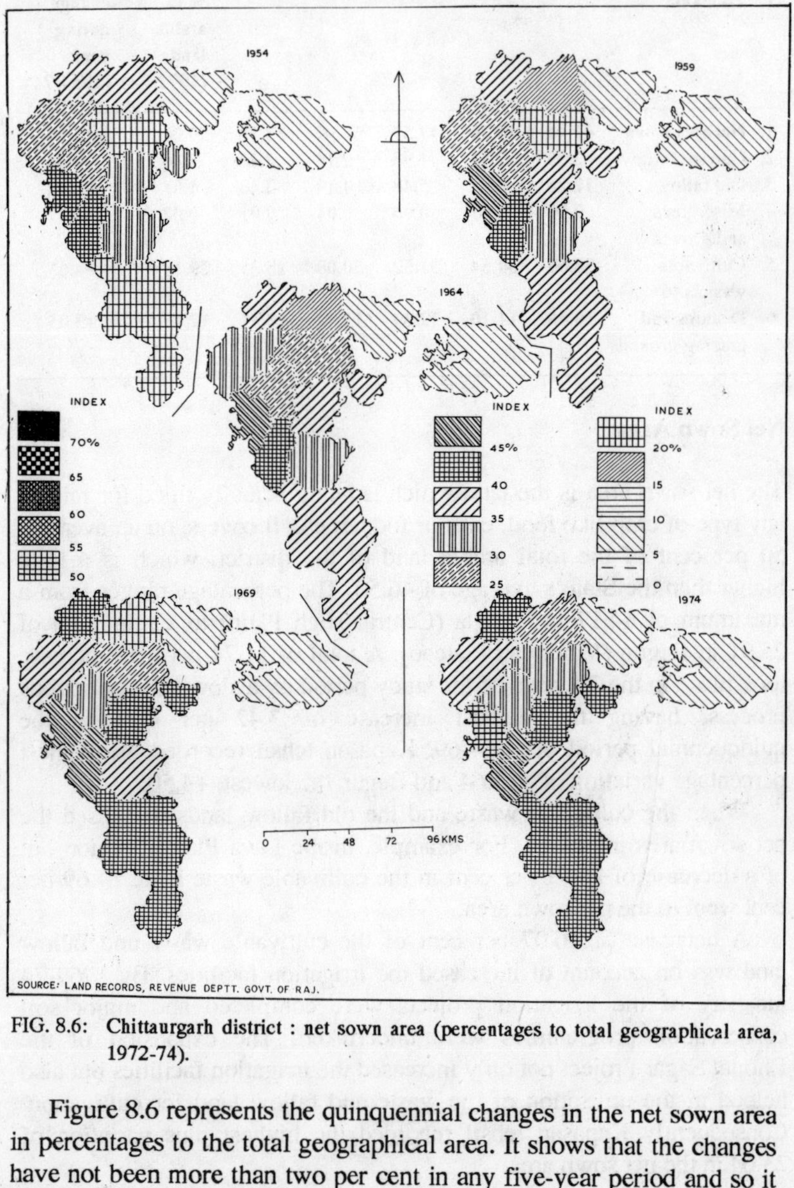

FIG. 8.6: **Chittaurgarh district : net sown area (percentages to total geographical area, 1972-74).**

Figure 8.6 represents the quinquennial changes in the net sown area in percentages to the total geographical area. It shows that the changes have not been more than two per cent in any five-year period and so it is insignificant. It is also characterised by lack of rainfall or heavy rainfall, causing greater erosion hazards.

On the basis of the three-year (1972-74) average percentages of the net sown area, the tehsils are grouped into classes as given in Table 8.6.

TABLE 8.6: Intensity groups of net sown area

Intensity Groups	Average percentages to arable land (1972-74)	No. of Tehsils	Name of Tehsils
High	55-65	5	Pratapgarh (lava Plateau), Nimbahera (Riverian Low Plain), Chhoti Sadri, Bari Sadri, Dungla (Central High Plain)
Medium	45-55	5	Chittaurgarh, Gangrar, Kapasan (Riverian Low Plain), Bhadesar (Central High Plain), Rashmi (N.Wn. High Plain)
Low	Less than 45	1	Begun (Vindhyan Plateau)

The tehsils of the low-intensity group in Table 8.6 are generally characterised by a highly rugged and hard slopy surface (Vindhyan Plateau). The soils are generally black and brown with shallow to moderate depths, moderate texture and moderate fertility. The steep scarps of Singoli and Kota Plateau are generally devoid of good soils. The Vindhyan hilly half of the Chittaurgarh tehsil is equally inhospitable for agriculture. The cultivated land, as a rule, is patchy and confined to the valleys and lowlands, where the rich transported soils containing high organic matter have accumulated and enough water is available from tanks, nallahs and wells dug at the foothill zones.

The tehsils falling in the medium group are Gangrar, Bhadesar, Rashmi, Kapasan and part of Chittaurgarh. These belong to the Uplifted Plain region. In the first three of these tehsils the surface is rocky, foothills are covered with weathered material and coarse textured soils. In the last two tehsils though the irrigation facilities are the most extensive in the whole district, yet the saline-alkali soils have reduced them to the medium-intensity group. In Rashmi the cultivated land is specially confined to the Banas Valley along which the soil is the deepest and the water strata the thickest.

Five tehsils fall in the high-intensity group (Fig. 8.6). These tehsils cover the Central High Plain and the Lava plateau. It has a gentle slope ranging from one to six degrees and 700 mm. of rainfall (Fig. 8.2). The rich fine textured black cotton soils with a high organic content, medium depth (nearly 50 to 90 cms.) and moderate fertility status (Fig. 8.2) in respect of nitrogen, phosphate and potash, cover the valleys, depressions and uneroded flat surfaces of the Lava Plateau (Pratapgarh). The eastern half of this tehsil possesses the continuous tracts of cultivated land. The western dissected part is generally covered with forests and grasses. So the cultivated land in this part is confined to the valleys and at places even deforested hilly steep slopes. Lying to the north of this plateau is the Central High Plain which is covered with brown and yellow soils with medium fertility and good facilities for well irrigation.

Current Fallow

Only 3.4 per cent of the whole district is left as current fallow and it ranges from 1.50 per cent in Nimbahera to 4.52 per cent in Kapasan. The common practice is to leave the land at rest after one or two ploughings, generally for one year, locally known as "Savana", and at times for three continuous years to regain the lost fertility. The fertilisers are applied by comparatively well-to-do farmers only. Otherwise, the land is left without crops or sown with leguminous and other green manuring crops such as guar, arhar, sannhemp, urad, etc., which are nitrogen fixers. These crops are then ploughed in while green.

On the whole, out of the 11 tehsils, only seven have a change of less than one per cent, showing that even after a long period after independence, the district is quite sluggish regarding the improvements in the farm practices. Nine tehsils fall into low-to-medium percentages groups. The percentages range from less than two to four. It shows that the areas with good fertile soils or sufficient rainfall or having adequate water supply, generally, have lesser area under current fallow.

It may be observed that the tehsils which have a low net sown areas, also have less areas under current fallow. It is because in areas where the cultivated land is limited, the practice of leaving the crop land idle cannot be afforded.

Old Fallow

The district has an average of 4 per cent of land under old fallow. But it records continuous decrease with a maximum of 5.43 per cent in the quinquennial period 1954–59. In this period all the tehsils showed a reduction in old fallow due to the good amount of rainfall (departure from normal, +7.8 cms) occurring till October. The overall variation in the old fallow during the 20 years is 8.42 per cent, the greatest decrease being in Chittaurgarh (–16.85%) and the lowest in Bari Sadri (–1.74%). The greatest positive change is recorded in Begun tehsil (Vindhyan Plateau) in the year 1964 with +6.75 per cent. The increased irrigation facilities are also responsible to bring a large portion of the old fallow land under the plough.

It shows that still there is good scope to increase the net sown area by utilising such-long-lying idle lands for cultivation.

Land under Miscellaneous Trees and Groves

The district possesses a negligible area (0.02 per cent of the total arable land) under this head. Every tehsil is far below the district average (0.46%). Pratapgarh tehsil (Lava Plateau) with 0.1 per cent under this head has orchards of guava, papaya and lemon while among the permanent trees the mango is widely grown. In Pratapgarh and Bari Sadri the "chironji" (a dry fruit) tree also grows.

Cultivable Waste

The land lying continuously idle at least for the past ten years, according to the *Patwaris*' records, is designated as cultivable wasteland. In comparison to the State with an average of 17 per cent of the cultivable wasteland, the district average of 29.11 per cent is quite high. The variation in this type of wasteland over the 20 years is only –7.74 per cent. The greatest variation of –3.92 per cent took place in 1964. The maximum variation of 26.79 per cent and the greatest quinquennial (1959-64) change of –11.8 per cent took place in Pratapgarh (Lava Plateau). Of this 10.20 per cent was transferred to the net sown area in the same period. The least change of –0.05 per cent was found in Rashmi tehsil of the Central High Plain (Fig. 8.7).

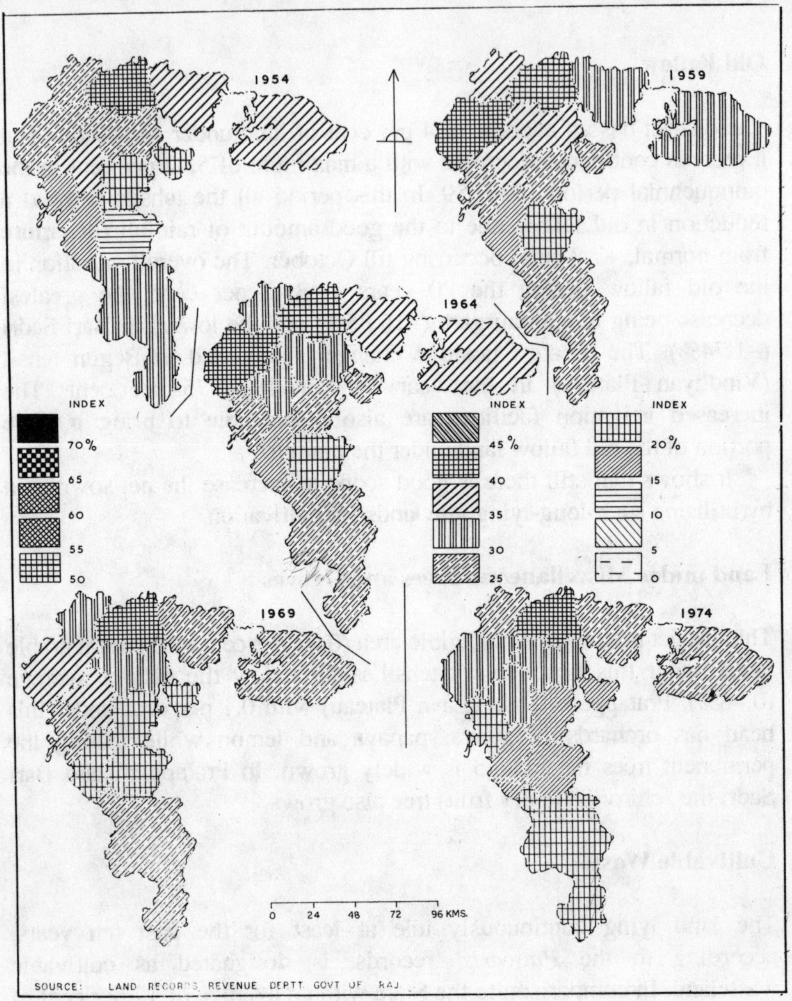

FIG. 8.7: **Chittaurgarh district : cultivable wasteland (percentage to total area).**

The data indicate a general trend of a constant decrease in the cultivable wasteland all over the district except in the tehsils of Kapasan, Chittaurgarh and Rashmi where it has been fluctuating. The decrease in cultivable wasteland is counter-balanced by an increase in the net sown area except in Begun and Pratapgarh where it has been utilised for other purposes also. For example, in Pratapgarh, out of a total decrease of 26.76 per cent of the cultivable wasteland, 15.56 per cent went to the net sown area, 4.20 to the pasture land and 7.0 to forest.

TABLE 8.7: Intensity groups of cultivable wasteland

Intensity Groups	Percentages to the total arable land	No. of Tehsils	Name of Tehsils
High	35-40	2	Gangrar (Riverian Low Plain), Begun (Vindhyan Plateau)
Medium	30-35	5	Nimbahera, Kapasan, Chittaurgarh (Riverian Low Plain), Rashmi (N. Wn. High Plain), Bhadesar (Central High Plain)
Low	Less than 30	4	Bari Sadri, Chhoti Sadri, Dungla (Central High Plain), Pratapgarh (Lava Plateau)

An analysis of the tehsilwise quinquennial differentials brings to light a fact that in 1964 and in 1974 the percentages of wasteland had significantly gone down.

The slow and steady decrease in cultivable wasteland from 1964 onwards like fallow land is directly related to the irrigation facilities. For example, in Nimbahera. Chittaurgarh, Begun, Kapasan and Pratapgarh the completion of the Gambhiri Canal System, Orai Dam and Gadola Dam have opened new vistas for the use of cultivable wasteland.

Pastures and Grazing Ground

This category of land includes all such lands which are under grass cover, whether government or private. These may be permanent pastures which are kept reserved as a village common grazing ground, or vast tracts of protected grasslands designated as *mahphoos charnot* (by the Settlement Department), not open for free grazing, and the unreserved grasslands where the cattle of any locality are allowed for stray grazing. Any ideal crop plan should provide land equally for both food for the farmers and feeds and fodders for his farm animals.

The district with 13.81 per cent under permanent pasture and grazing grounds to the total arable land is in much better position in comparison to 5.33 per cent of the State. There has been a constant increase in pasture lands, to the extent of 3.05 per cent during the study period. In the last quinquennial period the highest percentage increase of 24.69 is recorded in Begun (Vindhyan Plateau). In Gangrar, Kapasan and Rashmi, all located in the N. Wn. High Plain and the Riverian Low Plain, the pastures have decreased by overgrazing on hill slopes and by converting the common grazing grounds into farms.

The pasture lands are covering larger areas in those tehsils where terrain is hilly and undulating with shallow and coarse textured infertile soils.

In Bari Sadri, Chhoti Sadri, Nimbahera and Chittaurgarh tehsils, the Forest Department has converted the hilly undulated tracts into reserved pastures. The grazing is allowed after the monsoon period, but only for three months. After the maturing of grasses, these "beers" are either sold to the contractors or their grasses are collected by the Forest Department itself.

A study of the correlation of the total livestocks and total available pasture land would not be out of place. The district has 46 livestock per hectare of pasture land, while it is 22 in the State. Table 8.8 classifies the tehsils into high, medium and low classes on the basis of pressure of livestock per hectare of pasture land.

TABLE 8.8: The pressure of livestock per hectare of pasture land

Pressure	No. of livestock per hectare of pasture land	No. of Tehsils	Name of Tehsils
Higher than the district's average	102-120	4	Kapasan, Chittaurgarh, Gangrar (Riverian Low Plain), Bhadesar (Central High Plain)
Medium	40-80	6	Nimbahera (Riverian Low Plain), Rashmi (N. Wn. High Plain), Dungla, Chhoti Sadri, Bari Sadri (Central High Plain), Pratapgarh (Lava Plateau)
Lower than the District Average	–40	1	Begun (Vindhyan Plateau)

It is clear from Table 8.8 that only Begun has pressure lower than the district. The greatest pressure on pasture land is found in Bhadesar, with 107 livestock per hectare of pasture land. It shows that the district suffers from shortage of food and fodder for its present livestock population. Taking 0.5 hectares of pasture land per head of cattle as standard, fixed by the State Settlement Department, some 20 years back, none of the tehsils had the required area under pasture land. This is shown in Table 8.9.

TABLE 8.9: The percentage shortage of pasture land

Group	Shortage in percentage	No. of Tehsils	Name of Tehsils
High	above 85	4	Chittaurgarh, Gangrar, Kapasan (Riverian Low Land), Bhadesar (Central High Plain)
Medium	70-85	6	Rashmi (N. Wn. High Plain), Nimbahera, Dungla, Bari Sadri, Chhoti Sadri (Central High Plain) Pratapgarh (Lava Plateau)
Low (less than)	70	1	Begun (Vindhyan Plateau)
District average 80.39 per cent			

It is evident that all the tehsils, except Begun, have an acute deficiency of pasture lands. In Begun, the plateau surface is mostly covered with natural grasses and the livestock population is comparatively low.

The cultivable wasteland and old fallow land are the potential sources of herbage. Most of these are marginal lands often with steep slopes as in the hilly and undulated tracts of Chittaurgarh and Pratapgarh tehsils, or with shallow soils of low fertility, as in the plateau region of Begun, or with soils of poor moisture storage capacity, commonly seen in Gangrar, Bhadesar and Rashmi. The forage available from these marginal lands is inadequate in quantity and poor in quality. The grass lands in the district are at a low level of productivity and the natural grasses growing on them are of poor nutritive value and the fodder crops are not assigned their due share in the cropping pattern. Cultivated fodders are other sources of forage. But the percentage share of fodder crops to the total cultivated land is only 1.8 which is quite low.

In Pratapgarh tehsil the tribals have destroyed good pasture lands by burning the grasses to reclaim land for cultivation. This has enhanced the soil erosion hazard and depleted the permanent grasses.

Similarly in Begun tehsil, owing to the lack of vigilance due to shortage of staff, the Gurjars with their herds trespass the reserved areas. The grazing of the tender grasses is depleting the good grass covers. The different types of grasses and fodder crops in some regions are grown in relation to the ecological conditions. Such lands can be increased and improved by restricting cultivation on hill slopes.

To this study of landuse changes we may add a study of landuse combinations after Kostrowicki. It is "percentage method of landuse combination". The synthesis reveals that there is hardly any change in landuse combinations during the study period of 20 years (1954–74) as is evident from Table 8.10.

TABLE 8.10: **Landuse combinations (change over a 20-year period 1954—1974) (after Kostrowicki's percentage method)**

Tehsils	1954	1974
Chittaurgarh	A_4 n. cw. fl. + NA_2 b.	A_4 n. cw.+ NA_2 b.
Gangrar	A_5 cw. n.	A_5 n. cw.
Kapasan	A_0 n. cw.	A_5 n. cw.
Rashmi	A_5 n. cw.	A_5 n. cw.
Begun	A_4 cw. n. fl. p. + NA_2 b.	A_4 cw. p. n. + NA_2 b. f.
Pratapgarh	A_4 cw. n. + NA_2 b. f.	A_4 n. cw. + NA_2 b.
Nimbahera	A_4 n. cw. + NA_2 b.	A_4 n. cw. + NA_2 b.
Bhadesar	A_5 cw. n.	A_5 n. cw.
Dungla	A_5 n.	A_5 n. cw.
Bari Sadri	A_4 n. cw. p. + NA_2 b. f.	A_4 n. + NA_2 b. f. na.
Chhoti Sadri	A_4 n. cw. fl. + NA_2 b. f.	A_4 n. + NA_2 b. f.
District	A_4 n. cw. p. + NA_2 b. f.	A_4 n. cw. + NA_2 b. f.

Notation	Role	Rank	Percentage share
A = Arable	Pre-dominant	5	80<
NA = Non-arable	Dominant	4	60-80
n = net sown area	Preponderant	3	40-60
cw = cultivable waste	Equal	2	20-40
fl = fallow land	Accompanying	1	<20
f = forest			
na = land not available for agriculture.			
p = pastures			

It is observed with regard to the arable and non-arable classes that all the tehsils have retained the same status in 1974 as they had in 1954. Of course, in four of the tehsils the significance of particular use has changed. The cultivable wasteland which occupied the first rank in the arable class in four units in 1954 has given place to the net cultivated land in three tehsils in 1974. In Begun tehsil (Vindhyan Plateau) the barren land has given place to forests to some extent while in Pratapgarh (Lava Plateau) the forest land has decreased to be dropped from the combination in the 20 years. Similarly the pasture land has also decreased during this period to the extent that it is no more retained in the arable class of land. In Bari Sadri cultivable wasteland and pastures and in Chhoti Sadri cultivable wasteland and fallow land which were significant in 1954 were excluded from the combination in 1974. In the same way other changes may be noticed in Fig. 8.8.

The Cropping Intensity

By measuring the cropping intensity indices* of the district under study, efforts are made to find out the nature of the cropping pattern and the possibilities of intensification of agriculture by extending double cropping. As a general rule the cropping intensity indices are intimately related with the amount of rainfall, extent of irrigation facilities, soil types and input capacity of farmers.

The district as a whole has a cropping intensity index of 128. Referring to this average index, all the tehsils of the district have been put into three main intensity groups as shown in Table 8.11.

TABLE 8.11: Representing the cropping intensity groups

Intensity Group	Indices	Double cropped area % to G.C.A.	No. of tehsils	Name of tehsils
High	130 & above	30-35	3	Begun (Vindhyan Plateau) Chittaurgarh (Gangrar Riverian Low Plain)
Medium	110-130	20-30	7	Rashmi (N. Wn. High Plain) Dungla, Bari Sadri, Chhoti Sadri (Central High Plain) Kapasan, Nimbahera (Riverian Low Plain)
Low	110	20	1	Pratapgarh (Lava Plateau)
District Average	128	25 per cent		

$$\text{* Cropping Intensity} = \frac{\text{Gross cultivated land}}{\text{Net cultivated land}} \times 100$$

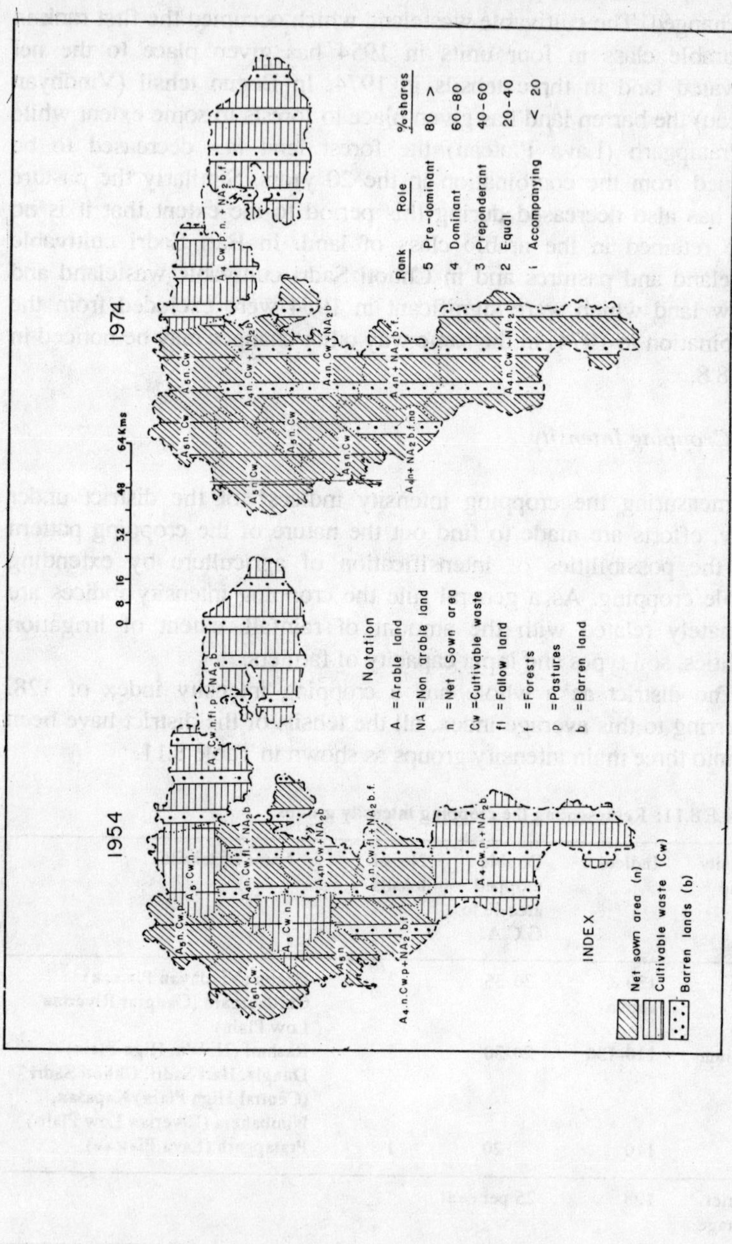

FIG. 8.8: Chittaurgarh district : landuse combinations (1954—74).

The high intensity group includes the tehsils of Riverian Low Plain and Vindhyan Plateau. This region has a moderate to high rainfall and high irrigation intensity, which encourages double cropping. Further, a large group of farmers in this region practise an economic crop-rotation through the use of early maturing varieties of seeds, which provide sufficient time to prepare the seed-beds for the Rabi crops.

The high density of population in these tehsils demands greater intensity of cropping to make up for food deficiencies. Thus the irrigation facilities, coupled with better crop rotation and high food demand have increased the area under double cropping.

The rest of the area in the Central High Plain and parts of the Riverian Low Plain falls in the medium intensity group. Low rainfall which, at times, limits the release of sufficient water for irrigation, causes a high percentage of seasonal fallowing. In spite of high-to-medium irrigation intensity, Rabi fallowing is quite prevalent in this region. In the North-Western and Central High Plains the hilly irregular surfaces, exposed rocks and the sandy-to-gravelly soils with low-to-medium fertility status limit the double cultivation. In Dungla the saline-alkali soils, on which inadequate irrigation is risky, the double cropping is not practised in larger areas.

The annual crops like sugarcane, cotton, arhar, or heavy feeding crops like opium, etc., which need seasonal fallowing are common in the cropping pattern of this region. In Bari Sadri the best lands have been reserved for *Ajwain* (Bishop's weed), a commercial crop of great economic importance.

Pratapgarh tehsil singularly forms the low cropping intensity region with a cropping intensity index of 107. Though soils in the large areas are fertile, shortage of irrigation facilities has restricted double cropping. The low-lying areas are generally left fallow during the Kharif season due to impeded drainage and waterlogging. Contrarily, on the plateau surfaces vast stretches are left as Rabi fallow in the areas where irrigation from wells is not available due to the low water table, and low water discharging capacity of the wells. At the top of all this, the area is tribally dominated and consequently the low economic standard, illiteracy, poverty and sluggish nature of the tribals handicap the adoption of improved techniques and better crop-rotation which help in increasing the extent of double-cropped areas.

If the change in the index of cropping intensity is studied over a period of years, it is found from Table 8.12 that the difference in the index values in the district is 9.0.

TABLE 8.12: **Quinquennial changes in the cropping intensity indices**

Tehsil	1954	1959	1964	1969	1974	Average Index No. (1972–74)	Change over 20 years (1954–74)
Chittaurgarh	131	135	138	129	146	133	+ 15
Gangrar	139	160	150	135	140	135	+ 1
Kapasan	135	170	120	120	135	128	0
Rashmi	120	132	116	115	125	121	+ 5
Begun	130	210	140	132	136	133	+ 6
Pratapgarh	111	115	109	110	112	107	+ 1
Nimbahera	118	160	119	120	161	128	+ 43
Bhadesar	132	140	133	130	135	130	+ 3
Dungla	122	160	118	117	123	121	+ 1
Bari Sadri	119	129	118	120	130	122	+ 11
Chhoti Sadri	119	130	120	122	130	123	+ 11
District	126	160	130	123	135	128	+ 9

The minimum difference of one is seen in Dungla, Pratapgarh and Gangrar due to the salinity hazards, backward tribal economy and gravelly soils, respectively. Bhadesar has a very restricted cultivated area in the inter-hill valleys.

In Rashmi, they have developed dry farming in the northwest and salt-resisting varieties on the saline soils of the southeast besides increasing the irrigation capacity of the existing tanks. In Begun and Chittaurgarh the expansion of canals from the Bassi tank and Orai dam has facilitated irrigation respectively. Figure 8.9 also represents the pattern of change in the cropping intensity quinquennially.

Standard Nutrition Unit

More useful is the study of Standard Nutrition Unit (SNU) which is the carrying capacity of land in terms of population in the component unit. The techniques of SNU by Shafi and Jasbir Singh are applied in Chittaurgarh district which give almost the same results as shown in Fig. 8.10. On the basis of SNU the area of high, moderate, low and very low efficiency have been deduced and shown in Table 8.13.

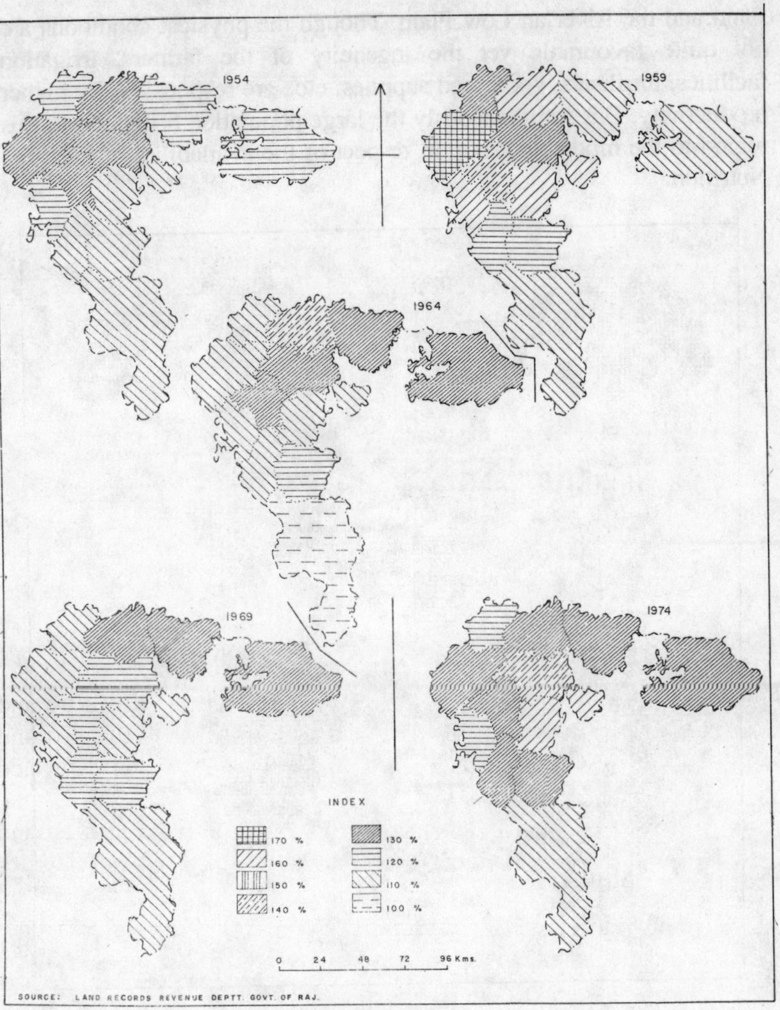

FIG. 8.9: Chittaurgarh district : cropping intensity.

It shows a high agricultural efficiency in the Central High Plain. Though the yields of food crops per unit area are not very high, yet it is a food surplus region due to low population density. As the available Standard Nutrition Unit is more than sufficient for the existing population, it is a potential area of the district.

The moderate efficiency region lies in the North-Western High

Plain and the Riverian Low Plain. Though the physical conditions are not quite favourable yet the ingenuity of the farmers, irrigation facilities, fertilisers, good seed supplies, etc., are responsible for higher productivity. But simultaneously the large population brings down this region to the moderate group in respect of the availability of Standard Nutrition.

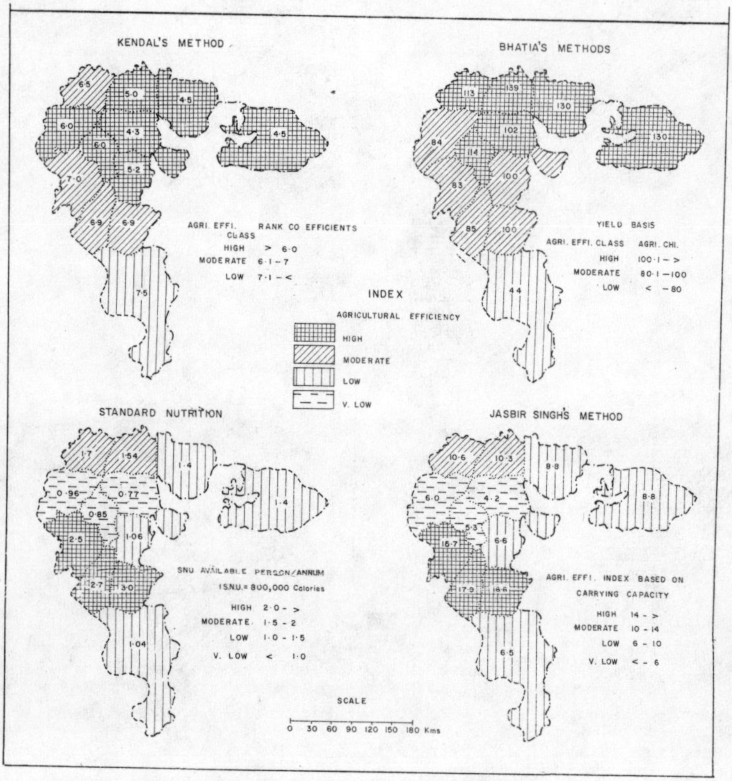

FIG. 8.10: Chittaurgarh district : levels of agricultural efficiency.

TABLE 8.13: Showing the agricultural efficiency regions with available SNU/person/annum and agricultural efficiency indices

Efficiency Group	Agricultural Index Values	Available SNU/Person/ Annum	No. of Tehsils	Name of Tehsils
High	More than 15	More than 2.0	3	Dungla, Bari Sadri, Chhoti Sadri (Central High Plain)
Moderate	10-14	1.5-2.0	2	Gangrar, Rashmi (N.Wn. High Plain)
Low	6-10	1.0-1.5	3	Nimbahera (Riverian Low Plain), Pratapgarh (Lava Plateau) Begun, (Vindhyan Plateau)
Very Low	6 & below	less than 1.0	3	Chittaurgarh, Kapasan (Riverian Low Plain), Bhadesar (Central High Plain)

The low agricultural efficiency is found in the Vindhyan and Lava Plateau regions. In the former the yield of food crops per hectare is, no doubt, high but the percentage of the net cultivated land is low. In the latter the tribals' rudimentary ways of cultivation do not give good returns. Further, in the eastern part of this plateau, in the fertile black soils, a good amount of land is devoted to spices, opium, etc. Thus the quantity and quality of the food production is low. Even a slight increase in the population will disturb the harmony.

In no way can it be inferred that these tehsils have low yields per hectare, but since the population is large in comparison to other tehsils they suffer from food shortage.

Conclusion

The aforesaid analysis of landuse for a period of 20 years is indicative of a very slow and insignificant change in the landuse pattern in Chittaurgarh district. In this duration the district shows an increase of only four per cent in the forest area. That too of a very low quality due to improper management. In comparison to the adjoining States of U.P., Haryana, Punjab and Gujarat it gives an impression of sluggish improvement. Destruction of trees and the neglect of cultivable wastes have increased the barren lands in some areas of the Riverian Plain. The construction of reservoirs, roads, factories and the expansion of townships have added 4.8 per cent to the land under non-agricultural uses during these two decades thus reducing not only the barren land and cultivable wastes but even the forests in some parts.

With regard to the change in the arable land during the study period the increase is only 1.36 per cent which is slow but continuous. The change of –0.64 per cent in current fallow during the same period is negligible. The current fallow shows maximum change in the west-central parts, while the minimum in the southern parts. The old fallow has increased only in the Vindhyan Plateau region.

Like the fallow lands the overall change of +0.07 per cent in land miscellaneous trees and groves also indicates a negligible change in the landuse pattern. The wastelands have decreased by 7.74 per cent with a maximum reduction in the Lava Plateau where the increased mechanisation of well irrigation has enabled the cultivators to utilise such idle lands.

The pastures which are significant in this district have decreased due to the harmful practice of overgrazing of the hill-slopes and due to the encroachment of cultivation on the village common grazing grounds.

The reduction of barren lands, cultivable wastes, current and old fallows is a sign of future economic development but its quantum hardly offers any promise for the future. The Rabi areas record an increase of only nine per cent in spite of so much development in irrigation, fertilisers and hybrid seeds while the Kharif has recorded only about half of the Rabi areas.

The cropping frequency graphs and the location quotients of important crops show that the cereals have always been holding a dominant place in the cropping pattern. In the latter years of the study period, gram, cotton, spices and narcotics overtook cereals while the area under oilseeds, pulses and fodder crops receded. Consequently the district provides only 1.23 Standard Nutrition Unit/person/annum. This shows that the availability of food is marginally sufficient and a little increase in population may bring about a catastrophe if efforts are not made to secure agricultural activity in Chittaurgarh district.

REFERENCES

Amani, K. Z., 1976 : *Agricultural Landuse in Aligarh District*, Kumar Publication, Aligarh.

Ayyar, N. B., 1976 : "The Geography of Nutrition", in *Essays in Applied Geography*, (eds.) Mishra, Ayyar and Pramila Kumar, University Press, Sagar, M.P., pp. 103-10.

Bhatia, S. S., 1965 : "A New Measure of Agricultural Efficiency in U.P.", *Economic Geography*, XLIII, pp. 244-60.

Census Operations, Rajasthan, Jaipur, 1971 : *District Census Handbook* — Chittaurgarh District, Series 18, Pts. XA and XB.

Deptt. of Agriculture, Rajasthan, Jaipur, 1974 : *Rajasthan Agricultural Diary*-1974.

———, 1970 : *Agricultural Chemistry — 20 years of Agricultural Research in Rajasthan*.

Gupta, N. L., 1964 : "Landuse of Udaipur Plateau" (unpublished Ph.D. Thesis), University of Udaipur, Udaipur.

Kostrowicki, J., 1976 : "On the Method of Determining Landuse, Crop and Livestock Combinations as used in the Polish Studies", in *Essays in Applied Geography*, Mishra, Ayyar and Pramila Kumar (eds.), University Press, Sagar, M.P.

Mohammad, N., 1978 : *Agricultural Landuse in India*, Concept, New Delhi.

Shafi, M., 1976 : "Approaches to the Measurement of Agricultural Efficiency — A Study in Agricultural Typology", in *Essays in Applied Geography*, Mishra, Ayyar and Pramila Kumar (eds.), University Press, Sagar, M.P.

Singh, Jasbir, 1972 : "A New Technique of Measuring Agricultural Efficiency in Haryana", *Geographer*, XIX, I, pp. 14-33.

Stamp, L. D., 1960 : *The Land of Britain : Its Use and Misuse*, Longmans, London, 2nd Edition.

Singh, Vijay Ram, 1968 : "Analysis of Calorific Intake in Rural Mirzapur", in *Applied Geography*, Singh, R.L. (eds.), National Geographical Soc. of India, p. 152.

9

TRANSFORMATION OF AGRICULTURE IN TRIBAL MADHYA PRADESH

C. K. JAIN

HERCULEAN EFFORTS have been made to increase the agricultural output after Independence in India. This has been made possible through the organised efforts of improving the physical base by utilising modern inputs, and through the improvement in agricultural practice and crop structure. These processes of planned change in agricultural economy may be called the transformation of agriculture. In the traditional agriculture pattern both seasonal and perennial crops played a part, with diversification emphasised rather than specialisation of a limited range of crops. Intercropping was also more characteristics than cropping in pure stands, cultivation by wooden plough was almost universal and the cropping practice was customary, crop yields were low. Fertility was largely maintained by keeping the land as fallow, by green manuring, and, within limits, by using cow dung and composite manures. In general, the peasant approached agricultural production with the aim of minimising risk rather than ensuring maximisation of production or profit. Transport was inadequate and inefficient. Marketing facilities were limited and storage facilities almost non-existent, while the use of mechanical power in farming operation was almost unknown. Holdings were fragmented and the system of land tenure unfavourable to increasing agricultural production. Such a system of farming becomes static and cannot keep pace with the rising demand of growing population. Such a situation necessitates improvement in agricultural system to meet the increasing demands. It led to the invention of new methods, techniques, and inputs to raise the production. Adoption of these developments leads to the transformation of traditional agriculture to modern farming (Singh, 1979).

The main characteristics of modern agriculture are wide spread use of these yield raising technologies. Among them, improved seeds, fertilisers, pesticides and machinery are worthmentioning. These are capital intensive inputs. Higher use of capital inputs caused commercialisation of farming. Application of these inputs as well as machine run by inanimate energy increases productivity of land and workers. Hence, modernisation implies transformation of traditional agriculture into a new and more efficient type of farming (Dayal, 1983; and Tiwari and Jain, 1989).

There has been a tendency of change in agricultural practice during the Plan period throughout the State. It has also been seen that the rate of change varies with the socio-economic condition of the farmers as well as with the physical resource base. Accepting the first hypothesis it is proposed to see the changing nature of agriculture in tribal tracts of Madhya Pradesh in post-Plan period. For this purpose fifteen major tribal districts having higher proportion of tribal population than the state average have been selected. They are Bastar, Rajanandgaon, Bilaspur of southern zone, Raigarh and Surguja of eastern zone, Sidhi and Shahdol of north-eastern zone, Mandla, Seoni, Betal and Chhindwara of south-central zone; Khandwa, Khargone, Dhar and Jhabua of western zone. In these districts, terrain is hilly, and highly dissected and physically is not suitable for cultivation. At the movement extension of cropland, irrigated area, High Yielding Variety (HYV) of seeds, use of chemical fertilisers, agricultural implements, use of power in agriculture and intensity of cropping have been analysed. The period of discussion is three decades commencing from Second Five Year Plan, upto the end of Sixth Plan (1984-85).

Growth of Net Sown Area

The period of three decades commencing with the inception of the First Five Year Plan, has seen several remarkable changes in tribal agriculture. Firstly, growing population and government efforts have caused increase in net sown area as far as possible. In 1984-85, total net sown area in these fifteen districts was 8.24 million hectares which is slightly higher than 37 per cent. This proportion is much lower than the State average of 42.9 per cent. There are only five districts where proportion of net sown area is higher than the State average. They are Rajanandgaon, Seoni, Khargone, Dhar and Jhabua. Low proportion of net sown area in south-central and eastern tribal zones is mainly due to hilly and dissected terrain and infertile soils. Not only that, growth in net sown area has been slower in these districts. It was 7 million hectares in 1955-56 which rose to 8.24 million hectares in 1984-85, thus recording an increase of 17.7 per cent in these fifteen districts. Average for the State is 23.0 per cent. The growth of net sown area varies from 3.0 per cent in Durg-Rajnandgaon to 47.0 per cent in Bastar districts. It is remarkable that four (Mandla, Sidhi, Surguja and Bastar) out of six districts recordingly higher growth than the State average are in the central part of the State. Another two are Jhabua and Dhar districts. Proportion of net cropped area was usually small in these districts in 1955-56, while extensive land were lying as cultivable waste and fallows. Now these potential lands have been cropped. It is also due to high pressure of rural population on land. Contrary to this, ten districts have low and very low growth in respect of net sown area. This slow growth is due to the fact that they had high developed their potential arable lands and there was least possibility of expansion of net sown area.

Expansion of Cultivation Land

Growth in net sown area depends on expansion of irrigation facilities, population growth and urbanisation. Other important factors include favourable climatic conditions, improvement of transport, reclaimation of fallow lands and intervention of government. There is wide range of variation in growth rate of net sown area during 1956-84. It varies from 3.0 per cent in Rajnandgaon-Durg district to 47.07 per cent in Bastar district. There are only six districts out of fifteen major tribal districts, recording very high growth rate than the State's average (23.0 per cent during the period). These are Bastar (47.07%), Surguja (26.4%), Mandla (24.7%), Sidhi (39.7%), Dhar (24.4%) and Jhabua (46.0%). In these districts high and very high increase has been made possible by the reduction of the fallow lands and reclamation of culturable wastelands during plan periods. In this period, minor and medium irrigation projects have been commissioned which have facilitated the expansion of the net cropped area.

Nine districts have experienced slow growth (than the State's average 23.0%) of net sown area. The districts lie in two tract — southeastern part of the State and western Malwa. In these areas, slow growth of net sown area is either due to undulating hilly and forested area, where expansion is physically not possible even after the development of irrigation (Betul, Shahdol, Seoni, Chhindwara, Raigarh) or they have highly developed their potential land (Bilaspur, Rajnandgaon-Durg, Khandwa, Khargone) and there is least possibility of expansion of net sown area.

The period of three decades commencing with the inception of the First Five Year Plan, has seen several remarkable changes in tribal agriculture. Fastly growing population and government efforts have caused increase in net sown area as far as possible, but it is much lower than the State's average of 42.9 per cent. There are only five districts where proportion of net sown area is higher than the State average. They are Rajnandgaon (55.4%) and Seoni (43.7%) in the southern part and Jhabua (51.1%), Dhar (61.0%) and Khargone (46.3%) in the western part of the State. Low proportion of net area in south-central and eastern tribal zones is mainly due to hilly and dissected terrain and infertile soils.

Changes in Intensity of Cropping

Intensification of cultivation, meaning the landuse efficiency, is definable as 'the extent to which the net area sown in cropped or resown'. The total cropped area as a percentage of the net sown area gives a measure of land-use efficiency, which really means the intensity of cropping (Singh, 1976; Sharma, 1988). Intensity of cropping largely depend on irrigation facilities, rainfall, pressure of population, extent of new farm-technology and socio-economic conditions of the farmers. In 1955-56, cropping intensity was 114 in major-tribal districts which has increased to 116 in 1984-85, recording total increase of 1.75 per cent. Four district, viz. Bastar, Bilaspur, Raigarh and Jhabua recorded minor growth. Only six district (Sidhi, Shahdol, Betul, Chhindwara, Khandwa and Khargone) have recorded higher growth than the State average (4.46%) during the period. In 1984-85, intensity of cropping varies from 104 in Bastar to 132 in Rajnandgaon district. Six districts (Bastar, Raigarh, Seoni, Khandwa, Khargone and Jhabua) have lowest intensity.

Change in Methods of Farming

After Independence, the Government of India decided that drastic economic changes were essential to survival. A new chapter of planned economic development began with the launching of the First Five Year Plan. The period between 1950 and 1970 saw profound changes in the agricultural scene of Punjab and this State also. These changes are largely reflected both in expansion and intensification of cropping which has been possible due to technological advances, such as irrigation, adoption of HYVs, use of chemical fertilisers and pesticides, and new iron based farm implements. These changes are also adopted by the farmers of tribal region.

Irrigation

Irrigation is one of the most important and basic ingredient in the process of transformation of agriculture (Sharma and Jain, 1988). Its use is conditioned by several variables. While low rainfall and its vagaries necessitate the development of artificial means of water supply, it is also essential to take maximum benefits out of the chemical fertilisers, and to some extent, high yielding variety of seeds.

Irrigation has been practised since quite long time in few parts of the region. The Chhattisgarh region of Central Province (C.P.) and Berar are worthmentioning, where irrigation potential were created during 1920-40. But in other parts, development of irrigation is the phenomenon of the plans. In 1955-56, total irrigated area was 279.8 thousand hectares which was only 3.5 per cent of the total cropped areas in the region. There is a wide range of variation in growth rate of irrigated area during 1955-85. Growth rates varies from −17.8 per cent in Bastar to 2545.2 per cent in Sidhi district. Very high growth occurred where initial proportion of irrigated area was extensively low i.e Mandla (0.20 per cent to total cropped area), Shahdol (0.26%), Surguja (0.81%), Jhabua (1.1%), Dhar (3.1%) and Sidhi (0.20%). In ten, out of fifteen districts, growth rate was higher than mean (230.3%), five of them (Surguja, Raigarh, Sidhi, Shahdol and Mandla) had less than one per cent. Jhabua Rajnandgaon-Durg, Seoni, Bilaspur and Betul had higher proportion of their cropped area under irrigation in 1955-56. Sidhi, recording highest increase, had only 3.7 per cent of their cropped area under irrigation in 1984-85. The phenominal increase in least irrigated districts is the result of very small denominator. Even at present there are only three districts (Khargone, Bilaspur and Rajnandgaon) where proportion of irrigated area is higher than the State average. All these districts (except above three) with initial poor irrigational base are still poor, despite of high and very high growth rates. In these districts dissected terrain dominated by hills and plateaus imped the extension of irrigation facilities. At present in Bastar (1.9 per cent of the total cropped area), Surguja (3.0%), Mandla (2.3%), Sidhi (3.7%) and Shahdol (2.2%) districts irrigation has not started in real sense.

Mechanisation of Farming

One of the major technical developments of the Post-Independence period has been the application of mechanical power to agriculture. Mechanisation in Madhya Pradesh and elsewhere in India, simply means the use of tractors, and power-operated pump-sets, tube-wells and wheat thresheres. Their impact in improving farm production performance per unit of agricultural land has been considerable (Singh, 1979). Mechanisation may be defined as the use of improved types of iron-based land tools, draught driven implements and power driven equipments (Singh, 1976). It would not be correct to say that only the power driven equipments have glamour; even the manually operated and animal driven implements can be used by a number of farmers, particularly by the sub-marginal and small farmers in this region.

Tractors

The use of tractor is not very common in this region due to the dominance of rice cultivation, small land holding, undulating hilly terrain and poverty of farmers. The number of tractor per thousand hectares of cropped area was 0.13 in 1974-75, whereas it is 0.48 tractor per thousand hectares of cropped area in the state. In 1984-85, there was a marked increase in the use of tractors in cultivation. Density of tractors per thousand hectares of cropped area increased from 0.13 to 0.48 during this period. Overall increase in the number of tractors during 1974-75 to 1984-85 is 274.8 per cent. In 1974-75, only four districts viz. Khandwa, Khargone, Dhar and Bilaspur, had higher density of tractor than the regional average.

Density of tractor per thousand of cropped area varies from 0.10 in Mandla to 0.89 in Khandwa and Raigarh district in 1984-85. The mean of these 15 districts figure arrive at 0.48. There are only 7 districts having higher density of tractors than regional mean i.e. 0.48. Three of them are in the eastern part (Raigarh, Sidhi and Bilaspur); one is in central tribal zone (Chhindwara) and other three are in the western zone (Khargone, Khandwa and Dhar). Density is low and very low in more than half of the districts which cover south-eastern parts of the State. In this area, animal force continues to be major source of power for agricultural operations. Thus it seems that the low proportion is due to undulating terrain, small and very small size of holdings.

Oil Pumps

Till the end of 1960 oil engine pumps were more common than the tractors in this region. The number of oil engine pumps per thousand hectares of cropped area was 2.54 in 1974-75. Number of oil pumps increased markedly during 1974-84. Their number crossed over 50 thousand and their proportion of per thousand hectares of cropped area become 5.90. Over all increase in the number of oil engine pumps during the period 1974-75 to 1984-85 was 145.3 per cent. Corresponding increase in the number of electric pumps was 286.1 per cent, indicating that, with the increase of rural electrification the electric pumps are becoming more popular. Only five districts fail to increase their growth of oil engine pumps above the State average (126.7%). Only Chhindwara recorded minus growth, in their oil pumps. Six districts (Rajnandgaon, Durg, Raigarh, Sidhi, Shahdol, Bilaspur and Jhabua) recorded more than double growth than the State average in oil pumps. On an average, there are 5.9 (States 6.8) oil pumps per thousand hectare of cropped area in 1984-85. The density is as high as 19.40 in Khargone and as low as 1.22 in Bastar. The districts with higher number than mean are mostly in western zone (Khargone, Khandwa, Dhar, Jhabua and Betul). Out of these five districts only Khargone has higher percentage of gross irrigated area. The six districts (Bastar, Rajnandgaon, Shahdol, Mandla, Chhindwara and Bilaspur) have not even three oil pumps for one thousand hectare of cropped area and all are in southern-central zone of tribal area. They are poor in respect of electric pumps also and do not have facilities of canal irrigation as well. Consequently, they have very poor and highly variable developed water resource base.

Electric Pumps

Electric pumps are more common than the diesel pumps. There are 110,611 electric pumps working in the State in 1984-85 as against

518,79 oil pumps. In 1974-75, 28,648 electric pumps are used for irrigation purpose (only 3.4 electric pumps per thousand hectare of cropped area). In this year, more than half districts, had not even one electric pumps for one thousand hectares of cropped area. Only five districts (Chhindwara, Dhar, Khargone, Khandwa and Betul) have more than the State average (4.2 electric pumps per thousand hectare) of the electric pumps. Over all increase in the number of electric pumps during the period 1974-75 to 1984-85 was 286.1 per cent. In this period only six districts (Bilaspur, Chhindwara, Mandla, Shahdol, Dhar and Khargone) have recorded increase in electric pumps more than the State average (260.5%). Only Bilaspur and Khargone district have a higher proportion of irrigated area. In 1984-85, number of electric pumps per thousand hectares ranges from 0.47 in Bastar to 40.6 in Chhindwara. Mean of the region is 12.6. All five districts with higher density than the mean are in the western zone in the tribal Madhya Pradesh. These districts are Chhindwara, Dhar, Khargone, Khandwa and Betul, while deficient district are mostly in the southern, eastern and central tribal zone of Madhya Pradesh.

Use of Power in Agriculture

The urgency of raising agricultural production has led to the expansion and intensification of cropped area. Expansion of cropped area has been done through the annexation of cultivable wastelands, which requires comparatively much larger amount of labour. Most of them, such ardous activities are expected to be performed with the machines motivated by the inanimate energy, particularly in areas where manual labour is comparatively scare. Another direction of change is the increasing application of yield-raising technology with the provision of ensured water supply. It leads to the intensification of farming which requires longer man-hours, as well as mechanical aids. For instance, farmer has to be more careful and active if he selects to irrigate his field in preference to dry farming. With application of modern inputs as high yielding variety of seeds, fertilisers, pesticides and insecticides, requirement of man-hours increases proportionately. To assess the use of power in agriculture, energy contributed by man as cultivators and agricultural labourers, animal and machinery engaged in farming and using inanimate energy have been taken into account after converting them into horsepower equivalent. For the conversion of energy contributed by different sources into horsepower units, equivalents are suggested by Sharma (1986). Use of power per hectare of cropped area has been changing with the changes occurring in the methods and techniques of farming.

Use of power per thousand hectares of net sown area ranges from 254 hp in Rajgarh to 513 hp in Khargone district in 1982-83. The districts can be grouped in four class in use of power per thousand hectares of net sown area (Table 9.1). There are only 5 districts with higher use of power than the State average (396 hp), while 10 districts have lower value. Availability of power for agricultural purpose is highest in Khargone followed by Khandwa, Chhindwara, Dhar and Bilaspur districts. It is due to higher proportion of diesel and electricity pumps. In absence of canals, commercial crops of this tract entirely depend on well irrigation, requiring a large number of pumps for lifting water. In Bilaspur district, power is derived from animate sources. Number of agricultural workers per unit of cropped area is very high in this district. In Chhindwara district, major contributors are irrigation pumps. Availability of power is lower in 10 districts. In these districts, farmers are basically dependent on animate power. Raigarh (254 hp) is at the bottom in use of power in the region. As a rule, proportion of irrigated area is low and very low in these districts.

TABLE 9.1: Use of power in agriculture in major tribal districts

Horsepower per hectare of N.S.A.	Level of use of power		Name of district
Above 0.487	Very high	(1)	Khargone
0.406 – 0.487	High	(4)	Chhindwara, Bilaspur, Khandwa, Dhar
0.325 – 0.406	Low	(5)	Shahdol, Sidhi Surguja, Ragarh, Betul, Jhabua
Below 0.325	Very Low	(5)	Seoni, Mandla, Rajanandgaon, Bastar

Use of Chemical Fertilisers

Fertilisers

Chemical fertilisers play a crucial role in agricultural production. An intelligent and sufficient application of the proper types of fertilisers is, therefore, of strategic importance to increase agricultural production. Manures can be organic or inorganic. Both types are necessary for good agriculture. The availability of organic manures is dependent on the extent to which green manuring crops can be grown in the fields. The cultivation of green-manuring crops has to fit in with the crop pattern and is possible under irrigated conditions. As their cultivation hampers the cultivation of at-least one crop, it is difficult to persuade the cultivators to take up the raising of green-manuring crops in a big way. The only way by which the manurial requirement can be met effectively, therefore, lies in the application of chemical fertilisers to the crops. Soil tests indicate that almost all soils in India must have additional nitrogen; about 85 per cent of the soils, need additional phosphorus and nearly 63 per cent of them require potash, if crop yields are to be increased by 50–100 per cent (Randhawa, 1976).

Use of chemical fertilisers to replenish fertility of soil cannot be said as a popular measure in this area. On an average, 12.1 kg of chemical fertilisers are applied per hectare of cropped area which is lower than the State average (19.1 kg/ha) in 1984-85. Poverty of farmers, tribal dominated population, lack of ensured timely water supply and subsistance nature of cropping may be reasons. In 1974-75, total fertiliser consumption in this area was 27.97 thousand tonnes which increased to 115.9 thousand tonnes in 1984-85. In last decade consumption of fertiliser increased by 311.1 per cent. There is a wide range of variation in growth rates. Varies from 4.5 per cent in Rajnandgaon-Durg to 1089.2 per cent in Sidhi districts. Eight districts have recorded higher growth in the use of fertilisers than the State average (295.4%). They are Sidhi, Surguja, Raigarh, Bilaspur, Mandla, Bastar, Jhabua and Dhar. In these districts (except Bilaspur and Dhar) irrigated area is much lower than the State average. There are eleven districts where the use of fertilisers is less than the State average (19.1 kg/ha). It is as low as 2.4 kg/ha in Mandla, 3.3 in Shahdol, 2.7 in Bastar, 5.7 in Jhabua, 4.0 in Sidhi districts. Khandwa (21.5), Khargone (30.2), Bilaspur (27.5) and Dhar (27.5) districts have higher use of fertilisers. It is due to the cultivation of commercial crops such as cotton, Soyabean, Moong etc. in these districts. They give comparatively high return with the use of fertilisers. It is noticed here that fertiliser are not popular due to several reasons. Farmers prefer to use manures to maintained their agricultural productivity.

High Yielding Variety of Seeds

No doubt provision of ensured timely water supply raise the

productivity of land, but the real benefit comes only when improved seeds and fertilisers are used alongwith the irrigation. The 'Green Revolution' was primarily based on high yielding variety of seeds. In the last twenty years of the third plan, several dwarf varieties of wheat as well as rice were evolved and introduced in the country and this area also (Sinha, 1975; Jain, 1988). It is known fact that only improved variety of cereals were particularly distributed. Therefore, rice, wheat, maize, bajra and jowar are considered for the present purpose. Total area under high yielding variety of seeds was 634.5 thousand hectares in 1974-75, which was 15.5 per cent of total cropped area of five crops. With the high quality of improved seeds area under high yielding variety of seeds increased 150.8 per cent from 634.5 thousand hectares in 1974-75 to 1591.2 thousand hectares in 1984-85. Growth rate varies from zero in Sidhi district to 655.1 per cent in Bastar district. In ten districts growth rates were higher than State average (131.1%) and regional average (150.8%). These districts are Bastar, Rajgarh, Shahdol, Mandla, Seoni, Betul, Chhindwara, Khandwa, Khargone and Dhar. Three of them (Bastar, Shahdol and Khargone) recorded more than double growth than mean during the period. But seven of them (Bastar, Chhindwara, Seoni, Mandla, Shahdol, Dhar and Khargone) had less than regional average (15.5%) and three of them (Raigarh, Betul, Khandwa) had higher proportion of their cropped area under high yielding variety of seeds in 1974-75. Bastar recorded highest increase, and had only 2.0 per cent of cropped area under high yielding variety of seeds. At present 34.1 per cent of the respective cropped area is under HYV of seeds. Nevertheless, more than half of the major tribal districts have lower proportion of area under high yielding varieties of seeds than the State average of 38.4 per cent. Only seven districts have higher proportion. These are Raigarh, Bilaspur, Surguja in the east, Chhindwara, Betul in the south-central, Khandwa and Khargone in the western tribal zone in the State. Three districts (Bastar 12.5, Sidhi 14.2 and Jhabua 16.5) have less proportion than the 50 per cent of the State average. Thus, use of high yielding variety of seeds could not get much favour in these districts.

Conclusion

The preceeding discussion on the changing agricultural practice in tribal tracts of Madhya Pradesh reveals depressingly low degree of the adoption of modern form technology. The traditional technology is yet more acceptable to the cultivators with marginal and small holding and limited economic means which make them more tradition oriented, and poverty stricken. There are not much changes in tribal agriculture. But, no doubt, efforts have been made for the upliftment of tribal agriculture. Inputs and credit facilities have been made available. Majority of tribal districts are still far behind in the race of development. There are no correlation with the density of tribal population and use of fertilisers (–0.40), extent of irrigation facilities (0.06), density of tractors (–0.05) and high yielding variety of seeds (–0.33) and density of pumps (–0.19). Nevertheless, districts of the south-eastern parts are in bare need of irrigation. First priority should be given to irrigation. Others (fertilisers, HYV and Mechanisation) are associated with it. They are very suitable to these areas because of several physical facts, such as undulating terrain with impervious rock-base and high rainfall. Improved seeds should follow it. Besides, fertilisers, pesticides and credit facilities are also to be provided.

REFERENCES

Dayal, E., 1983 : Modernisation of Agriculture in Uttar Pradesh. In Pandey (Ed.), *Modern Geographical Trends*, New Delhi : Today & Tomorrow Publ., pp. 475-89.

Jain, C. K., 1988 : *Patterns of Agricultural Development in Madhya Pradesh*, New Delhi : Northern Book Centre.

Mohammad, N. (ed), 1981 : *Perspectives in Agricultural Geography*, Vol. I, Concept, New Delhi.

Randhawa, M. S., 1986 : *A History of Agriculture in India*, Vol. IV, New Delhi, ICAR, pp. 291-92.

Sharma, S. K., 1986 : Use of Power in Cultivation and its Relation with Agricultural Productivity in Madhya Pradesh, *Hill Geographer*, Vol. 5 (1 & 2), pp. 13-14.

———, 1989 : *Resource Development in Tribal India*, New Delhi : Northern Book Centre, pp. 78-79.

Sharma, S. K. & C. K. Jain, 1980 : Use of Water Resources for Irrigation and Agricultural Development, in Noor Mohammad (Ed), *Perspective in Agricultural Geography*, New Delhi : Concept Pub. Co., pp. 463-83.

Singh, G. B., 1979 : *Transformation of Agriculture : A Case Study of Punjab*, Kurukshetra : Vishal Publication, p. 314.

Singh, Jasbir, 1976 : *An Agricultural Geography of Haryana*, Kurukshetra : Vishal Publication, p. 188, 235.

Sinha, B. N., 1975 : Modernisation of Indian Agriculture : High Yielding Varieties and Green Revolution, *Research Bulletin No. 1*, Eastern Geographical Society, Bhubaneshwar, p. 9.

Tiwari, P. D. & C. K. Jain, 1989 : *Modernisation of Agriculture and Food Availability in India*, New Delhi : Northern Book Centre, pp. 106-107.

10

CHANGING PATTERN OF AGRICULTURAL LANDSCAPE IN DROUGHT PRONE AREAS

N. B. K. REDDY AND Y. V. RAMANAIAH

IN THE recent times modernisation, development of irrigation and improved socio-economic conditions have brought about a revolutionary change in the scenario of Indian agriculture. But the most striking characteristic of the present-day agriculture in the drought prone areas is its high degree of instability. Yet another disturbing feature relating to the Indian agriculture is the adverse consequences due to erratic behaviour of the monsoon and frequent occurrence of drought conditions. The development of an agriculture in these areas where it is the main occupation of the people is imperative to achieve social objective.

Objectives of the Study

The main objectives of the present study are:

(i) to examine the dynamics and trends of crop landuse, irrigated land and land under fallows in Anantapur district,

(ii) to findout the mutual transference of land between net area sown and fallow land,

(iii) to examine the spatio-temporal distribution of HYV cultivation,

(iv) to measure the agricultural landuse efficiency by employing modified standard coefficient method, and

(v) to bring out the spatio-temporal variations in agricultural landuse efficiency.

Locational and Agro-Climatic Setting

Anantapur district is situated in the south-western part of Andhra Pradesh. It is the largest district but the most vulnerable to drought in Andhra Pradesh. The frequent occurrence of drought, long dry spells and crop failures have had a devastating effect on the agricultural economy of the district.

Anantapur district has an undulating terrain with ridges and clusters of rocky hills scattered over the region. The forest cover is thin and scanty. The river of some significance is 'Pennar', but its usefulness for irrigation in the district is not much. It is chronically drought prone and does not get enough rainfall to sustain its dry crops which account for 83 per cent of the gross cropped area. The normal rainfall of the district is 544 mm. The frequent dry spells of three to eight weeks between mid-June and the first week of September, led to the withering of crops, or reduced yields. It has the highest variability of rainfall and the co-efficient of variation varies from 20 to 30 per cent. Out of 26 years, 13 years were drought years in the district (Subramanyam, 1979). On an average, the district has been experiencing a drought once every two years. The Irrigation Commission (1972) identified all taluks of Anantapur district as drought prone taluks.

Methodology

In the present study, the changing pattern of the agricultural landscape is examined between three points of time namely 1962-65, 1972-75 and 1982-85. In order to correct the distortions, a three year average is taken for the purpose of analysis. Taluk is taken as the basic unit of analysis. Here the study is confined to an examination of the dynamics and changes of net area sown, land under fallows, irrigated area and the area under important crops like groundnut, millets, pulses and paddy. Computer analysis is made to findout the trends of change. The linear regression equation $Y = a + bx$ is employed, where y is the index number of the element and X is the time. The goodness of fit of a trend line is measured to know how well the regression line fits the data and the extent to which the regression accounts for the variation in the observed values of the dependent variable. This is quantified by calculating the ratio between regression variance (S^2y) and observed variance (S^2y) and this ratio is known as coefficient of determination (r^2). Correlation analysis is employed to findout the mutual relationship between net area sown and fallow land.

An endeavour is also made to examine the spatio-temporal variations in HYV cultivation and differential adoption of HYVs between 1972-75 and 1982-85. An attempt is also made to measure the agricultural landuse efficiency by employing modified standard coefficient method and analysing its spatio-temporal variations in the district.

Changing Pattern of Net Area Sown

The net area sown includes the land which is actually cropped during the current agricultural year. Obviously its distribution is profoundly influenced by physical factors especially isohytal, edaphic and terrain conditions. However, the socio-economic factors also exert significant influence on it. The net area sown accounted for 46.3 per cent of the geographical area of the district (Table 10.1).

TABLE 10.1: Changing distribution of net area sown in Anantapur district

Pattern	Per cent to total Geographical area			Variation between
	1962-65	1972-75	1982-85	1962-65 & 1982-85
Average	51.9	49.2	46.3	-5.6
Lowest	34.2	33.2	33.1	-1.1
Highest	85.0	79.5	71.6	-13.4

The variation in the percentage of net area sown in Anantapur district showed a significant decrease of 5.6 per cent between 1962-65 and 1982-85. In hectarage, the area under net area sown has shrunk from 1005 thousand hectares in 1962-65 to 877 thousand hectares in 1982-85 showing a net decrease of 128 thousand hectares. The net area sown has decreased in all the taluks except Madakasira in the 20-year period. In 1962-65, the net area sown accounted for 51.9 per cent of the geographical area of the district. It was due to a fairly good amount of rainfall i.e. 900 mm which helped to bring much cultivable land under the plough. In 1972-75 the proportion of net area sown has decreased to 49.2 per cent due to the low rainfall of 600 mm. In 1982-85, the net area sown has further decreased to 46.3 per cent. The chronic drought conditions for five to six years have badly affected the crop farming resulting in steep decrease in the net area sown. In view of the predominance of rainfed farming (83%), the fluctuations in the yearly and seasonal rainfall affect the spatio-temporal distribution of the net area sown in the district.

Trend

From 1964-65 onwards, the trend of net area sown in the district has been marginally negative. The goodness of fit of the regression trend i.e. coefficient of determination states that the present negative trend is to the tune of 0.5 per cent only. It means that the negative trend of the net area sown is now exactly perfect and true. It is highly fluctuating from year to year depending on the variation of rainfall. The trend analysis has revealed that six taluks have shown negative trend while the remaining five taluks positive trend. The negative trend in the net area sown in Anantapur, Tadpatri and Gooty taluks is determined by coefficient of determination value of 50 per cent. In contrast, Penukonda, Kadiri and Madakasira taluks witnessed a positive trend in the net area sown by confirming to 20 per cent of coefficient of determination.

Normally more than 40 per cent of the area under plough has been considered as the basic requirement (Singh, 1975) and Anantapur district fulfils this condition. But the proportion of land under plough has shown wide fluctuations from year to year on account of rainfed farming in a drought-prone region.

Changing Pattern of Fallow Land

Land under fallows indicates the availability of land for areal expansion of crop farming. Extension of crop landuse through scientific utilisation of fallow land is important for the all round development of agriculture although modernisation and intensification of agriculture are more important. In the drought prone Anantapur district where competition for arable land is low due to low density of population and adverse climatic conditions, the land under fallows is a common phenomenon. Land under fallows accounted for 18.5 per cent (1982-85) of the geographical area of the district which is indeed a high proportion.

In the 20 year period, the proportion of fallow land increased by 5.5 per cent amounting to about 106 thousand hectares. The increase is found at all levels and in all taluks of the district barring a small decrease in Dharmavaram and Penukonda taluks. The increase in the proportion of fallow land in Anantapur District is double than the increase in current fallows (Table 10.2).

TABLE 10.2: **Changing distribution of fallow land in Anantapur district**

Pattern	Per cent to total Geographical area			Variation between 1962-65 & 1982-85
	1962-65	1972-75	1982-85	
Average	13.0	10.7	18.5	5.5
Lowest	3.3	2.1	9.9	6.6
Highest	10.5	19.6	27.8	7.3

Trend

An increasing trend in the area under fallows is noticed in Anantapur district. This positive trend is accounted by 15 per cent of coefficient of determination. A trend of increase in the area under fallows is observed in eight taluks out of which, Gooty, Tadpatri, Kadiri and Hindupur taluks have shown a trend of high increase. In contrast, a decreasing trend is noticed in three taluks. Dharmavaram taluks has shown a trend of high decrease in fallow land.

The high temporal variations in the area under fallows are due to the vagaries of monsoons and lack of practice of dry land farming technology. There is an ample scope for better utilisation of fallow lands for developing farm forestry and animal husbandry.

Mutual Transference of Net Area Sown and Fallow Land

Owing to the changing agro-climatic conditions, socio-economic and technological processes, the arable land is bound to play a dynamic role. Let us examine the direction in which mutual transference of arable landuse has taken place. Usually the net area sown increases at the cost of fallow land and vice versa. Here an endeavour is made to make correlation analysis to understand whether there exists a mutual relationship between these two landuse types.

The correlation analysis has revealed a high negative correlation between net area sown and fallow and the coefficient of correlation is — 0.85 in the district. All the taluks in the district have shown negative correlation. It shows that these two landuse types are inversely correlated and functionally related.

Intensity of Cropping

Intensity of cropping refers to the multiple use of agricultural land, that is, double or tripple cropping systems in the same field and in the same year. The percentage of gross sown area to net sown area gives a measure of index to intensity of cropping (Table 10.3). The degree of intensity of cropping and its spatio-temporal variation is related to the intensity of irrigation, rainfall distribution and soil fertility.

TABLE 10.3: Changing distribution of intensity of cropping in Anantapur district

Pattern	Gross area sown as percentage of net area sown			Variation between 1962-65 & 1982-85
	1962-65	1972-75	1982-85	
Average	103.3	103.5	103.0	-0.3
Lowest	100.3	100.3	100.2	-0.1
Highest	109.3	10.6	110.1	0.8

The intensity of cropping in Anantapur district was insignificant and accounted for 103.0 per cent in 1982-85. Low intensity of cropping all over the district is due to lack of perennial and protective irrigation facilities, poor edaphic and unfavourable terrain conditions. The assumption that increase in the intensity of cropping is mainly achieved by increasing irrigation is valid as is evident from the fact that the positive correlation coefficient between the two variables is as much as 0.99.

The intensity of cropping has marginally decreased in the district from 103.3 per cent in 1962-65 to 103.0 per cent in 1982-85. In terms of the actual area, the double cropped area has decreased from 33,582 hectares in 1962-65 to 26,748 hectares in 1982-85, showing a net decrease of 6,834 hectares. The severe and continuous drought conditions during 1982-85 badly affected the ground water resources and lowering of the water level in wells, tanks and reservoirs. The positive trend is accounted by 20.6 per cent of coefficient of determination. It may be due to ceaseless efforts of farmers and the Government in tapping the groundwater and the development of canal irrigation.

Irrigated area

A study of the conjuctive use of surface and ground water for irrigation provides an overall view of the status of irrigation in the district. Added to low rainfall, there are poor irrigation facilities in the district. The gross irrigated land in the district from all the sources forms about 17.6 per cent (1982-85) of the gross cropped area (Table 10.4). Out of the gross irrigated area of 159 thousand hectares, canals accounted 28 per cent, tanks 14 per cent, wells 52 per cent and other sources 5 per cent.

TABLE 10.4: Changing distribution of irrigated area in Anantapur district

Pattern	Gross irrigated area as percentage of gross sown area			Variation between 1962-65 & 1982-85
	1962-65	1972-75	1982-85	
Average	8.6	20.7	17.6	9.0
Lowest	0.4	3.9	9.2	8.8
Highest	18.3	36.5	30.5	12.2

The percentage of gross irrigated area has increased from 8.6 per cent in 1962-65 to 17.6 per cent in 1982-85. In hectarage, gross irrigated area has increased from 89 thousand hectares in 1962-65 to 159 thousand hectares in 1982-85, showing a net increase of 70 thousand hectares.

There have been fluctuations in the irrigated area from one period of time to another period of time which is evident from the fact that the gross irrigated area which was 8.6 per cent in 1962-65 increased to 20.7 per cent in 1972-75 and decreased to 17.6 per cent in 1982-85. The regression trend analysis too has shown a positive trend of irrigated area but with low value of 12.4 per cent of coefficient of determination. This confirms the fluctuating trends of irrigated area. The intensity of irrigation i.e. the gross irrigated area as percentage of net irrigated area has increased in absolute terms but not in relative terms. The area under irrigated land more than once has increased from 24,835 hectares in 1962-65 to 25,293 hectares in 1982-85, showing a net increase of 458 hectares.

Changing Pattern of Crop Farming

An attempt is made here to review the changes that occurred in the cultivation on four important crops since they account for 88.4 per cent of the gross cropped area of the district. They are groundnut, millets, pulses and paddy.

Groundnut Cultivation

An interesting feature of the present cropping pattern of the district is the predominant cultivation of groundnut which accounted for 48.0 per cent of the gross cropped area of the district (Table 10.5).

TABLE 10.5: Changing distribution of groundnut cultivation in Anantapur district

Pattern	Per cent to the gross cropped area			Variation between 1962-65 & 1982-85
	1962-65	1972-75	1982-85	
Average	20.5	26.2	48.0	27.5
Lowest	10.0	11.9	23.3	11.4
Highest	38.4	55.8	78.5	40.1

It is significant to note that there has been a spectacular spatial spread in groundnut cultivation between 1962-65 and 1982-85. In 1962-65, groundnut cultivation accounted for 20.5 per cent of the gross cropped area of the district which has increased to 26.2 per cent in 1972-75. Groundnut cultivation has received much preferential treatment, as a result of which it has further increased to 48.0 per cent in 1982-85. In terms of hectarage, more than a two-fold increase was observed i.e. from 212 thousand hectares in 1962-65 to 434 thousand hectares. The spatial spread of groundnut cultivation is observed in all the taluks of the district.

Trend

The regression trend analysis has also revealed that there has been a high positive trend in groundnut cultivation in the 20-year period. The regression trend is significantly accounted by 45.0 per cent of coefficient of determination.

The study has revealed that groundnut cultivation in highly favoured by the farmers of Anantapur district because (i) it is a cash crop, (ii) it fetches good income, (iii) it is a short growing crop, (iv) the cost of its cultivation is moderate, (v) it has a good market, (vi) it requires less water, (vii) its cultivation is easy, (viii) it is less disease-prone, (ix) the red sandy soils are more suitable, and (x) it can be cultivated under rainfed conditions. Groundnut cultivation has tended to replace millet crops and some of the minor millets are getting eliminated from the cropping pattern.

Millet Cultivation

Millet cultivation is by far the most important and extensively grown foodgrain crop in Anantapur district accounting for 26.3 per cent (1982-85) of the gross cropped area. Millets are predominantly grown as rainfed crops in the poorer and light soils. Millets provide an ideal, economic and productive crop in the chronically drought prone areas (Table 10.6).

TABLE 10.6: Changing distribution of millet cultivation in Anantapur district

Pattern	Per cent to the total gross cropped area			Variation between 1962-65 & 1982-85
	1962-65	1972-75	1982-85	
Average	47.1	40.7	26.3	-20.8
Lowest	37.4	24.3	11.7	-25.7
Highest	56.9	55.0	40.0	-16.9

In 1962-65 the proportion of millet cultivation was high covering 47.1 per cent of the gross cropped area of the district. The concentration of millet cultivation gradually declined between 1962-65 and 1972-75. A further decline has occurred during 1972-75 and 1982-85. The proportion of millet cultivation accounted for 40.7 per cent in 1972-75 and 26.3 per cent in 1982-85. In terms of hectarage, the area under millet cultivation has recorded a rapid decline from 489 thousand hectares in 1962-65 to 238 thousand hectares in 1982-85. It shows a net decline of 251 thousand hectares or 20.8 per cent. The trend analysis has also revealed that there has been a high decreasing trend in the millet cultivated area. This trend of decrease is confirmed by a high value of 84.0 per cent of coefficient of determination. The declining trend of millet cultivation may be due to its declining preference as a foodgrain and changing dietary habits of the people. It is observed that the area under millet cultivation has shrunk proportionately with the spatial spread of groundnut cultivation.

Pulses and Gram Cultivation

Pulses and gram are often largely cultivated as mixed crops along with groundnut and millet crops in the district (Table 10.7). Cultivation of pulses and gram does not require much care and investment and they are grown in poorer soils. Pulses accounted for 8.0 per cent (1982-85) of the gross cropped area of the district.

TABLE 10.7: **Changing distribution of pulses and gram cultivation in Anantapur district**

Pattern	Per cent to gross cropped area			Variation between 1962-65 & 1982-85
	1962-65	1972-75	1982-85	
Average	10.6	11.8	8.0	-2.6
Lowest	2.9	4.4	4.4	1.5
Highest	19.4	21.6	16.2	-3.2

In 1962-65, the area under pulses and gram cultivation accounted for 10.6 per cent of the gross cropped area of the district. The crop has slightly gained in its areal extent 1972-75 accounting for 11.8 per cent. But in 1982-85, the concentration of pulses declined to 8.0 per cent. The area under pulses cultivation decreased from 110 thousand hectares in 1962-65 to 73 thousand hectares in 1982-85, showing a net decrease of 37 thousand hectares. The decrease in pulses cultivation is noticed in all taluks of the district. The regression trend too has shown that there has been a negative trend in the area under pulses cultivation. This negative trend is accounted by 34.0 per cent of coefficient of determination. But in many taluks, the trend of pulses cultivation is not constantly negative but has shown fluctuations from year to year.

Paddy Cultivation

The spatio-temporal distribution of paddy cultivation indicates its close relation to soil and water. In order of importance paddy crop takes the fourth position. With a total cultivated area of 54,771 hectares, paddy cultivation accounted for 6.1 per cent of the gross cropped area of the district in 1982-85 (Table 10.8).

TABLE 10.8: Changing distribution of paddy cultivation in Anantapur district

Pattern	Per cent to gross cropped area			Variation between 1962-65 & 1982-85
	1962-65	1972-75	1982-85	
Average	8.0	9.8	6.1	-1.9
Lowest	0.5	2.1	2.0	-1.5
Highest	16.4	18.4	12.9	-3.5

Not even a single taluk in the drought prone Anantapur district grows paddy in any significant degree as is evident from the fact that the highest percentage of paddy cultivation has been 12.9 per cent in Anantapur taluk. The biggest constraint for paddy cultivation is non-availability of adequate amount of water and rainfall.

Only 8.0 per cent of the gross cropped area of the district was under paddy cultivation in 1962-65. It increased to 9.8 per cent in 1972-75, but the proportion of paddy cultivation had declined to 6.1 per cent in 1982-85. Paddy area has decreased from 83 thousand hectares in 1962-65 to 55 thousand hectares in 1982-85 thus registering a net decrease of 28 thousand hectares. The regression trend analysis has revealed that there has been a low negative trend in the area under paddy crop as shown by the low value of coefficient of determination. A high decrease in paddy crop area is found in southern taluks of the district namely Kadiri, Hindupur, Madakasira, Penukonda and Dharmavaram where sericulture is the most competitive crop for paddy cultivation. Many of the paddy fields under well irrigation have been replaced by mulberry crop and fruit farming.

Spatio Temporal Distribution of HYV Cultivation

The innovations and diffusion of the high yielding varieties and new technology have transformed the agricultural scenario. The cultivation of HYV crops requires heavy investment and protective irrigation (Table 10.9). The farmers cannot afford to take the risk of heavy investment due to uncertainity of rainfall and low irrigation facilities. In 1982-85, the area under HYV cultivation accounted for 3.9 per cent of the gross cropped area of the district.

TABLE 10.9: **Changing distribution of HYV cultivation in Anantapur district**

Pattern	HYVs as percentage of gross cropped areas		Variation between 1972-75 & 1982-85
	1972-75	1982-85	
Average	2.6	3.9	1.3
Lowest	0.6	1.8	1.2
Highest	5.2	12.3	7.1

Not even a single taluk in the district has contributed significantly to HYV cultivation. The area under HYVs increased from 24,248 hectares in 1972-75 to 34,875 hectares in 1982-85 showing a net gain of 10,627 hectares in the 10-year period or 1.3 per cent increase in their concentration. The increase in the concentration of HYV is observed in all taluks of the district except Hindupur. In Hindupur taluk, the HYV paddy could not stand competition from mulberry cultivation.

Differential Adoption of HYVs

The HYV paddy, groundnut, jowar and bajra together accounted for 83.9 per cent of the total area under HYVs in the district. The HYV paddy has shown a significant shrinkage in its share in the total area under HYVs i.e. from 60.9 per cent in 1972-75 to 31.5 per cent in 1982-85. In contrast, a high increase is noticed in HYV groundnut cultivation where it increased from 1.0 per cent to 27.9 per cent. The percentage of HYVs to their respective total crop areas showed a small increase between 1972-75 and 1982-85 (Table 10.10).

TABLE 10.10: **Changing distribution of the differential adoption of HYV crops in Anantapur district**

Name of the crops	Different HYV crops as percentage of the total area under HYVs		Different HYVs as percentage of their respective total crop areas	
	1972-75	1982-85	1972-75	1982-85
Paddy	60.9	31.5	15.9	20.0
Jowar	12.4	14.4	1.7	5.7
Bajra	16.2	10.1	7.2	9.2
Groundnut	1.0	27.9	0.1	2.2
Wheat	3.3	3.1	30.4	73.3
Others	6.2	13.0		

The success of the Green Revolution and the various developmental programmes in agriculture have not yet achieved any significant results in the district. The district has not much benefitted by HYVs because of limited irrigation facilities, low rainfall conditions and lack of awareness among the farmers on dry land farming technology.

Changing Pattern of Agricultural Landuse Efficiency

An attempt is made here to measure the agricultural landuse efficiency in Anantapur district by employing the modified standard coefficient method (Reddy and Ramanaiah, 1985). Here, the instead of assigning ranks, the actual percentages of the variables of a particular component areal unit are added and averaged. The mean percentage value of the variables of a component areal unit thus obtained is directly proportional to the degree of agricultural landuse efficiency in which 100.0 per cent is the highest efficiency value.

In the present study, six positive variables namely, (i) net area sown, (ii) area sown more than once, (iii) irrigated area, (iv) net sown area as percentage of arable land, (v) area under HYVs, and (vi) commercial cropping system are chosen as the parameters since they are considered as the essential determinants or indicators of agricultural landuse efficiency (Table 10.11).

TABLE 10.11: **Changing distribution pattern of agricultural landuse efficiency in Anantapur district**

Sl. No.	Name of the taluk	Mean per cent value of Agricultural Landuse Efficiency			Variation between 1962-65 and 1982-85
		1962-65	1972-75	1982-85	
1.	Anantapur	29.1	36.8	35.4	6.3
2.	Tadpatri	31.8	31.3	32.0	0.2
3.	Gooty	33.6	33.6	36.1	2.5
4.	Uravakonda	36.3	33.5	33.6	-2.7
5.	Dharmavaram	27.7	31.5	39.6	11.9
6.	Kalyandrug	28.4	31.5	32.9	4.5
7.	Rayadrug	32.5	34.7	32.8	0.3
8.	Penukonda	28.8	32.7	35.2	6.4
9.	Kadiri	31.2	36.3	32.9	1.7
10.	Hindupur	33.5	36.6	30.0	-3.5
11.	Madakasira	26.2	34.7	34.3	8.1

The agricultural landuse efficiency of Anantapur district, as could be expected, is low i.e. 33.5 per cent in 1982-85. There has not been much spatial variation in the agricultural landuse efficiency in the district.

There has been some marginal improvement in agricultural landuse efficiency in the district i.e from 30.1 per cent in 1962-65 to 33.5 per cent in 1982-85. In fact, all the taluks except Hindupur and Uravakonda have improved their agricultural landuse efficiency during the period. A high mean increase of over 10.0 per cent is found in Dharmavaram taluk. The development made in irrigation, hybridisation and better cropping system could not overcome the severity of drought impact on agriculture in the district. Much could be done to improve the performance of agriculture and to mitigate the sufferings of the farmers. The conjuctive utilisation of ground and surface water, better land and water management, and improvement in socio-economic conditions of the people well lead to better socio-economic development of the district. An application of dry land farming techniques will greatly help to develop agriculture in the district.

Conclusion

An examination of the changing patterns and trends in the agricultural landscape of Anantapur district has revealed that the erratic behaviour of the monsoon and severe drought conditions have had a debilitating effect on the agricultural economy of the district. The moisture deficiency caused by low rainfall with high variability, scanty irrigation facilities, poor red soils, undulating terrain with rocky exposures, low farming technology as well as poor socio-economic conditions of the people are the detrimental factors which come in the way of agricultural development. Undoubtedly, protective irrigation and suitable dry land farming practices are the prime need for stable and prosperous agriculture in the district.

A welcome feature in the cropping system of Anantapur district has been the shift from inferior millet crops to superior variety of crops like groundnut, sericulture, fruit farming, etc. Now what is required is to bring about better yields in the existing crops through dry land farming practices. The development of inter-cropping techniques and mixed farming practices will help to improve the performance of agriculture. Better if the fallow lands are brought under farm forestry and fodder crop cultivation.

REFERENCES

Bagchi, K., 1979 : Agriculture in Drought Prone Areas, *Geographical Review of India*, pp. 293-96.

Ebdon, D., 1983 : *Statistics in Geography,* Basil Blackwell, Oxford, England.

Government of India, 1973 : *Climate of Andhra Pradesh,* Indian Meteorological Department, New Delhi.

———, 1976 : *Report of the National Commission on Agriculture*, Part IV: Climate and Agriculture, Ministry of Agriculture and Irrigation, New Delhi.

Ramanaiah, Y. V., 1984 : *Pattern of Landuse and Cropping in Andhra Pradesh,* An unpublished, Ph.D. Thesis, Department of Geography, S.V. University, Tirupati.

Reddy, N. B. K. (ed), 1979 : *Proceedings of All India Symposium on Drought Prone Area of India,* Rayalaseema Geographical Society, S.V. University, Tirupati.

Reddy, N. B. K. and Reddy, D. N., 1981 : Agriculture in Arid and Semi-arid Regions of India, *Transactions of the Institute of Indian Geographers,* Vol. 3, No.2, pp. 115-35.

Reddy, N. B. K. and Ramanaiah, Y. V., 1985 : Changes in Agricultural Landuse Efficiency in Andhra Pradesh, *Landuse Policy,* Vol. 2, No.3, pp. 210-16.

Singh, V. R., 1975 : Landuse Orientation in Central India, *Geographical Review of India,* Vol. 37, No.4, pp. 356-63.

Subramanyam, V. P. and Hema. Malini, 1979 : "Drought prone areas of Andhra Pradesh", N. B. K. Reddy (ed.) *Proceedings of Drought Prone Areas of India,* Rayalaseema Geographical Society, S.V. University, Tirupati.

11

CROPPING INTENSITY IN UTTAR PRADESH

O. COUTINHO AND T. C. SHARMA

IN THE quest for a strategy for maximising agricultural production, multiple cropping i.e., raising more than one crop in a sequence in an agricultural year on a unit of cropland (National Commission on Agriculture, 1976) holds a vast potential in most rural areas of India in view of the prevailing abundance of farm labour, a favourable climate and only a small proportion (25%) of the net sown area being sown more than once compared to 60 per cent in Bangladesh and 90 per cent in China. The disparity is also wide within the country. The future strategy should emphasise on a faster growth in the incidence of multiple cropping in areas where it remains low for which purpose regional diagnostic exercises are essential.

The present inquiry aims at analysing the growth patterns in multiple cropping and the role of some important factors responsible for the disparity in it in Uttar Pradesh for the period from 1966-67 to 1980-81 with the presumption that the rise in yields of most crops has served as an added incentive and there has been a considerable rise in it since the launching of the High-Yielding Varieties Programme in the earlier year and that the trends and stabilised by the latter. As the trends since then have been similar, the results of the study remain relevant. The exercise covers the 48 districts and five agro-economic regions of Uttar Pradesh which formed the study area in 1966-67. The area was called Uttar Pradesh (Plains) during the first triennium of the study period. It is divided into five agro-economic regions: Western, Central, Eastern, Bundelkhand and Hills. Only Nainital and Dehrąydun districts from the Hills region are included as data for the other districts is not available. The extreme disparity in the agricultural development within the State and its large cropped area were the two weighty reasons behind the choice of Uttar Pradesh for the study.

Methodology

Multiple cropping is conventionally expressed as cropping intensity in per cent:

$$\frac{\text{Total Cropped Area (TCA).100}}{\text{Net Sown Area (NSA)}}$$

Indices of cropping intensity have been computed for the first and last trienniums of the study period for all the districts, regions, Uttar Pradesh and all-India.

A simple correlation matrix with 8 explanatory (independent) variables as x1, x2, x3, ... x8 and cropping intensity as a dependent variable has been attempted to assess their role in the spatial disparity in the index values of the latter for the 48 districts during the last triennium. The independent variables included are average size of the land holdings, rainfall (actual), gross irrigated area, irrigation intensity (gross irrigated area as percentage of net irrigated area), tubewell-irrigated area, fertiliser consumption, sugarcane cropped area and density of rural population. Irrigation is a *sine qua non* for growing crops in the Indian conditions in the *rabi* and *zaid* (summer) seasons. All the three irrigation factors i.e., gross irrigated area, irrigation intensity and tubewell irrigated area play an important role in multiple cropping often in a complementary manner (Narain and Roy, 1980). The size of land holdings decides the financial status of the farmers upon which, in turn, depends their capacity to invest in irrigation or other infrastructural development and inputs. Sugarcane area as proportion of total cropped area is important because of its direct contribution to multiple cropping the crop being taken as two and as it acts as a surrogate for all types of modernisation and development (National Commission on Agriculture, 1976). Rainfall has a complementary role along with irrigation in multiple cropping in many areas (Sawant, 1975). Multiple cropping absorbs labour in abundance which is, in fact, often the main push behind it in densely populated rural areas with little or no alternate employment opportunities (Boserup, 1965; Hayami and Ruttan, 1971).

Results and Analysis

The cropping intensity index rose in Uttar Pradesh during the study period from 136.37 per cent to 148.12 per cent compared to the increase from 117.63 per cent to 124.64 per cent in the case of India as a whole. Its growth in Punjab and Haryana for the period was 132 per cent to 160 per cent and 140.5 per cent to 151 per cent respectively indicating clearly that though Uttar Pradesh has done better than all-India this State is way behind the two progressive states (Punjab and Haryana) both in temporal growth and the present position. Also the growth was quantitatively/financially less productive in Uttar Pradesh in view of the relatively lower yields in all crops and poorer cropping patterns with a large area under coarse cereals. A comparison of the index values for the two periods suggests a rapid rise in the variation among the regions and districts.

The total cropped area in Uttar Pradesh increased during the study period by 8 per cent (17,39,886 hectares) of which nearly 74 per cent was contributed by expansion in rabi crops (mostly wheat) and sugarcane; 20 per cent by kharif crops; and 15 per cent by *zaid* crops. As a result, kharif and rabi crops now account for 50 per cent and 48 per cent respectively of the total cropped area compared to 53 per cent and 46 per cent respectively in the first triennium. The *zaid* crops occupy only a small area. Within the State the cropping intensity index value showed a spatial variation during the last triennium from 104.62 per cent in Hamirpur district to 204.76 per cent in Muzaffarnagar and the simple correlation matrix (Table 11.1) and the stepwise regression model (Table 11.2) revealed the highest degree of positive correlation (r=0.79) between cropping intensity index and sugarcane area followed by gross irrigated area (r=0.71), irrigation intensity (r=0.69), tubewell irrigated area (r=0.66) and fertiliser consumption (r=0.63). Under the new technology, a high degree of complimentarity exists among these factors and it reflects clearly in the case of Uttar Pradesh in the area under sugarcane. A collinearity, therefore, exists among all these four factors specially in the Plain districts. Rural population density showed a low degree of positive correlation (r=0.28) because of little variation in rural population density in the Plain districts which occupy over 90 per cent of the total cropped area of the State. The other two factors, land holding size and actual rainfall, showed a weak negative correlation because of an inverse relationship between these and multiple cropping. Bundelkhand has larger holdings but low cropping intensity. An exercise for the study region excluding these districts shows a positive correlation (*r*=0.46). The study indicates that rainfall has no impact on cropping intensity in the region.

TABLE 11.1: **Simple correlation matrix for all variables**

(*n* = 48)

	Y	X_1	X_2	X_3	X_4	X_5	X_6	X_7	
Y	1.00								
X_1@	–0.12	1.00							
X_2	–0.31	–0.28	1.00						
X_3	0.71**	–0.04	–0.52**	1.00					
X_4	0.69**	0.16	–0.29	0.68**	1.00				
X_5	0.66**	–0.16	–0.53**	0.74**	0.29	1.00			
X_6	0.63**	–0.28	–0.24	0.61**	0.47**	0.58**	1.00		
X_7	0.79**	–0.16	–0.35**	0.51**	0.54**	0.49**	0.48**	1.00	
X_8	0.28	–0.65*	–0.09	0.38*	–0.10	0.56**	0.50**	0.14	1.00

Note: **1-tailed significance; *.01 **.001**

@ r=0.46 for 43 districts excluding Jhansi, Jalaun, Hamirpur, Banda and Mirzapur

Y = **Cropping Intensity**
X_1 = **Land Holding Size**
X_2 = **Rainfall (Actual)**
X_3 = **Gross irrigated area**
X_4 = **Irrigation intensity**
X_5 = **Tubewell irrigated area**
X_6 = **Fertiliser consumption**
X_7 = **Sugarcane area**
X_8 = **Rural population density**

TABLE 11.2: **Stepwise multiple regression model of significant variables regressed against cropping intensity**

Variables	Multiple R	R_2	B	Standard Error	*t*-value	*F*-value
Sugarcane area	.79	.63	1.40	.15	8.90	79.24
Gross irrigated area	.87	.76	.54	.15	4.93	71.85
Land Holding Size	.89	.80	–9.57	3.23	–3.01	59.55
Irrigation intensity	.91	.83	.42	.14	2.85	53.94
Tubewell irrigated area	.92	.85	.35	.09	3.90	51.46

Of the five agro-economic regions, the cropping intensity index for the last triennium was the highest (166.02%) after an increase of 22.54 per cent over the first triennium in the Western region. The index value rose to 204.76 per cent, the highest, in Muzaffarnagar district closely followed by Meerut district with 202.71 per cent, 78.11 per cent and 88.0 per cent respectively of the total cropped area is irrigated in these two districts. They also have a high irrigation intensity and a large area under sugarcane cultivation. Besides these two districts, the regional average is exceeded in 7 other districts. These include Saharanpur, Bulandshahr and Aligarh where over 60 per cent of the total cropped area is irrigated, and Bijnor, Moradabad, Pilibhit and Rampur, where a moderate irrigated proportion (51-57% of TCA) is complemented by a higher rainfall. All have over half of the total cropped area under irrigation and a large area under sugarcane, specially Bijnor. The land holding size is above the State average in all and the rural population density is quite high (above 400 persons per km^2) in most districts. The fertiliser consumption per hectare is also above State-average in all. The cropping intensity index is much lower in the other 9 districts of this region. The group includes the western districts of Mathura, Agra, Etah, Mainpuri, Farrukhabad, Budaun and Etawah where though the irrigation extent is often quite high (40-60% of TCA), a low rainfall necessitates irrigation also during kharif season and hence a low irrigation intensity; and Barellly and Shahjahanpur due to low irrigation extent (47-53 per cent of TCA). All these districts have a negligible area under sugarcane and the rural population density is also the lowest in the region. An important observation is that the index value has increased in all the districts, except Budaun, and the growth was above 15 per cent in as many as 10 of these (above 20 per cent in Saharanpur, Muzaffarnagar, Meerut, Bijnor, Budaun, Moradabad and Pilibhit).

The Central region had a cropping intensity index value of 139.15 per cent during the last triennium (compared to 127.94% in 1966-68) which is lower than the State-average and much below the Western region. Of the nine districts of the region, only two, Barabanki and Kheri, had a value above the State-average (154.77 per cent and 151.4 per cent respectively) and a relatively high rainfall alongwith a moderate gross irrigated area and irrigation intensity (in all of which the former district is better placed) contributed to their better performance. Kheri has a large area under sugarcane and Barabanki a sufficiently high rural population density. A poor irrigation development (22-50% of TCA) as also a smaller size of holding (below one hectare) in a majority of districts seem to be the major lacunae, and more irrigation, especially from private tubewells, is necessary for expanding the multiple cropping incidence in the whole region. Though all districts, except Rae Bareli, had an increase in the index; its extent was only marginal (below 10%) suggesting a stagnant position. Rae Bareli district had a low decline.

In the Eastern region, the multiple cropping incidence (147 per cent and 136 per cent respectively for the two trienniums) though slightly below the State-average is sufficiently high, and it appears to be mainly due to a high rural population density, it being above the State-average (314 persons per km^2) in all the districts (except Mirzapur and Bahraich), exceeding 500 persons per km^2 in as many as 7 and Deoria recording the State's highest (609 persons per km^2), where it appears to be the most dominant factor. The cropping intensity index in all these 7 districts is above 150 per cent. Irrigation development is inadequate in all the districts. Bahraich having only 13.16 per cent and Gonda 26.14 per cent of their respective cropped areas under irrigation and it being below the State-average in as many as six districts. Besides Bahraich, the situation is particularly bad in Mirzapur, Gonda and Sultanpur which have the poorest development. Except Mirzapur and Allahabad, where the physical conditions are unfavourable, all districts have a much smaller land holding size than in the Western region or even the State average, it being 0.7 hectare or less in as many as 10 districts. Tubewell irrigation, which is quite important in the region, does not seem to provide any significant advantage due to an overall irrigation paucity, only 40 per cent of the gross sown area being under it and the irrigation intensity also being low — in 12 districts below 120 per cent. Also most of the tubewells are public tubewells with a low efficiency. The fertiliser consumption is below 32 kg per hectare in Mirzapur, Basti and Bahraich. In view of the poor resource base of the farmers, only large-scale subsidies could perhaps make any significant improvement in this region of tiny holdings and poor yields. In only four districts (Jaunpur, Ghazipur, Gorakhpur and Azamgarh) the growth was above 16 per cent and one district (Deoria) had, in fact, a decline.

The two Hill districts of Nainital (Plains) and Dehradun have a cropping intensity index (182.67 per cent and 169.08 per cent respectively for the two trienniums) on par with the Western districts due to a high land proportion under irrigation coverage (61 per cent of TCA) in the former and a high rainfall in the latter compensating substantially its poor irrigation coverage (39 per cent of TCA). The small land holding size in Dehradun (0.6 hectare compared to 1.9 hectare in Nainital) accounts for the lower intensity compared to Nainital. During the study period the index value rose by 13.59 and 17.10 per cent in the two districts respectively.

The poorest development in multiple cropping (index 112.15 per cent in the last triennium and 111.62 per cent in the first) obtains currently in the 4 districts of Bundelkhand. Hamirpur has the State's lowest index of 104.68 per cent; and it is 106.4 per cent in Jalaun, 116.8 per cent in Jhansi, and 119.6 per cent in Banda. The situation remains stagnant in Jhansi, Hamirpur and Banda with an increase of 0.53-5.7 per cent whereas Jalaun had a decline. Only 21.36 per cent of the total cropped area is irrigated in the region and only 4.19 per cent of it by tubewells. Fertiliser consumption is below 10 kg per hectare in Jhansi, Hamirpur and Banda. A low rural population density, poor soils and large holdings with low yields and hence large-scale fallowing also, perhaps, explain the poor response in multiple cropping in the region.

Conclusions

Though the incidence of multiple cropping in Uttar Pradesh is lower than in Punjab, it is much higher than the all-India average mainly due to a high rural population density, subsistence nature of farming and absence of alternate employment opportunities. However, within the State, the demographic pressure seems to explain the disparity to some extent only between Bundelkhand and the other regions. Within the Plains, excluding Bundelkhand, there is little variation in rural population density and hence the low positive correlation between it and cropping intensity.

Irrigation emerges as the most important factor affecting the spatial distribution of cropping intensity in Uttar Pradesh. Gross irrigated area, tubewell irrigated area and irrigation intensity, all the three irrigation factors, show a high positive correlation with the cropping intensity index. Sugarcane cropped area also has a positive correlation with cropping intensity because of its direct contribution and as it acts as a surrogate for irrigation development. Sugarcane cultivation has acted as a catalyst factor in the success of the high-yielding varieties in crops like wheat and rice. Areas with sugarcane concentration provide the ideal conditions and enjoy a much higher cropping intensity and yields than the other regions. Land holding size explains part of the disparity only within the Plain regions (excluding Bundelkhand districts and Mirzapur).

Below State-average cropping intensity occurs in 19 districts of which 8 are in the Eastern region, 7 in the Central region, 4 in the Western region, and all 4 in Bundelkhand. The Western region districts and Nainital Plains not only have an above State-average intensity but also the gap between these and the remaining districts has widened over the years due to a much faster growth in this region than in the rest of the State. The Bundelkhand region and Mirzapur district with the lowest level in cropping intensity have inadequate irrigation development due mainly to a rugged relief. In most other below-average districts, it is the small size of holdings generating little income and hence the farmers inability to sink tubewells which have greatly boosted irrigation intensity in many other areas. In either case the problem requires State intervention. Bundelkhand has perhaps the lowest public investment in irrigation and other infrastructure development. Time has come to focus attention on this neglected region in the State. Loans and subsidies on a liberal scale alone can solve the problem of the marginal and small farmers specially in the Eastern and Central regions where their number is overwhelmingly large. Measures are also immediately called for the reclamation of ravines, badlands, waterlogged and alkali-affected soils and to prevent excess of irrigation which is the root cause of degradation of the last mentioned two types.

REFERENCES

Boserup, Easter, 1965 : *The Conditions of Agricultural Growth*, Aldene, Chicago.

Hayami, Y. and Ruttan, Vernon W., 1971 : *Agricultural Development : An International Perspective*, The John Hopkins University Press, Baltimore.

National Commission of Agriculture, 1976 : *Report of the National Commission on Agriculture, Part VI : Crop Production. Sericulture and Apiculture*, Ministry of Agriculture and Irrigation, New Delhi, p. 33.

Narain, Dharm, and Roy, Shyamal, 1980 : *Impact of Irrigation and Labour Availability in Multiple Cropping : A Case Study of India,* Research Report 20, International Food Policy Research Institute, Washington, D C.

Sawant, S. D., 1975 : "Extent of Multiple Cropping in Irrigated and Unirrigated Areas of India : Some Implications for Usefulness of Irrigation Statistics, *Indian Journal of Agricultural Economics*, Vol. 30, No 2, pp. 47-53.

12

AGRICULTURAL LANDUSE AND PRESSURE OF POPULATION

K. K. L. DAS AND K. N. DAS

THE REGION designated as North Bihar Plain, lying wholly to the west of the river Kosi is hemmed in between the parallels of 25°15′15" and 27°31′0" N and meridians of 83°50′0" and 87°15′0" E. Landlocked from all sides this is a compact area, covering 35,822.6 km^2 and had, according to the 1971 census, a total population of 2,00,12,288. The region is separated from Nepal in the north by the Bihar-Nepal international border, while it is broadly delimited by the river Ganga in the south. In the west, the regional boundary is coterminus with the eastern extension of Uttar Pradesh while the river Kosi sets its eastern limit. It comprises, wholly or in part, some 14 districts: West Champaran, East Champaran, Gopalganj, Siwan, Saran, Vaishali, Muzaffarpur, Sitamarhi, Madhubani, Darbhanga, Samastipur, Begusarai together with Khagaria (Monghyr) and Naugachhia (Bhagalpur) subdivisions (Fig. 12.1). In all, there are 180 anchal-cum-development blocks for which population and landuse data have been compiled and computed for analysis. In this paper, we endeavour to discuss the physical setting that has conditioned agricultural landuse and peopling since several hundreds of years. The area is representative of one of the early settled portions of the Middle Ganga Plain in which the pressure of population on land resources is high.

Physical Setting

With the exception of the Himalayan foothills corresponding to the Siwalik strata exposed as north-west bulge of the Someshwar and Dun ranges in West Champaran, almost the entire North Bihar Plain is an area of great physiographic homogeneity throughout its length and breadth. The relief varies from 30 metres in south-east to over 150 metres in north-west. The general lie of the country is from north to south as well as from west to east. This is revealed by contours and stream courses which make the topographic trend from north-west to south-east. The flow of rivers bears testimony to an arcuate bend. With the exception in the extreme north-west where it is a bit rugged, the

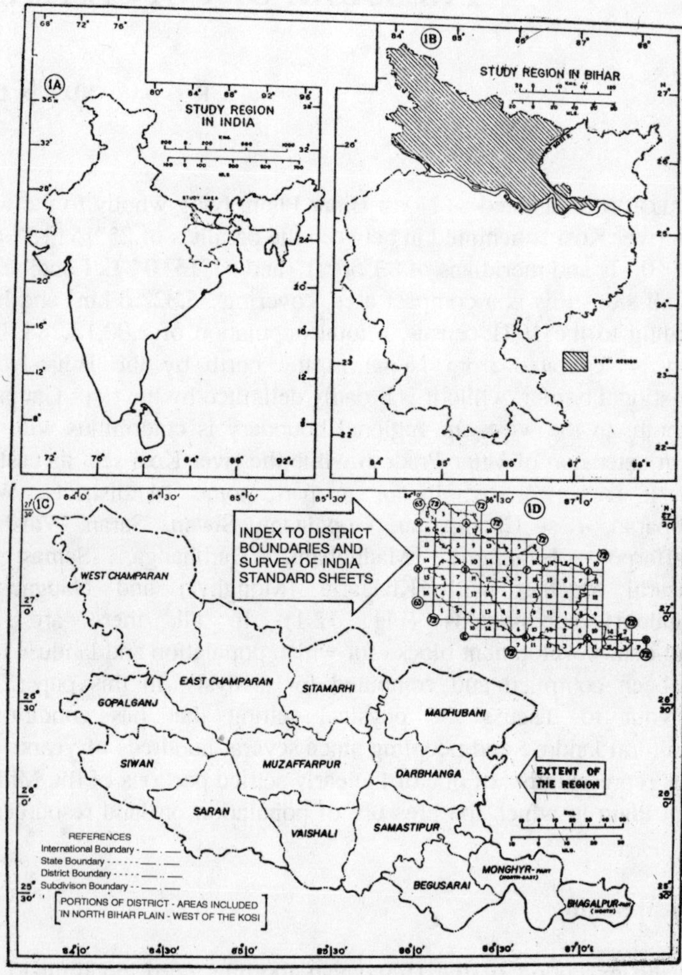

FIG. 12.1: North Bihar Plain—west of the Kosi location.

entire area is a remarkably featureless plain composed of river alluvia. This has resulted from the aggradational action of many big and small perennial Himalayan affluents debouching onto the Bihar Plain. Naturally this has remained the playground of rivers, e.g. Ghaghara, Gandak, Burhi Gandak, Baghmati (Dewakar, 1959). In addition, there are the Adhwara group of minor streams, e.g. Kamala-Balan, Tiljuga, Bhutahi Balan, Kosi and a host of minor streams which make their way to the Ganga in meandering, braided and frequently changing courses.

As a result of a sudden break in the gradient of streams downward from the *tarai* hills, almost an imperceptible relief, a huge load of detritus, sand and silt perpetuating the choking of the stream beds having conspicuously shallow channels that have given rise to such topographic prominences as alluvial cones, floodplains, levees, ox-bow or elongated lakes, ***chaurs*** (long semi-circular and marshy lowlands), *tals* (natural depressions filled with water during rains), meander bends and cut off channels besides dead and defunct courses or rivers (Das, 1968). These are clearly the result of alluviation and diluviation as erosion and scouring are common features co-existing side by side, at some stretch of the streams, with aggradational and building up action. The fluvial action, river-bank erosion, congestion and stagnation of beds, anastomosing channels separated by shoals and bars are a function of hydraulic variables to which the rivers of the plain are generally vulnerable. In turn, these give rise to flood hazard and therefore a study in fluvial morphology of the Kosi and other rivers is full of absorbing interest (Das, 1968). Geddes has recognised conspicuous presence of the Gandak cone in the west and the Kosi cone in the east and in between them lie a number of intercones that are manifest in the Burhi Gandak and Baghmati plains (Geddes, 1960).

The morphological characteristics of the plain are exhibited in its alluvial landscape. This has imparted to the plain some salient features that are marked by the formation of fertile soils, viz. clay loam, silt clay loam, sandy loam in varying proportions of fineness of clay, loam and sand. Aided by spill-flooding along the river banks as well as in interstream areas these soils prominently stand out as the valuable land resource base capable of sustaining millions of mouths (Fig. 12.2). Apart from the edaphic conditions, the seasonal rhythm of climate, viz. hot weather (mid-February to mid-June), general rains (mid-June to mid-October) followed by winter season (mid-October to mid-February) with adequate sunshine, abundant warmth and plentiful monsoon showers, has profoundly influenced the agrarian life and economy. The climate has allowed, in these parts, three to four cycles of plant growth in a year. The annual rainfall provides an important base for agriculture, an all too paramount pre-occupation of an over-whelming majority of the people. The amount slightly decreases from 2,053 mm in Bhainsalotan in north-west to 929 mm in Raghopur in south-west. The regional average of total annual rainfall stands at 1,205 mm Memoirs of Indian Meterological Department (M.I.M.D., 1962).

An abundant rainfall coupled with favourable natural conditions were responsible for luxuriant growth of forests in the Bihar Plain in the past which might be evident from thick vegetal cover in Champaran (O'Malley, 1930), derived from *Champa+aranya*,, i.e. forest of Champa trees in the west extending to Purnea, occasioned by *pur+aranya*, i.e., completely forested tract in the Kosi floodplain in proto-historical times (Roy Chaudhary, 1963). Centuries of settlement, development of the art of cultivation, population expansion and human interference in changing their physical landscape meant complete extirpation of natural vegetation from nearly all of the plain surface. Forest worth the name is now limited only to Dun-Sumeswhar hills and the *tarai* belt.

Population Characteristics

As available historical evidence shows, the Aryan civilisation grew along the Ganga plain some 15,000 years ago and, in harmony with favourable soils and salubrious climate, cultivation of crops had its origin roughly 10,000 years back (Mukherjee, 1938). There is ample evidence that peopling firstly took place along the river Ganga due to ground moisture conditions and potable water facilities available but later it was pushed to the north as forest clearings made room for expansion of settlement sites together with raising of food crops. Under favourable natural conditions population grew slowly but steadily. The growth has been influenced by an amalgam of ecological and cultural factors which we may refer to as agricultural-economic environment. The favourable ecological, economic and social factors have combined together to make the region agriculturally important in the production of a great variety of food, subsidiary food and cash crops which apart from being fundamental to the sustenance and survival of the population have influenced spatial setting and demographic structure, e.g. age and sex composition, occupational structure, literacy, population movement, etc.

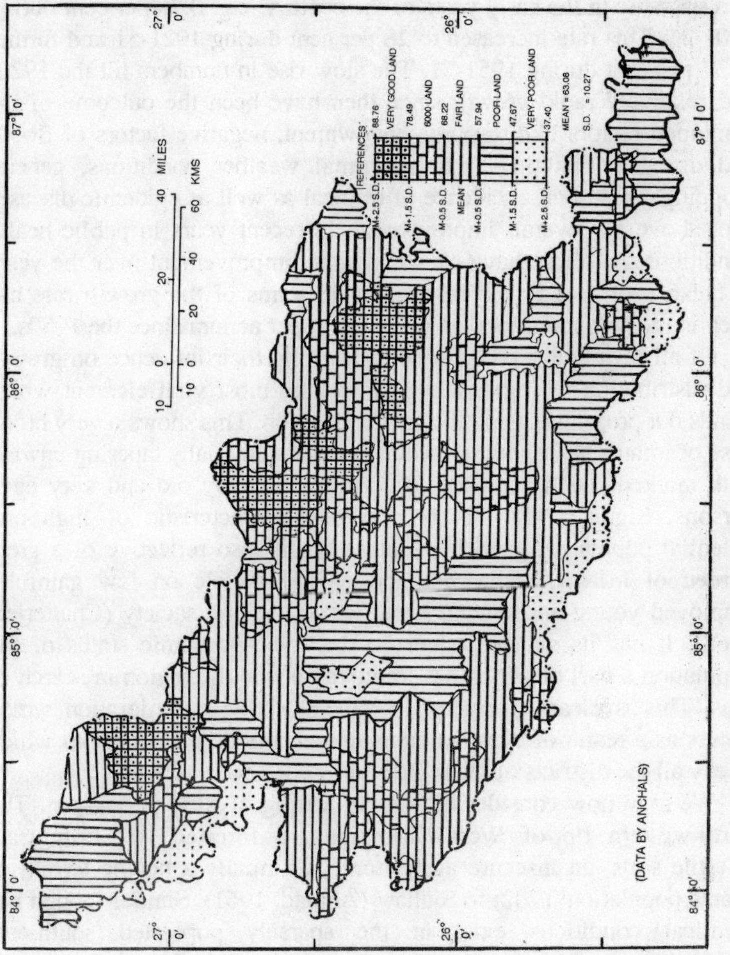

FIG. 12.2: North Bihar Plan, west of the Kosi — land capability classes.

The growth of rural population has grown from 1,07,43,758 in 1901 to 1,90,08,134 in 1971, a total gain of nearly .83 million over the past seven decades. The net accretion has been 77 per cent. The growth rate was sporadic in the early years of the century, e.g. 0.75 per cent during 1901-21. This rate increased to 28 per cent during 1921-51 and further to 37 per cent during 1951-71. The slow rise in numbers till the 1920s and relatively rapid growth since then have been the outcome of the combined factors like resource endowment, negative factors of floods and droughts resulting from abnormal weather conditions, general cropping conditions, incidence of tropical as well as epidemic diseases and, above all, overall improvement, in recent years in public health conditions. All these have caused relative improvement over the years in balance of natural increase which in terms of the growth rate has been in the neighbourhood of 1.8 per cent per annum since the 1950s.

Of all demographic components exerting their influence on growth and distribution of regional population the most vital element which stands out prominently is the age composition. This shows a very broad base of infants and extremely young persons gradually tapering upward with markedly an attenuated top characterised by old and very aged persons. Significantly this is not only characteristic of high and potential population growth in future but is also reflective of a great burden of infants, adolescent and elderly people on few gainfully employed young and mature bread-earners of the society (Chatterjee, 1968). It has its sinister effect on the socio-economic status of the population a part of which has to move outside the region in search of jobs. This is clearly manifest in male-selective out-migration which occurs as a result of agrarian crisis and economic distress from which nearly all the districts of the region badly suffers.

We may now consider the spatial setting of rural population. The north-western tip of West Champaran, a forested and hilly tract infertile soils, an insecure agriculture, is virtually a hostile terrain of sparse population[1] (202) in Sidhaw (Ahmad, 1961). Similar though not identical conditions exist in the sparsely populated south-east sub-region, as, for example, Gopalpur (292). This forms a trough of lowland onto which drainage courses converge, making the area a perennial chunk of flood hazards and inhospitality. The area had altogether a rural population of 1,90,08,134 accounting for over 35 per cent of the State's share in about 21 per cent of its land area. The spatial pattern is, by all means, uneven but average per sq. km. density works out to 535 as against the State's figure of 324 which is indeed high considering the agriculture-based economy which is unable to offer fairly good living conditions. It would not be out of place to mention the influence of river regime of the Gandak, Burhi Gandak, Baghmati, Baya, Jamuari, Kamla Balan, Bhutahi Balan, Tiljuga besides the all-important Ganga and Kosi—all of which in their building up activity have left good promise for crop husbandry that exists in the plain. In years of peak flows and high floods they however, pose a serious threat to organised cultivation. It naturally appears that density pattern varies in the northern tract compared to its southern counterpart fringing along the Ganga *diara*. The distribution in certain pockets is conspicuous by heavy concentration, very high to high, moderate and even sparse, depending on an interplay of resource endowment, favourable farming, nature of cropping, immunity from floods, droughts, etc., which vary a good deal from one part to another. For example, Bidupur and Vaishali (893) followed by Hajipur (800), Jandaha (724) in Vaishali apart from Patori (848), Dalsingsarai (749) in Samastipur and Keoti Ranway (752) in Darbhanga represent tracts of overcrowding. Elsewhere in the districts like East Champaran, Muzaffarpur, Sitamarhi, Madhubani and Begusarai density varies from 635 to 435.

The areal differentiation in density of population has been explained by variations in cumulative land capability scores, proportion of multiple cropping, proportion of cultivated land in total area together with percentage of cash crops and irrigated land in total sown area. The postulate that density of population is dependent on the above variables has been tested by calculating multiple correlation coefficient which reads as 0.68^2 and is highly significant at one per cent level of F-test of significance (Snedecor and Cochran, 1968 and Kenny and Keeping), 1954 (Table 12.1).

$$|W| = 0.25535409$$

$$|W_{33}| = 0.7958123$$

$$R^2{}_3.12456 = 1 - \frac{W}{W_{33}}$$

$$= 1 - \frac{0.2553541}{0.7958123} = 0.679128$$

$$F = \frac{R^2}{1 - R^2} \cdot \frac{n-k-1}{k} F k, n-k-1$$

$$= 61.0259$$

where *Fk, n–k–1* denotes the Snedecor's *F* distribution with *K* and *n–k–1* d.f.

Among the various demographic features, movement of population from one district to another, within the region as well as outside it, is more the result of economic push factor than solely the matrimonial and cultural ones. It becomes abundantly clear that an agrarian society, with very low levels of literacy (16.3, 26.4 and 6.1%) respectively for total, males and females, absence of high levels of urbanisation and industrialisation, can hardly afford to sustain all its primary workers within the region itself. Almost the entire area is syndrome of poverty and rural backwardness to which we will turn later to elucidate the main points of our contention.

TABLE 12.1: Determinant of correlation matrix | W |

	X_1	X_2	X_3	X_4	X_5	X_6
X_1	1	0.18780	0.25375	–0.19821	0.22876	0.22693
X_2	0.18780	1	0.80588	0.08169	0.00907	0.04607
X_3	0.25375	0.80588	1	–0.04880	0.05784	–0.03568
X_4	–0.19821	0.08169	–0.04880	1	–0.06503	0.01823
X_5	0.22876	0.00907	0.05784	–0.06503	1	0.01865
X_6	0.22693	0.04607	–0.03568	0.01823	0.01865	1

where X_1 — Cumulative Land Capability Scores
X_2 — Proportion of Multiple Cropping
X_3 — Density of Rural Population
X_4 — Percentage of Cash Crops in Total Sown Area (TSA)
X_5 — Proportion of Cultivated Land in Total Area
X_6 — Percentage of Irrigated Area in Total Sown Area

Agricultural Landuse

In an area of 3,542,650 hectares of land corresponding roughly to 35,510 km² pattern of land utilisation is dominated by agricultural landuse (Table 12.2). Data based on three yearly averages from 1970-71 to 1972-73 show that nearly 70 per cent of the total land is devoted to net sown area (NSA). If fallows including other fallows were added to it the picture that emerges roughly means a preponderantly high proportion of 79 per cent of cultivated or arable land. Pastureland, orchards plus cultivable wasteland taken together account for a little less than 4 per cent which truly speaks of an imbalance in vital uses of land. Almost the same proportion of land is barren and uncultivable (3.59%) as the above, while the area under forests is too small to merit separate treatment. The other most important use after cropland is the land put to non-agricultural uses (14%) viz. settlements, homestead and *barilands,* roads, canals, cultural features like community halls, playgrounds, parks, etc.

TABLE 12.2: Pattern of land utilisation (average of 1970-71 to 1972-73)

Category	Area in '000 hectares	Percentage of Total Area
Net Sown Area	24533.2	69.3
Current Fallow	2873.3	8.1
Other Fallows	577.9	1.6
Orchards and Pastures	1013.3	2.9
Cultivable Waste	241.6	0.7
Barren and Uncultivable Waste	1269.7	3.6
Non-agricultural Use	4797.0	13.5
Forests	120.5	0.3
Total Area	35426.5	100.0

Source: Government of Bihar, Directorate of Statistics and Evaluation : *Annual Season and Crop Report, 1970-71, 1971-72 and 1972-73.*

A cursory glance at landuse figures would readily reveal that cropland appears to be the most dominant feature of land utilisation which results from early occupancy of the area, staggering rise in population and food needs together with socio-economic conditions obtaining almost uniformly everywhere so that these interactions have produced a socio-economic milieu in which we find an overemphasis on the production of cereals, viz. paddy, wheat, maize, *marua*, etc. (76%), subsidiary food crops, viz. sweet potatoes, *cheena*, inferior millets (2%), pulses and oilseeds (12%) which are next in importance, apart from cash crops (10%) that are relegated to a third place with animal husbandry occupying a negligible rank in the general hierarchy of diversified cropping pattern. There is relevance in growing subsistence crops for direct human consumption and raising subsidiary food, pulses and oilseeds by taking advantage of three cycles of plant growth dependent upon seasonal rhythm and available surface as well as sub-surface ground moisture conditions.

An analysis of Fig. 12.3 showing crop association would clearly reveal how ecological factors and food needs of the people greatly motivate the farmers to grow subsistence crops or such crops that might supplement their needs. It would be important to note that with an increasing number of people dependent on farm production consequent upon steady contraction in per capita cropland, there has arisen a trend for intensive rather than extensive landuse. It is noteworthy that thousands of years of continuous sedentary farming on account of favourable geographical conditions have combined to influence cultivation that has been extended to farthest limits. In the face of an ever-rising population and their food needs, multiple rather than single cropping system is gradually gaining momentum. This means that the farmers are trying hard to eke out, from the same unit of land, as many harvests as possible. Naturally, the same unit of the NSA is devoted to raising two and, under favourable conditions, even three crops in an agricultural year. Table 12.3 indicates the number of harvests, their share in the TSA and percentage of area cropped more than once in the NSA.

Table 12.3 clearly shows that Aghani, Rabi, Bhadai and Garma are the principal harvests in order of importance. Aghani (rice, sugarcane, pulses, etc.) and Rabi (wheat, barely, khesari, gram, oilseeds, etc.) between them account for 70 per cent of the total sown area. Next in importance are Bhadai crops (maize, marua, inferior millets, minor pulses, etc.,) which roughly cover 24 per cent while Garma crops, locally important though, account for a small percentage of the TSA. Area sown more than once varies a good deal but taking the regional picture percentage of land cropped more than once in the NSA stands at 46 which clearly reveals that multiple cropping has continued to develop as a measure to counter the ill effects of population pressure on land to which we will turn later.

TABLE 12.3: **The Agricultural harvests and intensity of cropping (average of 1970-71 to 1972-73)**

Principal Harvests	Area in '000 ha.	Percentage of TSA
Bhadai	8428.9	23.5
Aghani	12636.3	35.3
Rabi	12408.4	34.6
Garma	2349.2	6.6
	35822.8	100.0
Area sown more than once in NSA	11289.7	46.0

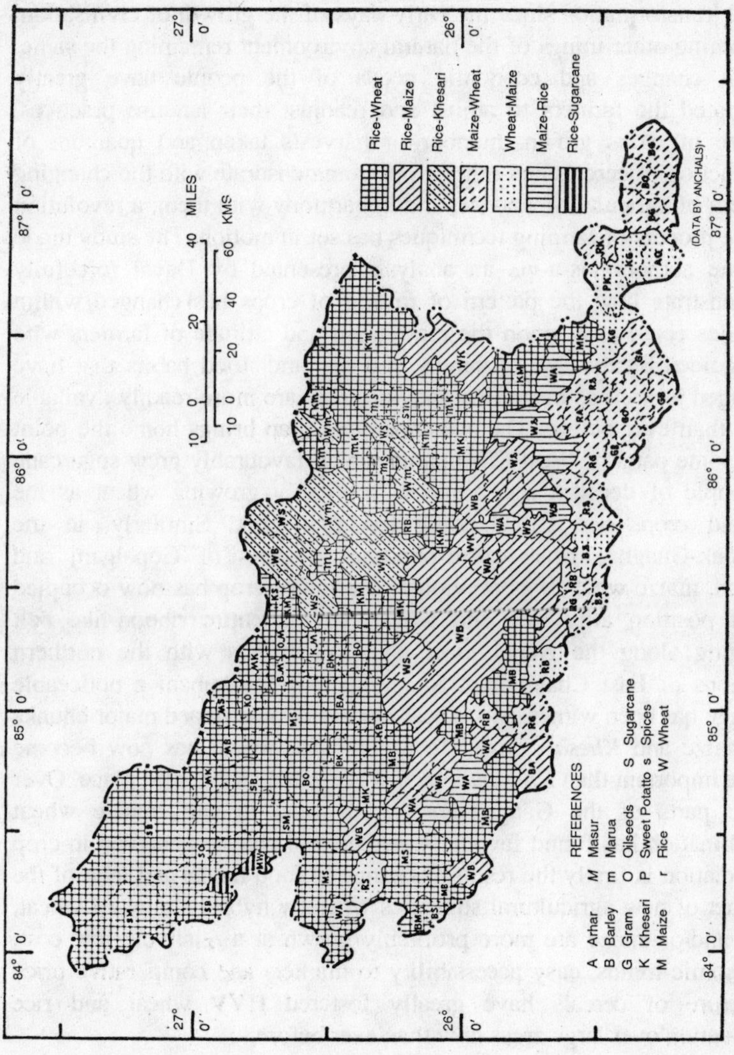

FIG. 12.3: North Bihar Plain, west of the Kosi—crop association regions.

We have earlier discussed that the area is one of early settled parts of the middle Ganga plain. The agricultural landscape has undergone rapid transformation since the early days of the growth of civilisation. Assuming other things of the natural environment remaining the same, social changes and economic needs of the people have greatly prompted the farmers to adjust and readjust their landuse practices, nature of crops grown, number of harvests taken and quantum of production of cereals and other crops commensurate with the changing pattern of agricultural landuse, and in harmony with them, a revolution in the process of farming techniques has set in motion. The study made by the authors vis-a-vis an analysis presented by Dayal forcefully demonstrate how the pattern of raising of crops has changed within decades consequent upon the motivation and attitude of farmers who are guided by the law of supply and demand, food habits that have changed as well as marketing facilities that are more readily available now than ever before. The accompanying map brings home the point that some parts of West Champaran which favourably grew sugarcane a couple of decades ago are now profitably growing wheat as the second crop, rice still holding the first rank. Similarly, in the Gandak-Ghaghra Doab corresponding to parts of Gopalganj and Siwan, maize which was previously the major crop has now occupied third position after rice and wheat. In the entire ribbon-like belt fringing along the Nepal *tarai* and co-terminus with the northern portions of East Champaran, Sitamarhi and Madhubani a noticeable change has been witnessed—wheat has easily outstripped major chunks of maize and *Khesari* lands. In Madhubani, wheat has now become more important than *Khesari* which is relegated to the third place. Over large parts of the Ganga *diara* from west to east, maize wheat combination has found favour. A perceptible change observed in crop association is firstly the result of change in food habits and then of the impact of new agricultural strategies whereby hybrid strains of wheat, rice and/or maize are more profitably grown at a relatively low cost. Economic trends, easy accessibility to markets and comparative price structure of cereals have greatly fostered HYV wheat and rice cultivation over large areas now than ever before.

In precise terms, agricultural landscape is presently much different and more diversified from what it was a century ago. It is not only important to point out how, and to what extent, socio-economic needs of the people inhabiting this old settled region have acted as a pull to farming, but what is more crucial today is innovation in science and technology of which some progressive and enthusiastic farmers are definitely reaping the fruits — by adopting new agricultural strategies. This fact gets support from another point of view, i.e. if an area of cultivation is not capable of being increased in consonance with population growth nor is there any chance of increase in per capita land, the population of the region, despite all natural hazards, is surely on march. This means that intensification and diversification at the one end of the scale and higher productivity accruing from proportionately higher inputs at a reasonably low cost of production, at the other, are a pointer towards increasing agricultural efficiency (Shafi, 1972). Changes in regard to the development of infrastructure in farm technology, i.e. more abundant use of agricultural water supply, improved variety of high-yielding strains in lieu of indigenous ones, plentiful use of both organic and inorganic manures and fertilisers, and above all, rural banking and credit facilities flowing from nationalised and cooperative banks are largely responsible for higher returns from the land now than what it was in the past (Banerjee. 1974) These are solely instrumental for an increase in the subsistence level of the mass although much still remains to be desired for a great bulk of them who live below the poverty line.

Measurement of Population Pressure

An exact definition of the pressure of population seems desirable so that identification of the regions of population pressure is made. However, the matter at issue is not so simple as it appears at the first sight, and the whole question is beset with a number of methodological and conceptual pitfalls. Nevertheless, it is not impossible to locate some basic benchmarks from which significant measurements can be made. There is, as we have seen earlier, uneven distribution of population in the study area which, on the whole, may be considered as one of the most densely populated parts of the country. Agriculture is the chief mainstay of an overwhelming majority of rural masses (over 90%) but owing to traditional agriculture resulting from low output from cropland it has failed to support people depending on it. Population pressure is high and even critical in areas of high concentration. This has its bearing on the cropping pattern. As the pressure of population on land increases, the area sown more than once in the NSA appears to increase correspondingly. Based on Pearsonian Product Moment Correlation Coefficient and putting the relevant data on the yardstick of statistical measurement it has been found out that density of population and proportion of multiple cropping is vcry strongly positively correlated (r = 0.81). The analysis of variance shows that 66 per cent variation in the concentration of population is wholly dependent on the abovenoted factors. The remaining 34 per cent is explained by other factors like industries, trade and commerce, which also strongly influence spatial distribution of rural population. With the population shooting upward the available land, in order to support an overwhelming majority of the people (over 90%), goes on steadily shrinking. It would be revealing to examine population-land relationship with a view to examining the case for appraisal of the pressure of population on land.

Man-land ratio provides an important element in the study of a measure for population pressure. This concept, which relates to the number of persons per unit of area, is particularly important in an agricultural area such as this. This can be expressed in various ways viz. crude or overall arithmetic density, man-soil density (Kuperus, 1938) or physiological density (Trewartha, 1953), agricultural density. nutritional density, caloric density and economic density. The crude density which is also termed as general density has been considered in terms of population divided by area. It, thus, depicts only very general conditions of population pressure. As neither all rural populations are dependent for their livelihood on land nor is all the land available in a particular area cultivated, the picture that emerges from dividing population by total land area is misleading. Man-soil density or physiological density is expressed as population divided by cultivated land; though it gives rather a more concrete picture, it again fails to convey a true picture of population pressure. It is so as it considers population in relation to cultivated land. In an overwhelmingly agricultural tract such as the area under study, pressure of population is brought out in a concrete form. But it is pertinent to note that all people even in the countryside are not wholly dependent on agriculture for their sustenance. This obviates the need for considering agricultural density which is worked by dividing total farm population by cultivated land (NSA plus fallows). It becomes more glaring to show the actual relationship between agricultural population as related to cropland. This provides a sound basis for calculating a direct measure of population pressure on the cultivated land since the area is purely agricultural where over 90 per cent of its rural population derives livelihood from the primary sector. Other parameters for measurement are found to be either non-workable in our case or else they have little relevance in pointing a way toward population pressure.

Agricultural Density of Population

The agricultural population constitutes over 90 per cent of the rural population. The average agricultural density works out at 6.3 persons per hectare of the cultivated land which varies from a minimum of 2.23 in Sidhaw *anchal* (West Champaran) in the extreme north-west to a maximum of 10.3 in Bidupur *anchal* (Vaishali) in the south. It is worth-while to note that as the percentage of agricultural population is uniformly high in all the *anchals* regional variations in agricultural density depend essentially upon natural as well as ecological factors which have governed the mode of agriculture to a large extent. The pressure of farm population is more acute in parts of south-west, south and north-east where they form compact blocks. It varies between 7.2 and 10.4 persons per hectare of cultivated land. The relatively high through very high to extremely high densities of farm population in these areas are owing to its optimum farm activities. Favourable drainage net, security from floods and at least one bumper harvest, particularly in the south and south-west, have provided an ample scope for cultivation. A large number of persons in these areas are solely dependent on farming. This tends to increase the pressure of agricultural population on cropland. The sub regions, on the whole, are endowed with bounties, viz. fertile land and better agronomic conditions.

The relatively low pressure of agricultural population is found over the entire north-west and mid-west comprising the whole of West and East Champaran and parts of Muzaffarpur, major portion of the south-east and extreme north-east, apart from a few others scattered here and there. The extremely low agricultural density (2.8 to 3.9 persons per hectare of cultivated land) in the case of north-west may be ascribed to it growing significance as an area where considerable proportion of population is engaged in other pursuits, e.g. commerce and trade, manufacturing, services etc. It may further be added that a high proportion of its population is constituted by the Scheduled Castes and Scheduled Tribes who do not own any significant chunk of land are mostly engaged as labourers on farm, forestry, commerce, service, etc. In the south-east, the low density figure of farm population does not imply any relief in population pressure because it co-exists with poor land and inferior cropping conditions which are a consequence of frequent occurrence of floods. They not only damage the standing *bhadai* crops but often convert cropland into fallows by aggravating waterlogging.

Pressure of agricultural population is moderate (7.1 to 8.2) over large areas in the eastern and south-west sections. These together constitute an arcuate-shaped area which is much wider east than in south-west. A slight moderation in the agricultural density is dependent chiefly upon the cultivation factor. While unfavourable ecological conditions arising out of the hazards posed by the Adhwara group of rivers together with the Kamla-Balan floods has led to a deterioration in agronomic conditions in north and east coupled with the alkalinity problem in the south-west whereby contrast appears to be sharp. These coupled with frequent failures of crops on account of lack of facilities for irrigation, agricultural prospects have, in the past few years, been bleak. However, comparatively moderate density of farm population does not mean any relief in pressure as their natural conditions create relatively unfavourable agronomic conditions. Thus, moderate pressure of agricultural population co-exists with unfavourable farming. The agricultural density, however, is too high and even the increased yield accruing from better natural and improved agronomic conditions is not capable of offsetting population pressure upon land.

Population vis-a-vis Agriculture

There appears to be a growing imbalance between the population number and natural resources, particularly those connected with the land. We have seen that increase in population means a drop in per capita cultivated land and total sown area (TSA) and consequently a corresponding rise in the pressure of population. Per capita availability of land, is thus, another significant determinant of population pressure in the region. Mukherjee had considered that for a country like India one acre (0.4 hectare) of cropland would be indispensable for the nourishment of one person (Mukherjee, 1938). It may be noted, however, that in the region under study, yield from one acre, as at present, is adequate for more than one person. Therefore, Mukherjee's contention made decades ago now appears to be too generalised and does not hold good in the changed circumstances as exist now. We will take recourse to some other indicators to highlight out findings as to the population pressure as related to agriculture as the sequel will show later.

The regional average of the per capita availability of cultivated land works out to only 0.14 hectare. An increase in population has led to a diminution in the average size of agricultural holdings, i.e. unit of land per rural household or family, per capita cultivated land and reduced supply of cereals to each person. The average size of holdings per rural household, for example, varies from 2.11 to 0.47 hectare regional average being 0.83 hectare. In consonance with this decline per capita cultivated land has consequently dropped to a significant level. Further, a low per capita figure of the TSA (0.19 hectare) and its spatial distribution similar to the per capita cultivated land further establishes areal distribution of population pressure. A low figure of 0.19 hectare of the TSA per head of population is by all means hopeless when considered in terms of Mukherjee's earlier contention as also due to prevailing food deficit in the area. There is, therefore, a conclusive evidence of an intensive pressure of population on available land resources.

The pressure of population on land has further been established by working out regional distribution of food surplus/deficit areas which overwhelmingly are sustained by agriculture. An increase in population has led to a sizable reduction in the supply of foodgrains to each person resulting in a widening gap between the available and needed quantities of food. The surplus/deficit foodgrain-producing areas have been found out on the assumption that on an average one person requires 0.45 kg of cereals (Das, 1972) and 0.85 grams of pulses (Aykroyd, 1962) in his daily diet to meet the minimum calorific need. Out of 535 grams of minimum quantum foodgrains, regional average works out to only 325 grams per head per day. This obviously shows an overall deficit in supply of foodgrains in which there is an acute shortage or pulses which might make up protein deficiency. As a matter of fact, consumption of fish or meat at irregular intervals, i.e. once in a week or fortnight, implies that an overwhelming majority of the mass are vegetarian in their dietary habits. The regional average in respect of pulses is in the neighbourhood of 20 grams against a daily minimum requirement of 85 grams. No part of the region has even near sufficiency in the availability of protein-rich diet, and with an abundance of carbohydrates there is an awful imbalance in the daily food intake of the rural people.

A statistical computation of data showing the regional pattern of surplus/deficit food production enable us to divide the region into five types of areas: (i) Food surplus areas; (ii) Near optimum food-growing areas (deficit below 10%); (iii) Marginal food-deficit areas (10-25%); (iv) Small food-deficit areas (25-50%); and (v) Severe food-deficit areas (over 50%). According to Sukhatme one in every four of India's people appears to be underfed and two out of every four are malnourished (Chopra, 1975). It holds equally good for the study area in which approximately more than half of the regional population is well below the poverty line. However, this does not all mean that there is mass starvation in areas of small to severe deficiencies of foodgrains particularly in the extremely congested rural parts of the region. The actual consumption of an individual or a family or a section of the population depends not so much on average availability but on the purchasing power. Due to high pressure of population, particularly in the parent districts of Saran, Muzaffarpur and Darbhanga, there is mass exodus of the male working population in search of jobs elsewhere. They provide purchasing power for meeting the needs of their families, for buying the food from the Government ration shops that are very many in rural villages of the area. Further, sweet potato whose yield per hectare of cultivated land is very high is grown in this area in bulk. It provides an important subsidiary food in the area which has reasonably been termed as "famine food" (Aykroyd, 1962) apart from mango, litchi, banana, jackfruit, etc., in which the region occupies a place of pride in the State. Areas of surplus or optimum availability of foodgrains co-exist with low-to-moderate densities and are found in the north-west and in its neighbourhood.

From what has been elaborated above there is conclusive evidence of an intense pressure of population on available land resources. The pressure, however, is not uniformly distributed over the region as the sequel will show. The concept of "relative coefficient of overpopulation, (Das, 1969), "intensity of population pressure" (Singh, 1971), and a technique of measuring the "degree of population pressure" (Krishan, 1971) are important steps towards studying the pressure of population on land resources. Following Krishan, areal differentiation of population pressure in the study region has been worked out, hypothetically, assuming that the region is overpopulated. It has been logically deduced from a study of the above determinants of population pressure together with the prevalence of most of the symptoms of population land relationships maladjustment in the region.

The picture of relative population pressure has been obtained by computing population pressure scores for the following determinants/ indicators for all the 180 *anchals* of the region. The principal determinants selected for the purpose were : (i) Agricultural density of rural population, (ii) Per capita cultivated land, (iii) Surplus/deficit foodgrain production, (iv) Sex ratio, and (v) Persons per rural household. Out of these five determinants selected, the first three assess population pressure in relation to agricultural economy, the fourth is reflective of out-migration indicting population movement prompted by stress and strain encountered by a part of the earning hands in eking out means of livelihood for their families and the last gives general impression about the population pressure. Four broad tracts of very high (over 11), high (10-11), medium (8-9) and low (7 and under) overpopulation may be recognised in the region (See Fig. 12.4). The spatial distribution of population pressure broadly follows the pattern of areal distribution of agricultural density of rural population and thus does not need further elaboration.

Evaluation and Remedial Measures

Now that the tracts of over-population have been identified the uphill task remains for us to explore possibilities about the potential carrying capacity of land (Land Capability Classes, Fig. 12.2) and to suggest remedial measures in order to counter the ill-effects of pressure of population on land. The carrying capacity of an area is "an estimate of the number of people that the area will support in perpetuity, under a given system of land usage, without deterioration of land resources" (Allan, 1972). Optimum carrying capacity of land, on the other hand, means "the number of people it can adequately feed to good health and activity and it is a function of agricultural efficiency (Singh, 1971). Shafi is much optimistic as he opines that the agricultural land of India is capable of bearing five times her existing population provided suitable infrastructure is built for that (Shafi, 1969).

The area is characterised by the traditional system of agriculture, marked absence of irrigation facilities and consequent upon it frequent failure of crops, low input-output ratios and less productivity despite bounties of nature. Under the present circumstances, it has been established that the area is characterised by overpopulation. The concept of measurement of potential carrying capacity of the land is, therefore, a basic requisite and has assumed great regional significance, particularly in an area like this where there is heavy reliance of population on subsistence crops. The need for such a study is of special interest in the present context as the scope for extension of cultivation is negligible. The one pragmatic way to meet the increasing

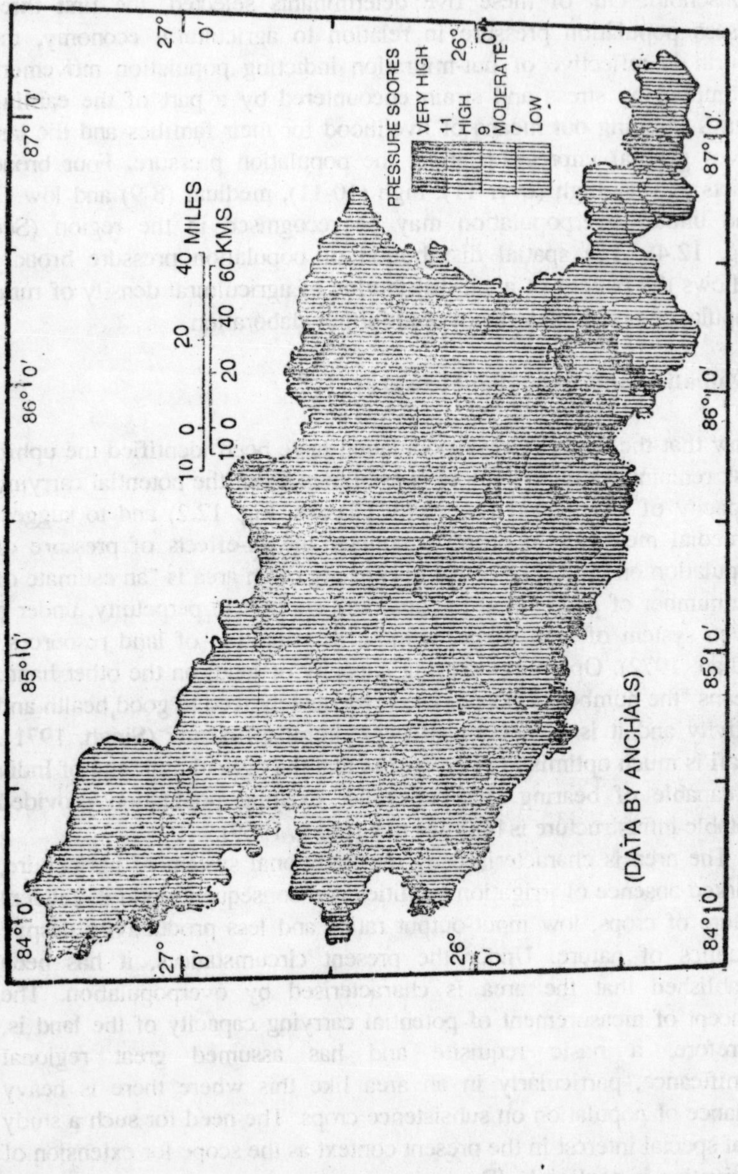

FIG. 124: North Bihar Plain, west of the Kosi—pressure of rural population on land.

burden of population appears to be an improvement of landuse, irrigation and per capita total sown land. Further, possibilities for agrarian reforms, lessons from other Indian as well as State areas for revolutionising traditional agriculture, general consensus among the bulk of the farmers to have the fruit of the much publicised green revolution, changes in farm technology and new innovations in the agrarian field, viz. opening of new canals, tube-wells apart from electric and diesel-operated pumps, liberal use of HYV seeds and hybrid strains of rice, wheat and maize, application of manure and NPK fertilisers as recent trends in farming technology and an overall view of agrarian revolution as a continuing process for higher productivity from the land as basic to the socio-economic needs, have to go a long way in increasing the inherent carrying capacity of the land. It is in the fitness of things that the Bihar Plain in general and the study area in particular have immense scope for the development of irrigation facilities by harnessing numerous perennial rivers of the area. It is, therefore, desirable that an index of land-efficiency as related to carrying capacity has to be obtained so as to find out which area is more capable of feeding a greater number of people at a reasonable level of consumption and which area, if not endowed with natural attributes, is able to support a comparatively small number of people. The whole object is to find out the potential carrying capacity of the land by dividing the farm produce by an individual requirement whereby the quotient will indicate an index of how a unit of land can carry and sustain the number of persons per annum, who may be dependent on it.

It has been estimated that in the North Bihar Plain west of the Kosi one hectare of very good to good land under good farm management can support as many as 25 persons, fair land roughly 18 persons, poor land 10 per cent and very poor land having *usar* and waterlogged conditions only five persons (see. Table 12.4).

The potential carrying capacity of land is everywhere larger than the existing one. The explanation of regional differences is to be sought in association with corresponding map Fig. 12.4. Based on Table 12.4 and the map prepared four types of areas have been identified:

(i) Areas of very high potential carrying capacity (20 to 25 persons per hectare of cultivated land),

(ii) Areas of high potential carrying capacity (15 to 20 persons per hectare of cultivated land),

(iii) Areas of moderate potential carrying capacity (10 to 15 persons per hectare of cultivated land), and

(iv) Areas of low to very low potential carrying capacity (five and below persons per hectare of cultivated land).

TABLE 12.4: Extent of carrying capacity of agricultural land in north Bihar plain west of the kosi

Land Capability Class	Farm Produce per hectare/ year in kg	Carrying Capacity* (No. of Persons per ha.)
I. Very good to good land	5,500 to 6,000	20 to 25
II. Fair Land	4,000 to 4,500	15 to 20
III. Poor land	2,000 to 2,500	10 to 15
IV. Very Poor land	700 to 1,000	5

* The Carrying Capacity has been estimated in terms of foodgrain requirement of population. Cf. Land Capability and Pressure of population on Land maps.

A spatial analysis of potential carrying capacity of land reveals two important conclusions: firstly, a visual comparison of the two maps showing pressure of rural population on land and another showing Land Capability Classes makes it clear that there seems to be no reason to become scared of the growing population pressure as radical changes for rural uplift are already taking place, and if they are based on a scientific appraisal of existing landuse and potential capabilities, they are likely to be more effective. It is expedient to think of increasing the carrying capacity to at least thrice the present level by bringing an agronomic revolution so that the future population may be supported at a reasonable level. Secondly, a comparison of the two maps shows that while the potential carrying capacity of land is high to exceptionally high mostly in the north, pressure of population, on the other, is the highest in the south and south-west. This naturally leads us to conclude that a further interval redistribution of population will take place. This will also occur in harmony with the transformation of the agricultural-economic environment for development that has now set in motion at the *Panchayat* and block levels.

NOTES

1. Figures in brackets for an individual *anchal* refer to per sq. km. density (1971).
2. The authors have made a detailed and painstaking study of various parameters of

land and people and relevant data have been elaborately computed on micro-wave computer and compiled for use here. Unfortunately, methodological approach to statistical correlation is so elaborate and detailed that scanty space does not permit their inclusion in the text.

REFERENCES

Ahmad, Enayat, 1961 : "The Rural Population of Bihar", *Geog. Rev.,* Vol. 51(2), pp. 253-75.

Allan, W., 1972 : "How Much Land Does a Man Require?" in *People and Land in Africa : South of the Sahara* (Ed. R.M. Prothero), New York, p. 317.

Aykroyd, W. R., Gopalan, C. and Subramaniyan, S. C., 1962 : "The Nutritive Value of Indian Food and the Planning of Satisfactory Diets", *ICMR,* New Delhi, p. 32.

Banerjee, Bireswar (Ed.), 1969 : *Essays on Agricultural Geography : 4 Memorial Volume to Dr. B.N. Mukerji,* Calcutta.

Banerjee, Bireswar, 1974 : "Green Revolution in India : A Geographical Analysis", *Science and Culture,* Vol. 40, pp. 2-3.

Chatterjee, S. P. (Ed), 1968 : *Developing Countries of the World,* cited from India by P. Dayal Calcutta, p. 347.

Chopra, P. N. (Ed.), 1975 : *The Gazetteer of India,* Vol. III : *Economic Structure and Activities,* New Delhi, p. 349.

Das, K. N., 1968 : "Soil Erosion and the Problem of Silting in the Kosi Catchment", *Jour of Soil and Water Cons. in India,* Vol. 16 (3 and 4), pp. 61-65.

———, 1968 : "Westward Shift in the Courses of the Kosi", *The Nat. Geog. Jour. of India,* Vol. XIV (1), pp. 28-30.

———, 1969 : *Population and Landuse Changes in the Kosi Region Bihar,* Bhagalpur Univ. Ph.D. Thesis (Unpublished), pp. 132-37.

———, 1972 : "Landuse and Level of Nutrition in the Kosi Region, Bihar", *The Geographer,* Vol. XIX, pp. 55-85.

Das, K. K. L., 1979 : *Population and Land Resources in North Bihar Plain, West of the Kosi* (Ph.D. Thesis submitted to the University of Bhagalpur), pp. 34-319.

Diwaker, R. R. (Ed), 1959 : *Bihar Through the Ages,* Cited from *Geography of Bihar,* Calcutta, pp. 24-25.

Geddes, A., 1960 : "The Alluvial Morphology of the Indo-Gangetic Plain : Its Mapping and Geographical Significance", *IBG. Trans, and Papers,* Vol. 28, pp. 263-65.

Harrison, A. J., 1977 : *Economics and Landuse Planning,* London.

Kenny, J. F. and Keeping, E. S., 1954 : *Mathematics of Statistics,* London.

Krishan, G., 1971 : "A Measure of Population Pressure in Developing Areas : A Case Study of Uttar Pradesh", *Jour. Ind. Inst. of Econ., Geog.,* Vol. 2 (I and II), pp. 53-57.

Kuperus, G., 1938 : "Relation Between Density of Rural Population and Utilisation of Soil in Jawa", *Proceedings of International Geographical Congress,* 11, Section — 3, Amsterdam, p. 466.

Memoirs of the Indian Meteorological Department, 1962 : *Monthly and Annual Normals of Rainfall and of Rainy Days,* Vol. XXXI (III), Delhi, pp. 87-90.

Mohammad, N. and Amani, K. Z., 1970 : "Crop Combination in Trans-Ghaghara Plain", *Geog. Rev. of India.* 32(1), pp. 47-59.

Mohammad, N., 1978 : *Agricultural Landuse in India,* New Delhi.

Morgan, W. B. and Munton, R. J. (Eds), 1974 : *Agricultural Geography,* London.

Mukherjee, R. K., 1938 : *Food Planning for Four Hundred Millions,* London, p. 36.

———, 1938 : "The Changing Face of Bengal : *A Study in Rural Economy"*, Calcutta, p. 239.

Myrdal, Gunnar, 1977 : *Asian Drama* : *An Inquiry into the Poverty of Nation* (Abridged in one volume by Seth S. King), Harmondsworth.

O'Malley, L. S. S. 1930 : *Champaran District Gazetteer,* p. 1.

Roy Choudhary, P. C., 1963 : *Purnea District Gazetteer,* p. 12.

Shafi, M., 1972 : "Measurement of Agricultural Productivity of the Great Indian Plains", *The Geographer,* Vol. XIX(1), pp. 4-13.

———, 1969 : "Can India Support Five Times Her Population?", *Science Today,* May, p. 24.

———, 1960 : *Land Utilisation in Eastern Uttar Pradesh,* Aligarh.

Singh, Jasbir, 1971 : "Optimum Carrying Capacity of Land, Caloric Density and Intensity of Population Pressure Changes in Punjab, 1951-61, *National Geographical Journal of India*, Vol. XVII (I), pp. 43-45.

———, 1974 : *An Agricultural Atlas of India : A Geographical Analysis,* Kurukshetra.

Stamp, L. D., 1960 : *Applied Geography,* Harmondsworth.

Snedecor, G. W. and Cochran, W. G., 1968 : *Statistical Methods,* Calcutta.

Trewartha, G. T., 1953 : "A Case for Population Geography", *Annals of the Assoc. of Amr. Geographers,* Vol. 43, pp. 94-95.

Tudge, Colin, 1977 : *The Famine Business,* London.

13

STATUS AND PROSPECTS OF SOMACLONAL VARIATION IN SUGARCANE AND OTHER CROPS

PANKAJ SHARMA AND A. Q. KHAN

CROP IMPROVEMENT is based on the creation of genetic variability from which superior genotypes are selected. In conventional plant breeding, the genetic variability is generated by means of sexual cycle employing suitable breeding methodology. However, it imposes limitations on the transfer of desirable traits from related species or genera due to crossability barriers. Mutation breeding was initially considered to bring about desired heritable changes in crop plants but the progress has been meagre because of randomness of the technique and generation of mostly uneconomical variability. However, the development of *in vitro* studies have opened up possibilities to overcome the limitations of both conventional and mutation breeding.

Exhilarating developments have taken place recently in plant cell and tissue culture initiated in 1900 by Gottlies Haberlandt. He worked on single isolated cells of the leaves and conceived the role of "growth enzymes", now known as growth hormones. In mid-1930s, the auxins were identified as natural growth regulators and the importance of B-vitamins in plant growth was also recognised. Gautheret, White and Nobicourt independently established continuous growing cultures of meristimatic tissues in 1939 which provided the impetus for further work in the field of tissue culture.

However, the induction of divisions in differentiated tissue had to await the discovery of cytokinins by Miller and coworkers. Further progress was made when the callus was broken into small aggregates and single cells in suspension cultures which were picked up mechanically and induced to divide by the "nurse culture technique".

However, it was Braun who demonstrated totipotency in crown gall teratoma for the first time which was confirmed by Vasil and Hildebrandt by raising whole plants from single cells of tobacco. Initially, the tissue and cell culture techniques were limited to physiological studies. Gradually, its sphere expanded to micropropagation, overcoming natural sexual barriers by embryo rescue of interspecific hybrids raising of haploid plants, and development of somatic hybrids. Recently, these techniques have also been exploited in vegetatively propagated crops like sugarcane. By chromosome manipulation at the cellular level, it may be possible to get sufficient genetic variability to rectify specific defects without completely disrupting the genetic architecture of a highly adapted variety.

Sugarcane (*Saccharum* spp) is a unique and most efficient photosynthetic plant which not only provides cheap food in the form of readily available sugars in the interior of the stalk but also yields a vast range of by-products leading to a concept of total utilisation of the cane plant. The cultivated varieties of sugarcane are complex polyploids and derived from mostly trispecific hybrids involving *S. officinarum* L., *S. Barberi* Jesu. and S. *Spontaneum* L. Besides, it imposes certain limitations in sexual breeding due to irregular chromosomal segregation, practical difficulties in hybridisation due to its peculiar floral biology and high sensitivity to environmental conditions for flowering. However, the established cultivars can be improved by elimination of a defect through tissue culture which is otherwise impossible through normal breeding techniques. Therefore, *in vitro* techniques now available have opened up new vistas to sugarcane breeders.

Tissue Culture in Sugarcane

The use of cell and tissue culture in sugarcane commenced at the Experiment Station of Hawaiian Sugar Planter's Association (HSPA) in 1961 to investigate the physiological and biochemical aspects of yield decline in sugarcane. Nickell was the first person to develop callus from the mature parenchyma of internodal tissue of clone H50 — 7209. Material from this original callus was grown as a suspension culture. Several years later, single cells were isolated from the suspension to produce a callus from each of them by the nurse culture technique. Of the five isolated cells, each produced tissue differing in appearance: one was white, two were beige, and to were yellow. The two beige ones grew at different rates, as did the two yellow ones. It appeared at that time that these were five different clones. Subsequent cytogenetic studies showed that each of these subclones had different chromosome number. Cytogenetical and morphological differences were observed among the clones developed from the same cultivar. Later, cytologists in Hawaii and Australia found that the commercial sugarcane cultivars were chromosomal mosaics, i.e., bearing different chromosome numbers in different somatic cells.

With the beginning of plant tissue and cell culture at HSPA, initial emphasis was for defining the nutritional requirements of the culture medium. Barba and Nickall stated that White's medium was superior for the growth of sugarcane tissue and cells, but not for the expression of totipotency. They concluded that growth in or on White's medium for more than one sub-culture leads to loss of the ability to differentiate when the auxins are removed. The reasons for this phenomenon was not known. Nickell and Maretzki developed a chemically defined medium which proved satisfactory for biochemical investigations. In general, sugarcane clones in suspension culture grow better in a medium supplemented with yeast malt extract than in a medium containing coconut milk.

A variety of carbohydrates can serve as energy sources although sucrose and glucose are the outstanding sources. The best medium for callus differentiation and plantlet formation is that of Murashige and Skoog or its modification (which uses 5 mg/1 IAA and 2 mg/1 kinetin as growth promoting substances. Barba and cowokers indicated that plantlets exposed to light intensities of 400 to 1000 footcandles produced a profusely branched root system, had a high proportion of dry matter, and gave a vigorous stand and a high survival rate when transplanted to soil. It was also observed that the growth of shoots differentiated from sugarcane callus was enhanced by naphthalene acetic acid (5mg/1) in the absence of coconut milk, while 2,4-D (0.2 mg/1) stunted their growth. In their absence, a crown developed at the base of the shoot on which new shoots and roots developed, grew and formed a clump of plantlets. Root growth could also be induced by trimming the leaves or by incubation at 25°C for 4 to 6 weeks. These results indicated that a medium which stimulates differentiation does not necessarily stimulate shoot growth; and the hormonal requirement for shoot differentiation differ from that of root induction in sugarcane callus.

Krishanmurthi observed that as soon as the callus is removed from the influence of 2,4-D, it redifferentiates into shoots. The onset of differentiation is preceeded by aggregation of cells, followed by the appearance of green pigment. In a modification of sugarcane plantlets from callus tissue, 2,4-D has been replaced with NAA (1mg/1) and kinetin (1mg/1). Irvine and co-workers tested a large number of chemical compounds for callus inducing ability and compared them with 2,4-D. Of the 79 compounds tested, 25 induced callus formation. Most of the effective compounds (about 95%) were in chemical groups with known auxin activity. They observed that simple structures, phenol derivatives or a halogenated hydrocarbon chain were sufficient to induce callus formation.

Apart from nutritional studies, initial workers also tried out various portions of the plant for establishment of callus cultures and regeneration of plants. While the first report came for development of callus from the mature parenchyma of internodal tissue, callus tissue was soon induced on parenchyma tissue of shoot apices, leaves and infloroscences using a basal nutrient medium supplemented with coconut milk (10% v/v) and 2,4-D. On transfer of callus tissue to a non 2, 4-D basal medium, the differentiation of plantlets could be induced. Liu and Chen induced calli of shoot apices, rolled young leaves and young inflorescences of 58 varieties. In 1977, Heinz and coworkers finally demonstrated that cultures can be established from almost any portion of the plant.

What is Somaclonal Variation?

Historically, the tissue culture cycle was considered solely as a sophisticated method of asexual propagation enabling a rapid rate of multiplication. Thus, it became the accepted dictum that all plants arising from tissue culture would be exact copies of the parental genotype.

Phenotypic variants have, however, been frequently observed among plants regenerated from cell cultures. Initially, these were considered to be merely artifacts of tissue culture arising from epigenetic events and unworthy of any use or of genetic interest. However, researches in Hawaii observed variation among plants regenerated from cell cultures of sugarcane as early as 1969 and viewed this as a potential tool in crop improvement. Skirvin and Janick observed considerable variation among calliclones regenerated from cell cultures of *Pelargonium* sp. (scented geranium) and concluded that tissue culture techniques may have special uses for the plant breeder, and should provide an additional method to increase intraclonal variation. These two instances of generating variation through tissue culture provided a new avenue for the improvement of asexually propagated species whose genetic improvement by conventional plant breeding programmes is difficult. Greater credence to this approach was subsequently provided by Shepard and coworkers who observed extensive agronomically useful variation among protoclones of potato (cv. Russet Burbank), such as growth habit. tuber colour, fruit production and tuber uniformity. In addition, some protoclones were identified with field resistance to early blight and late blight.

In 1981, Larkin and Scowcroft proposed a general term 'somaclones' for the plants derived from any form of culture of somatic cells, and somaclonal variation for the variation displayed among such plants. The term 'somaclone' encompasses all such terms as 'calliclone' (plant regenerated from callus tissue), 'mericlone' (plant regenerated from meristimatic tissue) and 'protoclone' (plant regenerated from protoplast).

Though initially efficient cell cultures were established in non-agriculturally important crops, now efficient cell cycles have been established with several important legumes and important grain crops. This work owes a lot to the persistent efforts by the workers of HSPA Experiment Station with sugarcane, Green and Phillips with maize, and Bingham and co-workers with alfalfa.

At present, a number of plant species have been studied and somaclonal variation appears to be the general rule. While earlier studies on somaclonal variation were limited to asexually propagated crops such as sugarcane and potato, now it is well documented that seed propagated species also exhibit somaclonal variation for a wide range of characters. Morphological variants have been reported in oats, barley, wheat, rice, sugarcane, maize, sorghum, garlic, potato, tobacco, *Medicago sativa*, *Pelargonium*, crysanthemum and *Fragaria*. Somaclones showing variation for various agronomic characteristics like tillering habit, floral and seed fertility, pigmentation, essential oil content, germination and yield have also been reported in oats, barley, wheat, rice, sugarcane, maize, sorghum, potato, tobacco, *Medicago sativa* and *Fragaria*. Chromosomal variants amongst somaclones have been reported in oats, barley, sugarcane and groundnut. Interestingly, somaclones showing mtDNA sequence rearrangements have been demonstrated in maize which clearly indicate that cytoplasmic gene changes take place during the tissue culture cycle. Perhaps the most important categories of somaclones are those showing disease resistance. These have been reported in sugarcane, maize, potato and *Brassica napus*.

Thus, somaclonal variation has been reported for a wide range of characters in almost all crop plants which have been investigated. The major hinderance in putting these variants to practical use has been the incomplete knowledge of the genetics of these variations.

Origin of Somaclonal Variation

Genetic analyses of plants regenerated from tissue culture was delayed for quite some time due to limitations imposed by the initial working material.

Tissue culture has resulted in somaclones showing both epigenetic and genetic variations. The epigenetic changes are not useful for crop improvement as they are not passed on to successive generations and have often hindered the accurate genetic analysis of asexually propagated crops by compounding the heritable variation. Many reasons have been attributed for the origin of somaclonal variation which are discussed below.

Pre-existing and Culture Induced Variation

It was proposed by Wenzel and coworkers and Brettell and Ingram that somaclonal variation merely reflects genetic changes occurring in the somatic cells of the explant prior to culture. But most of the existing evidence points to the fact that the majority of genetic changes leading to variation occur during the culture phase.

There have been numerous reports which demonstrate that the genotype of plants and the source of explant are very important for somaclonal variation. In potato, differences have been observed in the number of regenerated plants between two cultivars and also for the regeneration of protoclones of different cultivars. The most dramatic effects have been observed between species and varieties of *Triticum*, and between genotypes of *Pisum sativum*. The frequency of regeneration could be increased dramatically after successive cycles of recurrent selection in alfalfa while the frequency of polyploid regenerants and chlorophyll deficiency mutants was found to vary significantly between the indica and japonica varieties of rice. We have also observed differences in the frequency of regeneration between two cultivars of sugarcane.

The source of explant is also a critical variable for the recovery of somaclonal variants. In geranium, plants from stem cuttings were uniform whereas, plants from root and petiole cuttings were quite variable. Similarly, propagation by stem cuttings of sugarcane cultivars which are chromosome mosaics can result in plants with variable chromosome number. It is also possible to obtain different frequencies and spectra of somaclonal variants among plants from different explants subjected to different selective pressures.

Earlier, it has been generalised that somaclonal variation is limited to plants derived from long-term cultures or protoplasts. Recently, however, somaclonal variants have been reported in plants derived from somatic embryos in celery, rice, sugarcane, maize, *Triticum durum* and *T. aestivum*. Previously, mericlones were advocated for large scale commercial propagation. Somaclonal variants have recently been reported among mericlones of *Lotus comiculatus*, strawberry, thornless blackberries, and orchids. Thus somaclones have been detected in tissue culture systems traditionally viewed to be genetically stable such as somatic embryos and meristems.

It has been known for quite some time that growth regulator composition of the culture medium may influence the frequency of karyotypic alterations in cultured cells 2, 4-D has been most frequently considered to be responsible for chromosome variability as in barely and potato. In *Nicotiana*, variants could only be regenerated at high concentrations of 6-Benzyl Adenine.

While there is evidence that chromosomal stability can be achieved in long term cultures, it is widely accepted that most long-term cultures are chromosomally variable, and was first reported in *Daucus carota*. Polyploid plants have been recovered in geranium, ornamental *Nicotiana*, cultivated tobacco, tomato and alfalfa. Aneuploid plants have been recovered more often from plants of polyploid or hybrid origin like *N. tabaccum* and *Saccharum* hybrids where addition or loss of a few chromosomes can be tolerated.

Several workers have observed that variation increases with the increasing duration of the culture. It has also been reported that somatic hybrids are a richer source of variability than comparable sexual hybrids. Several unique variants have also been detected between clones of somatic hybrids in some cases.

Chromosomal Abnormalities

The frequency of gross karyotypic changes in somaclones is much reduced relative to that observed in cell cultures *per se*. This has been confirmed in the karyotypic studies of potato, *Pelargonium,* lucerne and maize where numerical chromosome changes are not very prevalent among somaclonal variants. Further, there seems to be a very little correlation between phenotypic variation among somaclones and changes in ploidy level.

Mitotic analysis of regenerated plants gives only limited information about cytological abnormalities. Detailed meiotic analysis of regenerated plants in barley, *Howorthia* and oats have revealed extensive chromosomal rearrangements, such as reciprocal and non-reciprocal translocations, inversions, sub-chromatid aberrations, and partial chromosome loss. It, therefore, seems that the occurrence of chromosome breakage and interchanges during cell culture is a relatively more frequent event than thought previously.

These somatic interchanges may favour transgression of desirable genes from alien species into commercial cultivars which is otherwise difficult in conventional interspecific hybridisation because of the lack of recombination between divergent genomes. This has been confirmed by the phenotype of plants regenerated from cell cultures of interspecific and intergeneric hybrids. Wide hybridisation is being increasingly viewed as a plant breeding option and thus cell culture holds tremendous potential in enhancing gene exchange.

Nuclear Gene Changes

In contrast to gross chromosomal changes, more subtle changes resembling single gene mutations have also been detected by several workers. Detailed genetic analysis has only been completed for tomato where a large number of single gene mutations have been identified. These include recessive mutations for male sterility, jointless pedicel, several fruit colours, lethal chlorophyll deficiency, virescence and mottled leaf appearance; a semidominant mutation controlling isoenzymer of alcohol dehyderogenase; and dominant mutations controlling fruit ripening, growth habit and disease resistance. Preliminary analysis has also been conducted in maize, *N. sylvestis*, rice, wheat and triticale.

In some cases, use of specific genetically marked strains has aided in evaluation of plants regenerated from cell culture. In *N. tabccum* Dulieu and Barbier using a_1 *y* and Loci and Lorz and Scowcroft using Su/su locus with chlorophyll deficiency present in heterozygous tissue detected genetic changes by the appearance of albino or dark-green regenerated plants.

It has been noticed that in addition to genetic variants common to all mutagens, several gene and chromosonal changes occur at higher than expected frequency as a consequence of somaclonal variations. Evidence also suggests that mitotic crossing over may also account for some somaclonal variation.

Cytoplasmic Gene Changes

In addition to single nuclear gene changes, cytoplasmic genetic changes have also been detected, at quite a high frequency. The most detailed work has been done on cytoplasmically controlled male-sterility in maize. Sensitivity to host-specific toxin of *Dreschslera maydis* race T, the causative agent of southern corn leaf blight, is associated with all genotypes containing Texas male-sterile (cms-T) cytoplasm. In seed-derived plants, these two characters are tightly associated, and both controlled by mitochondrial DNA (mt DNA). Gengenback *et al* selected *in vitro* for resistance to the toxin with the aim of combining it with male-sterility, but even in the regenerated somaclones resistance was associated with reversion to male-fertility. However, the restriction endonuclease patterns of mt DNA indicated that significant changes had occurred in the mt DNA of plants derived from cell culture. One mutation to male sterility and toxin insensitivity has been shown to be a frameshift mutation in mt DNA. Variations in mt DNA have also been detected in high frequency among potato somaclones.

There are fewer organelles in a cell in the developing shoot apex than in a mature cell. Hence if a mutation occurs in organelle DNA, it is more likely to become the distinct organelle type in subsequent somatic cell divisions. During time course experiments using wheat callus cultures, novel restriction fragments appeared in mt DNA and some of the novel fragments underwent amplification, diminution, or were lost.

Gene Amplification and Diminution

The quantity of a gene product can be increased or decreased simply by differential gene amplification or diminution. With appropriate selection, stable amplification can result. In plants, ribosomal RNA gene amplification and depletion has been documented in wheat, rye, hyacinth, maize, *Vicia*, melon and tobacco. The DNA of flax is known to change in response to environmental pressure. Reassociation of kinetics showed that one plant form possessed a class of moderately repeated DNA not present in the other form.

If such amplification (or depletion) of DNA sequences can occur in plant cell cultures, it may account, at least, in part for somaclonal variation. Amplification and reduction of repeated DNA have been reported in both callus and regenerated plants of flax. Similar amplification and reduction of specific repeated sequences have been reported for cereal cultures. While selecting for resistance to a competitive inhibitor of glutamine synthase in alfalfa, tolerant cell lines were identified containing 4 to 11 times the number of glutamine synthase gene sequences.

Activation of Transposable Elements

Transposable genetic elements are DNA sequences which can move from one position in the genome to another, independently of extensive sequence homology. Activation of transposable elements has also been detected in regenerated plants. In maize, activation of Ac was detected in 3 of 94 independently derived plants, suggesting a high frequency of recombination events. Similarly, the Spm (En) element of maize has been reactivated by tissue culture regeneration.

Somaclonal Variation in Sugarcane

Sugarcane was amongst the first crops where somaclonal variation was detected. In fact, it was detected as early as the tissue culture technique was first applied to this crop. Initiality, differences were observed in colour of the tissue produced; subsequent cytogenetic studies showed difference in three chromosome numbers. The early discovery of chromosomal mosaicism in sugarcane was the beginning of a considerable effort to produce and isolate sub-clones on the basis of their chromosome number for improved agronomic characteristics.

Variation in Chromosome Number

It was assumed that variation in agronomic characteristics was primarily due to variation in chromosome number. With this notion, attempts were made to double the chromosome numbers of sugarcane clones. The first success was achieved in *S. spontaneum* by seed treatment with colchicine. However, further attempts failed with the colchicine treatment of both seeds and lateral buds. Nevertheless, it was achieved in *S. spontaneum* using cell suspensions for colchicine treatment and in commercial clones. It was later established that greater variation was observed in plants derived from a chromosomally mosaic population than in plants derived from a chromosomally stable populations. A shift towards lower chromosome numbers in some regenerated sub-clones also observed. Subsequently, chromosome counts were done routinely in the somaclones and a large degree of variation in chromosome number was consistently observed.

By now it has been well established that there is a lot of variation among somaclones with regard to chromosome number. However, there appears to be no correlation to chromosomal changes with any agronomic characteristic. This has also been compounded by the occurance of different chromosome numbers in the same plant tissue.

Disease Resistance

Shortly after culture conditions for sugarcane were established and variation was observed among regenerated plants work was initiated in Fiji to isolate sub-clones of Pindar variety of *S. officinarum* which would be resistant to Fiji disease (a virus disease transmitted by a hemipteran bug. *Ferkinsiella Vitensis* and to downy mildew (caused by *Sclerospora sacchari*, causing eyespot disease of sugarcane) has been developed and utilised for selection of disease resistant somaclones. Somaclones have also been identified with increased resistance to *Ustilago scitamenea*, causing whip smurt of sugarcane.

Now it is clear that somaclonal variation is an effective means to increase the degree of disease resistance in a population; it can be more effective if suitable *in vitro* selection pressure is applied so that a higher frequency of desirable somaclones can be obtained.

Increase in Sucrose Content

As with disease resistance, somaclones have also exhibited variation with regard to sucrose content, a very important agronomic characteristic in sugarcane. Liu and Chen first reported increased sucrose content among regenerated plants and has been corroborated by subsequent studies. To utilise the full potential of somaclonal variation, an effective selection strategy should be devised for *in vitro* selection.

Apart from these characteristics, somaclones of sugarcane have been identified with higher salt resistance increased tolerance to herbicides, and changes in isozyme band patterns.

Though widespread, somaclonal variation in sugarcane still remains largely random and unrepeatable. If the genetic causes for variation are unravelled, it may be possible to produce desirable somaclones at a higher frequency and, perhaps, at will.

Present Problems and Future Prospects

The major drawback of somaclonal variation lies in its randomness with both uneconomic and desirable variants being generated. Therefore, it is necessary to generate and screen large number of regenerated plants to identify the desirable somaclones. Further, substantial varietal and crop-specific differences occur due to genetic makeup and composition of the culture medium. Thus, specific protocols need to be developed on a species or even variety basis to ensure appropriate level of genetic variation.

The recovery of stable genotypes after selection of desirable somaclones is also difficult. As large number of genetic changes are based on point mutations or chromosonal rearrangements, most of the regenerants are composed of segregating populations. Thus, selection of somaclones for quantitative traits such as yield and morphological characters becomes difficult. However, genotypic stability has been reported for disease resistance, sucrose content, fibre percentage and fresh weight in sugarcane.

There is also paucity of *in vitro* selection systems to be utilised with somaclonal variation, so that pre-selected plants could be transferred to the field. No selection systems are available for quantitative traits such as yield but lately some *in vitro* selection systems have been developed for simply inherited and easily detectable traits.

Despite the above drawbacks, somaclonal variation holds enough promise to stimulate considerable interest for plant improvement.

Selection can be applied during culture medium and plant regeneration protocol, and at greenhouse stage to recover desirable genotypes for rapid variety improvement. Further, as a consequence of the clonal origin of plants derived from tissue cultures, stable breeding lines can be established in a shorter period of time than is possible via conventional mutagenesis.

One of the major advantages of somaclonal variation is that new classes of genetic variation arise during tissue culture regeneration which offers the breeder an opportunity to generate specific variants. It is possible to couple the induction of variation with cell selection as is being done for most of the qualitative traits where correlation exists between the cellular and whole plant response to specific chemicals used as selective agents. Three stable plant varieties have been developed using variants from cell culture. These include two cultivars of tomato. i.e. DNAP 9, a solid tomato variety and DNAP17 expressing resistance to *Fusarium* Race 2, and Bell Sweet, a yellow bell pepper variety with much less seeds per fruit as compared to the parental variety.

In addition to the premise for crop improvement involving agronomic characteristics, somaclonal variation is beneficial for increasing the production of plant derived chemicals. The key to technical and economic feasibility rests on the ability to induce and select genetically stable whole plants or cell cultures that overproduce specific chemicals, and the development of scale-up technology that exploits the biological capabilities of plant cells and provides efficient chemical production. Cell lines have been isolated with increased concentrations of secondary products such as shikonin, ajmalacine, scopalamine and rosmarinic acid, which have a number of industrial and pharmaceutical uses. Also, media must be optimised to ensure successful production of chemicals.

Results of somaclonal variation may also have an impact on gene transfer experiments using recombinant DNA. An awareness of the role of medium components in somaclonal variation should lead to the development of optimum regeneration protocols for use in transformation experiments.

Conclusion

Somaclonal variation has stimulated considerable interest amongst plant breeders as a potent tool of creating novel genetic variability in a shorter time period. It has, however, to be emphasised that much will depend on the extent to which somaclonal variation is exploited by the plant breeders. For optimal utilisation of this variability, a close relationship between tissue culturists, geneticists and plant breeders is very essential.

REFERENCES

Ahloowalia, B. S., 1976 : Chromosomal changes in para-sexually produced ryegrass, In *Current Chromosome Research* (eds. K. Jones and P.E. Brandham), Elsevier North Holland Biomedical Press, Amsterdam, Holland, pp. 115-122.

———, 1982 : Plant regeneration from callus cultures in wheat, *Crop Sci.*, 22, pp. 205-410.

Bajaj, Y. P. S., Ram, A. K., Labana, K. S. and Singh, H., 1981 : Regeneration of genetically variable plants from the anther — derived callus of *Arachis hypogea* and *Arachis villosa, Plant Sci. Lett.*, 23, pp. 35-49.

Barba, R.C. and Nickell, L.G., 1969 : Nutrition and sugar differentiation in tissue culture of sugarcane — a monocotyledon, *Planta*, 89, pp. 299-302.

Barba, R.C., Zamora, A.B., Mallion, A.K. and Linga, C.K., 1977 : Sugarcane tissue culture research, *Proc. Int. Soc. Sugarcane Technol.*, 16, pp. -843-1864.

Barbier, M. and Dulieu, H. 1983 : Early occurrence of genetic variants in protoplast cultures, *Plant Sci. Lett.*, 29, pp. 201-206.

Bayliss, M.W., 1975 : The effects of growth *in vitro* on the chromosome complement of *Daucus carota L.*, suspension cultures, *Chromosome*, 51, pp. 401-411.

———, 1980 : Chromosome variation in plant tissues in culture, In *Perspectives in Plant cell and Tissue Culture*, Int. Rev. Cytol. *Suppl. 11 B* (ed. I.K. Vasil), Academic Press, New York, USA, pp. 113-144.

Bendich, A.J. and Gauriloff, L.P., 1984 : Morphometric analysis of cucurbit mitochondria: The relationship between chondriome volume and DNA content, *Protoplasma*, 119, pp. 1-7.

Bingham, E.T., Hurley, L.V., Kaatz, D.M. and Saunders, J.W., 1975 : Breeding alfalfa which regenerates from callus in culture, *Crop Sci.*, 15, pp. 719-721.

Braun, A.C., 1959 : A demonstration of the recovery of the crown — gall tumour cell with the use of complex of single cell origin, *Proc. Natl. Acad. Sci.*, U.S.A., 45, pp. 932-938.

Brettell, R.I.S. and Ingram, D.S., 1979 : Tissue culture in the production of novel disease-resistant crop plants. *Biol. Rev.*, 54, pp. 329-345.

Brettell, R.I.S., Thomas, E. and Ingram, D.S., 1980 : Reversion of Texas male — sterile cytoplasm maize in culture to give fertile, T — Toxin resistant plants, *Theor. Appl. Genet.*, 58, pp. 55-58.

Brettell, R.I.S., Denniss, E.S., Scowcroft, W.R. and Peacock, W.J., 1986a : Molecular analysis of a somaclonal nutant of maize alchol dehydro-genase, *Mol. Gen. Genet.*, 202, pp. 235-239.

Brettell, R.I.S., Pallotta, M.A., Gustafson, J.P. and Appels, R., 1986b : Variation of the Nor loci in triticale derived from tissue culture, *Theor. Appl. Genet.*, 71, pp. 637-643.

Bright, S., Jarret, V., Nelson, R., Creisson, G., Karp, A., Franklin, J., Norbury, P., Kuct, J., Rognes, S. and Miften, B., 1983 : Modification of agronomic traits using *in Vitro* Technolgy, In *Plant Biotechnology* (eds. Mantell, S.H. and Smith, H.), Cambridge Univ. Press, Cambridge, U.K., pp. 251-265.

Brook Houzen, P.D., Cook, J.P, Cowley, C.R. and Hollingworth, M.D., 1984 : Analysis of corn mutants derived from *in vitro* culture, *Agronomy Abstracts. Amer. Soc. Agronomy*, Las Vegas, Nev., U.S.A., p. 60.

Browers, M.A. and Orton, T.J., 1982 : Transmission of gross chromosomal variability from suspension cultures into regenerated celery plants, *J. Heridity*, 73, pp. 159-162.

Burk, L.G. and Matzinger, D.F., 1976 : Variation among anther — derived doubled haploids from an inbred line of tobacco, *J. Heredity*, 67, pp. 381-384.

Calos, M.P. and Miller, J.H., 1980 : Transposable elements, *Cell*, 20, pp. 579-595.

Carlson, P.S., Smith, H.H. and Dearing, R.D., 1972 : Parasexual interspecific plant hybridization, *Proc. Natl. Acad. Sci.*, U.S.A., 69, pp. 2292-2294.

Chaleff, R.S. and Parsons, M.F., 1978 : Direct selection *in vitro* for herbicide resistant mutants of *Nicotiana, tabacum. Proc. Natl. Sci.*, U.S.A., 75, pp. 5104-5107.

Chaleff, R.S., 1986 : Selection of herbicide resistance mutants, In *Handbook of plant cell culture*, Vol. 4, (eds., Sharp, W.R. and Evans, D.A.), Macmillan Press, New York, U.S.A., pp. 133-148.

Chen, Z.Y., 1986 : A clone of sugarcane with thick stalks and a high sucrose content regenerated from tissue culture, *Scientia Agricultura Sinica*, 3, p. 90.

Cooper, D.B., Sears, R.G. Gvenzi, A.C., Lapitan, N.L., Jones, B.L. and Lookhart, G.L., 1984 : Heritable somaclonal variation in wheat gliadin proteins. Agronomy abstracts, *Amer. Soc. Agronomy*, Las Vegas, New, U.S.A., p. 62.

Crocomo, O.J. Ochoa — Alejo, N. Macheda, I.S. and Murayama, M.Y., 1984 : *In vitro* selection of sugarcane (*Saccharum spp.*) variants tolerant to herbicides. In *Plant Tissue and Cell Culture Application to Crop Improvements*, (eds., Noviak, F.J., Havel, L. and Dolezel, J.), Czechoslovak Academy of Sciences, Prague, Czechoslovakia, pp. 353-354.

Cullis, C.A. and Goldsborough, P.B., 1981 : Variation in the ribosomal RNA gene number — a model system for DNA changes in flax genotrophs, In *The Plant Genome*, (eds., Davies, D.R. and Hopwood, D.A.), The Joha Innes Charity, Norwich, U.K., pp. 91-97.

Cullis, C.A. and Cleary, W., 1986 : DNA variation in flax tissue culture, *Can J. Genet.*, 28, pp. 247-251.

Cummings, D.P., Green, C.E. and Stuthmam, D.D., 1976, Callus induction and plant regeneration in Oats, *Crop Sci.*, pp. 467-470.

Cummings, D.P., 1984 : Somaclonal variation in corn: A field selection experiment, Agronomy Abstracts, *Amer. Soc. Agronomy*, Las Vegas, Nev., U.S.A., p. 63.

De Paepe, R., Bleton, D. and Gnangbe, F., 1981 : Basis and extent of genetic variability among doubled haploid plants obtained by pollen culture in *Nicotiana sylvestris Theor. Appl. Genet.*, 59, pp. 177-184.

Deambrogio, E. and Dale, P.J., 1980 : Effect of 2, 4 - D on the frequency of regenerated plants in barley and on genetic variability between them, *Cereal Res. Commun.*, 8, pp. 417-423.

Donn, G., Tischer, E., Smith, J.A. and Goodman, H.M., 1984 : Herbicide — resistant alfalfa cells: *J. Molec. Appl. Genet.*, 2, pp. 621-635.

Dulieu, H. and Barbier, M., 1982 : High frequencies of genetic variant plants regenerated from cotyledons of tobacco. In *Variability in Plants Regenerated from Tissue Culture* (eds., Earle, E.D. and Demarly, Y.), Praeger Press, New York, U.S.A., pp. 211-229.

Earle, E.D., Gracen, V.E., Best, V.M., Batts, L.A. and Smith, M.E., 1987 : Fertile revertants from S - Type male - sterile maize grown *in vitro*. *Theor. Appl. Genet.*, 74, pp. 601-609.

Edallo, S. Zuccinali, C., Perenzin, M. and Salamini, F., 1981 : Chromosomal variation and frequency of spontaneous mutation associated with *in vitro* culture and plant generation in maize. *Mavdica*, 26, pp. 39-56.

Evans, D.A., Sharp, W.R. and Flick, C.E., 1981a : Growth and behaviour of cell cultures: Embryogenesis and organogenesis, In *Plant Tissue Culture : Methods and Applications in Agriculture* (ed., Thorpe, T.A.), Academic Press, New York, U.S.A., pp. 45-53.

———, 1981b : Plant regeneration from cell cultures, *Hortic, Rev.*, 3, pp. 214-314.

Evans, D.A. and Gamborg, O.L., 1982a : Chromosomal stability of cell suspension cultures of *Nicotiana spp. Plant Cell Rep.*, 1, pp. 104-107.

Evans, D.A., Flick, C.E., Kut, S.A. and Reed, S.M. 1982b : Comparison of *Nicotiana tabacum* and *Nicotiana nesophile* hybrids produced by ovule culture and protoplast fusion, *Theor. Appl. Genet.*, 62, pp. 193-198.

Evans, D.A. and Sharp, W.R., 1983a : Single gene mutations in tomato plants regenerated from tissue culture, *Science*, 221, pp. 949-951.

Evans, D.A., Bravo, J.E., Kut, S.A. and Flick, C.E., 1983b : Genetic behaviour of somatic hybrids in the geneus *Nicotiana: W. otophora + N. tabacum* and *N. sylvestris + N. tabacum Theor. Appl. Genet.*, 65, pp. 98-101.

Evans, D.A., Sharp, W.R. and Medina - Filho, H.P., 1984 : Somaclonal and gametoclonal variation, *Amer. J. Bot.*, 71, pp. 759-774.

Evans, D.A., and Bravo, J.E., 1986a : Phenotypic and genotypic stability of tissue cultures plants. In *Tissue Culture as a Plant Production System for Horticultural Crops* (eds., Zimmerman, R.H., Griesbach, R.J., Hammerschlag, F.A. and Lawson, R.J.), Martinus Nijhoff Publishers, Dordrecht, Holland, pp. 74-94.

Evans, D.A., and Sharp, W.R., 1986b : Somaclonal and gametoclonal variation, in *Handbook of Plant Cell Culture*, Vol. IV (eds., Sharp., W.R., and Evans, D.A.), Macmillan Publishing Company, New York, U.S.A., pp. 97-132.

———, 1986c : Application of somaclonal variation, *Biotechnology*, 4, pp. 528-532.

———, 1988 : Somaclonal variation and its application in plant breeding, Newlett., Int. Asso. Pl, *Tissue Cult.*, 54, pp. 2-10.

Evans, D. A., 1989 : Somaclonal Variation — Genetic basis and breeding applications, *Trends in Genet.*, 5, pp. 46-50.

Fitch, M.M. and Moore, P.H., 1981 : The selection for salt resistance in sugarcane (*Saccharum Sp. hybrids*) tissue cultures (Abstract), *Plant Physiol.*, 64 (Suppl.), p. 26.

Fukui, K., 1983 : Sequential occurrence of mutations in the growing rice callus, *Theor. Appl. Genet.*, 65, pp. 225-230.

Gamborg, O.L., Shyluck, J.P., Brar, D.S. and Constab. J., 1977 : Morphogenesis and plant regeneration from callus of immature embryos of sorghum, *Plant Sci. Lett.*, 10, pp. 67-74.

Gengenbach, B.G., Green, C.E. and Donoran, C.M., 1977 : Inheritance of selected pathotoxin resistance in maize plants regenerated from cell cultures, *Proc. Natl. Acad. Sci.,* U.S.A., 74, pp. 5113-5117.

Gengenbach, B.G. and Umbeck, P., 1982 : Characteristics of T - cytoplasm Revertants from tissue culture, *Maize Genet. Coop. Newslett.,* 56, pp. 140-142.

Grassel, C.O., 1966 : Comments, Int. Sec. Sugarcane Technol. Sugarcane breeders' *Newslett.,* 18, pp. 29-30.

Groose, R.W., and Bingham, E.T., 1984 : Variation in plants regenerated from tissue culture of tetraploid alfalfa heterozygous for several traits, *Crop Sci.,* 24, pp. 655-658.

Green, C.E. and Phillips, R.L., 1975 : Plant regeneration from tissue cultures of maize, *Crop Sci.,* 15, pp. 417-421.

Green, C.E., 1977 : Prospects of crop improvement in the field of cell culture, *Hortic. Sci.,* 12, pp. 7-10.

Guha, S. and Maheshwari, S.C., 1964 : *In vitro* production of embryos from anthers of *Datura, Nature,* 204, p.497.

Henke, R.R., Mansur, M.A. and Constantin, M.J., 1978 : Organogenesis and Plantlet formation from organ - and seedling derived calli of rice (*Oryza, sativa*), *Physiol. Plant.,* 44, pp. 11-14.

Heinz, D.J. and Mee, G.W.P., 1969a : Plant differentiation from callus tissue of *Saccharum* species, *Crop Sci.,* 9, p. 346-348.

Heinz, D.J., Mee, G.W.P. and Nickell, L.G., 1969b : Chromosome numbers of some *Saccharum species* hybrids and their cell suspension cultures, *Amer. J. Bot.,* 56, pp. 450-456.

Heinz, D.J. and Nickell, L.G., 1969 : Separating and recovering genetic variability of a sugarcane clone sexually by use of cell and tissue culture techniques (Abstract), *XI Int. Bot. Cong.,* p. 88.

Heinz, D.J. and Mee, G.W.P., 1970 : Colchicine - induced polyploids from cell suspension cultures of sugarcane, *Crop Sci.,* 10, pp. 696-699.

Heinz, D.J. and Mee, G.W.P., 1971 : Morphologic, cytogenetic and enzymatic variation in *Saccharum* species hybrid clones derived from callus tissue, *Amer. J. Bot.,* 58, pp. 257-262.

Heinz, D.J., 1973 : Sugarcane improvement through induced mutations using vegetative propagules and cell culture techniques, in *Induced Mutations in Vegetatively Propagated Plants,* IAEA, Vienna, Austria, STI/PUB/399, pp. 53-59.

Heinz, D.J., Krishnamurti, M., Nickell, L.G. and Maretzki, A., 1977 : Cell, tissue and organ culture in sugarcane improvement, in *Applied and Fundamental Aspects of Plant Cell, Tissue and Organ Culture,* (eds., Reinert, J. and Bajaj, Y.P.S.), Springer — Verlag, Berlin, FRG, pp. 3-17.

Irvine, J.E., Fitch, M. and Moore, P.H., 1983 : The induction of callus in sugarcane tissue cultures by selected chemicals, *Plant Cell Tissue Organ Cult.,* pp. 141-150.

Irvine, J.E., 1984 : The frequency of market changes in sugarcane plants regenerated from callus culture, *Plant Cell Tissue Organ Cult.,* 3, pp. 201-209.

Jagathesan, D., 1966 : Comments, Int. Soc. Sugarcane Technol. Sugarcane Breeders' *Newslett.,* 18, pp. 30-31.

Kasperbauer, M.J., Buchnes, R.C. and Bush, L.P., 1979 : Tissue culture of annual ryegrass X tall fesue F hybrids: callus establishment and plant regeneration, *Crop Sci.,* 19, pp. 457-460.

Kemble, R.J., Brettell, R.I.S. and Flavell, R.B., 1982 : Mitochondrial DNA analysis of fertile and sterile maize plants from tissue culture with the Texas male sterile cytoplasm, *Theor. Appl. Genet.*, 62, pp. 213-217.

Kemble, R.J. and Shepard, J.F., 1984 : Cytoplasmic DNA variation in a potato protoclonal population, *Theor. Appl. Genet.*, 69, pp. 211-216.

Kemble, R.J., Yarrow, S.A., Wu, S.C. and Barsby, T.L., 1988 : Absence of Mitochondrial and chloroplast DNA recombinations in *Brassica napus* plants regenerated from protoplasts, protoplast fusions and anther culture, *Theor. Appl. Genet.*, 75, pp. 875-881.

Krishnamurti, M. and Tlaskal., J., 1974 : Fiji disease resistant *Saccharum officinarum* var. Pindar sub-clones from tissue culture, *Proc. Int. Soc. Sugarcane Technol.*, 15, pp. 130-137.

Krishnamurti, M., 1977 : Sugarcane improvement through tissue culture, *Proc. Int. Soc. Sugarcane Technol.*, 16, pp. 23-28.

———, 1981 : Sugarcane tissue culture — an example for crop improvement, *Proc. Int. Tissue Culture Workshop*, pp. 12-22.

———, 1982 : Disease resistance in sugarcane developed through tissue culture, in *Plant Tissue Culture* (ed., Fujiwara, A.) Jpn. Assoc. Plant Tissue Culture, Tokyo Japan, pp. 769-770.

Landsmann, J. and Uhrig, H., 1985 : Somaclonal variation in *Solamum tuberosum* detected at the molecular level, *Theor. App. Genet.*, 71, pp. 500-505.

Larkin, P.J. and Scowcroft, W.R., 1981 : Somaclonal variation — a novel source of variability from cell cultures for crop improvement, *Theor. Appl. Genet.*, 60, pp. 197-214.

———, 1982 : Somaclonal variation and crop improvement, In *Genetic Engineering of Plants* (eds., Hollaender, A., Meredith, C.P. and Kosuge, T.) Plenum Press, New York, U.S.A., pp. 289-314.

Larkin, P.J. and Scowcroft, W.R., 1983 : Somaclonal variation and eyespot toxin tolerance in sugarcane, *Plant Cell Tissue Organ Cult.*, 2, pp. 111-121.

Larkin, P.J., Ryan, S.A., Brettell, R.I.S. and Scowcraft, W.R., 1984 : Heritable somaclonal variation in wheat, *Theor. App. Genet.*, 67, pp. 443-455.

Lee. T.S.G. and Bacchi, O.O.S., 1983 : Newslett., Int. Assoc. *Pl. Tissue Cult.*, 40, p. 9.

Liu, M.C. and Chen, W.H., 1976 : Tissue and cell culture as aids to sugarcane breeding. I. Creation of genetic variation through callus culture, *Euphytica*, 25, pp. 393-403.

———, 1978 : Tissue and cell culture as aids to sugarcane breeding. II. Performance and Yield potential of callus derived lines, *Euphytica*, 27, pp. 273-282.

Liu, M.C., 1981 : *In vitro* methods to sugarcane improvement, In *Plant Tissue Culture, Methods and Applications in Agriculture* (ed. Thorpe, T.A.), Academic Press, London, U.K., pp. 299-323.

Liu, M.C. and Yeh, H.S., 1983 : Regeneration of Nacl - tolerant sugarcane plants from callus reinitiated from Nacl preselected plantlets, *In vitro*, 19, p. 250.

Liu, M.C., Yeh, H.S. and Chen, W.H., 1984 : Tissue and cell culture as aids to sugarcane breeding. IV, A high sucrose and vigorously growing calliclone 71 — 4829, *Taiwan Sugar*, 31, pp. 77-83.

Lorz, H. and Scowcroft, W.R., 1983 : Variability among plants and their progeny regenerated from protoplasts of Su/su heterozygotes of *Nicotiana tabacum. Theor. Appl. Genet.*, 66, pp. 67-75.

Lu., W.Z. and Wu. G.N., 1986 : Current development of biotechnology and its application of plant breeding in China, *J. Agric. Sci.*, China, *2 (Suppl)*, pp. 1-9.

Lupi, M.C., Bennici, A., Baroncelli, S., Gennai, D. and D. Amato, F., 1981 : *In vitro* regeneration of durum wheat plants. II. Diplomatic selection in aneusomatic plants, *Z. Planzenzuecht.*, 87, pp. 167-171.

Maddock, S.E., Lancaster, V.A., Risiott, R. and Franklin, J., 1983 : Plant regeneration from cultured inflorescence of 25 cultivars of wheat (*Triticum aestivum) J. Exp. Bot.,* 34, pp. 915-926.

McCoy, T.J., Phillips, R.L. and Rines, H.W., 1982, Cytogenetic analysis of plants regenerated from oat (*Avena sativa*) tissue cultures: High frequency of partial chromosome loss, *Can J. Genet. Cytol.*, 24, pp. 37-50.

Mel, G.W.P., Nickell, L.G. and Heinz, D.J., 1969 : Chemical Mutagens — their effects on cells in suspension culture, *Ann. Rep. Expt. Stat.,* Hawaii Sugar Planters Assoc., 7/8.

Meins, F. Jr., 1983 : Heritable variation in plant cell culture, *Ann. Rev. Plant Physiol.*, 34, pp. 327-340.

Miller, C.O., Skoog, F., Von Saltza, M.H. and Strong, F.M., 1955 : Kinetin, A cell division factor from deoxyribonucleic acid, *J. Am. Chem. Soc.*, 77, p. 1392.

Mix, G., Wilson, H.M. and Forough - Wehr, B., 1973 : The cytological status of plants of *Hordeum vulgare* L. regenerated from microspare callus. Z. Pflanzenziict., 80, pp. 89-99.

Morel, G., 1960 : Producing virus - free cymbidium, *Amer. Orchid Soc. Bull.*, 29, pp. 495-497.

Morrison, R.A., Whitaker, R.J. and Evans, D.A., 1988 : Somaclonal Variation: Its Genetic Basis and Prospects for Crop Improvement, in *Recent Advances in Phytochemistry. Opportunities for Phytochemists in Blant Biotechnology* (ed. Conn. E.E.), Plenum Press, New York, U.S.A., pp. 1-18.

Nyurm, W.H., 1953 : Cultural conditions favouring the isolation and growth of single cells from higher plants *in vitro*, Ph.D. Thesis, University of Wisconsin, U.S.A.

Muir, W.H., Hildebrandt, A.C. and Riker, A.J., 1954 : Plant tissue cultures produced from single isolated plant cells, *Science*, 119, pp. 877-878.

Murashige, T. and Skoog, F., 1962 : A revised medium for rapid growth and bioassays with tobacco tissue culture, *Physiol. Plantarum*, 15, pp. 473-497.

Nadar, H.M. and Heinz, D.J., 1977 : Root and shoot development from sugarcane callus tissue, *Crop. Sci.*, 17, pp. 814-816.

Nickell, L.G., 1964 : Tissue and cell cultures of sugarcane. Another research tool, *Hawaiian Planters' Record*, 57, pp. 223-229.

———, 1967 : Tissue and cell cultures for sugarcane research, *Proc. Int. Soc. Sugarcane Technol.*, 12, pp. 887-892.

Nickell, L.G. and Maretzki, A., 1969 : Growth of suspension cultures of sugarcane cells in chemically defined media, *Physiol. Plantarum*, 22, pp. 117-125.

Novak, F.J., 1980 : Phenotype and cytological status of plants regenerated from callus cultures of *Allium sativum L.Z. Planzenzuecht*, 84, pp. 250-260.

Ogihara, Y., 1981 : Tissue culture in *Haworthia*, Part 4 : Genetic Characterisation of plants regenerated from callus, *Theor. Appl. Genet.*, 60, pp. 353-363.

Oniuma, T. and Yoshida, T., 1974 : Genetic variation among doubled haploid lines of barley and tobacco varieties, *Jpn. J. Breeding*, 24, pp. 211-216.

Oono, K., 1981 : *In vitro* methods applied to rice, In *Plant Tissue Culture* (ed., Thorpe, T.A.), Academic Press, New York, U.S.A., pp. 273-298.

Orshinsky, B.R. and Tomes, D.T., 1984 : Comparison of plants derived from cuttings, node cultures, and ethylmethanesulfnate treated node cultures of bird's foot trefoil (*Lotus corniculatus*), *Can. J. Bot.*, 62, pp. 1501-1504.

Orton, T.J., 1980 : Chromosomal variability in tissue cultures and regenerated plants of *Hordeum. Theor. Appl. Genet.*, 56, pp. 101-112.

Phillips, A.Y., 1967 : Colchicine treatment of noble canes. 34th Ann. Rep. *BWI Cane Breeding Stat.*, pp. 12-14.

Prat, D., 1983 : Genetic variability induced in Nicotiana sylvestris by protoplast culture, *Theor. Appl. Genet.*, 64, pp. 223-230.

Reisch, B. and Bingham, E.T., 1981 : Plants from ethionine — resistant alfalfa tissue cultures: Variation in growth and morphological characteristics, *Crop Sci.*, 21, pp. 783-788.

Sacristan, M.D. and Melchers, G., 1969 : The caryological analysis of plants regenerated from tumourous and other callus cultures of tobacco, *Mol. Gen. Genet.*, 105, pp. 317-333.

Sacristan, M.D., 1982 : Resistance responses to *Phoma lingam* of plants regenerated from selected cell and embryogenic cultures of haploid *Brassica napus*, *Theor. Appl. Genet.*, 61, pp. 193-200.

Schimke, R.T., Alt, F.W., Kellens, R.E. and Bertino, J.R., 1977 : Amplification of dihydrofolate reductase gene in methotrexate - resistant variants of cultured murine cells, Cold Spring harbour Symp. *Quant. Biol.*, 42, pp. 649-657.

Scowcroft, W.R. and Larkin, P.J., 1985 : Somaclonal Variation, cell selection and genotype improvement, In *Comprehensive Biotechnology*, Vol. 4, (eds., Robinson, C.W. and Howell, J.A.), Pergamon Press, London, U.K., pp. 153-168.

Sears, R.G. and Deckard, E.L., 1982 : Tissue culture variability in wheat. Callus induction and plant regeneration, *Crop Sci.*, 22, pp. 545-550.

Secor, G.R. and Shepard, J.F., 1981 : Variability of protoplast - derived potato clones, *Crop Sci.*, 21, pp. 102-105.

Sharma, P., 1988 : Preliminary studies on tissue culture, evaluation of somaclones, and combining ability in sugarcane. (*Saccharum spp.* complex). M.Sc. Thesis, G.B.P. U.A & T., Pantnagar, India.

Shepard, J.F., Bidney, D. and Shahin, E., 1980 : Potato protoplasts in crop improvement, *Science*, 208, pp. 17-24.

Shepard, J.F., 1981 : Protoplasts as sources of disease resistance in plants, *Ann. Rev. Phytopathol.*, 19, pp. 145-166.

Singh, B.D., Kao, K.N. and Miller, R.A., 1975 : Karyotypic changes and selection pressure in *Haplopappurs gracilis* suspension cultures, *Can J. Genet. Cytol.*, 17, pp. 106-109.

Sisodia, N.S., 1965 : Techniques of colchicine application for chromosome doubling, 32nd *Ann. Rep. BWI cane Breeding Stat.*, pp. 30-31.

———, 1966 : Colchicine for chromosome doubling, 33rd *Ann. Rep. BWI Cane Breeding Stat.*, pp. 40-44.

Skirvin, R.M. and Janick, J., 1976 : Tissue culture - induced variation in scented *Pelargonium* spp, *J. Amer. Soc. Hortic. Sci.*, 101, pp. 281-290.

Smith, S.M. and Street, H.E., 1974 : The decline of embryogenic potential as callus and suspension cultures of carrot (*Daucus carota* L.) are serially subcultured, *Ann. Bot.*, 38, pp. 233-241.

Sreenivasan, J.V. and Jalaja, N.C., 1983 : Sugarcane varietal improvement through tissue culture, In *Plant cell culture in crop improvement* (eds., Sen, S.K. and Giles, K.L.), Plenum Press, New York, U.S.A., pp. 371-376.

Sun, Z.X., Zhao, C.Z., Zheng, K.L., Qi, X.F. and Fu, Y.P., 1983 : Somaclonal genetics of rice, *Oryza sativa L. Theor. Appl. Genet.* 67, pp. 67-73.

Sutter, E. and Langhans, R.W., 1981 : Abnormalities in *Chrysanthemum* regenerated from long-term cultures, *Artn. Bot.,* 48, pp. 559-568.

Swartz, H.J., Galletta, G.J. and Zimmerman, R.H., 1981 : Field performance and phenotypic stability of tissue — culture propagated thornless blackberries, *J. Amer. Soc. Hortic. Sci.,* 106, pp. 667-673.

———, 1983 : Resistant tobacco plants from protoplast-derived calluses selected for their resistance to *Pseudomonas* and *Alternaria* toxins, *Theor. Appl. Genet.*, 66, pp. 209-215.

Tlaskal, J., Hutchinson, P.B. and Roach, B.T., 1970 : Variation in chromosome numbers within tissues of sugarcane clones, Int. Soc. Sugarcane Technol, *Newlett.*, 25, pp. 20-24.

Tlaskal, J. and Hutchinson, P.B., 1974 : An objective method for counting chromosomes in sugarcane roots and meristems, *Proc. Int. Soc. Sugarcane Technol.*, 15, pp. 168-175.

Thomas, E., Bright, S.W.J., Franklin, J., Lancaster, V.A., Miflin, B.J. and Gibson, R., 1982 : Variation amongst protoplast — derived potato plants (*Solanum tuberosum* cv: Maris Bard), *Theor. Appl. Genet.*, 62, pp. 65-68.

Tisserat, B., Esan, E.B. and Murashige, T., 1980 : Semantic Embryogenesis in Angiosperms, *Hortic. Rev.*, 1, p. 78.

Urata, U. and Long, P.P., 1968 : A method for propagating grasses by tissue culture, Agronomy Abstracts, *Amer. Soc.* , Las Vegas, Nev., U.S.A., p. 26.

Vajrabhaya, T., 1977 : Variations in clonal propagation, In *Orchid Biology: Reviens and Perspectives*, Vol. I (ed., Arolittle, J.), Columbia Univ. Press, New York, U.S.A., pp. 179-201.

Van Harten, H.M., Bouter, H. and Broerties, C., 1981 : *In vitro* adventitious bud techniques for vegetative propagation and mutation breeding of potato (*Solanum tuberosum* L.), II. Significance for mutation breeding, *Euphytica*, 30 : pp. 1-8.

Vasil, V. and Hildebrandt, A.C., 1965 : Differentiation of tobacco plants from single, isolated cells in microculture, *Science*, 150, pp. 889-892.

Wenzel, G., Schieder, O., Przewozny, T., Sopory, S.K. and Melchers, G., 1979 : Comparison of single cell culture derived *Solanum tuberosum* L. plants and a model for their application in breeding programs, *Theor. Appl. Genet.*, 55, pp. 59-55.

Whitaker, R.J. and Hashimoto, T., 1986 : Production of secondary metabolites, in *Handbook of Plant Cell Culture*, Vol. 4, (eds., Evans, D.A., Sharp, W.R. and Ammirato, P.V.), Macmillan Press, New York, U.S.A., pp. 264-286.

White, P.R., 1939 : Potentially unlimited growth of excised plant callus in an artificial medium, *Amer, J. Bot.,* 26, pp. 59-64.

Wise, R.P., Pring, D.R. and Gengenback, B.G., 1987 : Mutation to male — sterility and toxin insensitivity in Texas (T) - cytoplasm maize is associated with a frameshift in a mitochondrial open reading frame, *Proc. Natl. Acad. Sci.*, U.S.A., 8, pp. 2858-2862.

Zehr, B.E., Williams, M.E., Duncan, D.R. and Widholm, J.M., 1987 : Somaclonal variation in the progeny of plants regenerated from callus cultures of seven inbred lines of maize, *Can. J. Bot.,* 65, pp. 491-499.

Zenkteler, M. and Melchers, G. 1978 : *In vitro* hybridization by sexual methods and by fusion of somatic protoplasts, *Theor. Appl. Genetl.,* 52, pp. 81-90.

14

DETERMINANTS OF PULSE CULTIVATION IN HARYANA

USHA TUTEJA

PULSES HAVE been the major source of protein in India. However, the review of pulse availability presents a very gloomy picture. The net per capita availability of pulses has declined from 61.6 gms per day in 1965 to 40 gms in 1989. The decline in availability of pulses has been partly due to increasing population and partly due to stagnation in the production of pulses. The compound growth rates of area and production of pulses in India between trieunium ending 1971-72 and 1988-89 are .05 and .16 per cent per annum respectively. The yield aspect is not encouraging too as per hectare productivity of pulses rose marginally i.e. at the rate of .10 per cent per annum during the last two decades. Particularly situation is grave in the Northern States where green revolution has spread widely. As a result of stagnation in the acreage and insignificant increase in the productivity, pulse production of 12.12 million tonnes during 1986-89 is slightly more than the 1969-72 output of 11.82 million tonnes. A peculiar situation has arisen when rising prices of pulses are not able to enthuse farmers to increase pulse production. Historically, pulse prices have risen at a higher rate than that of usual prices as the index numbers of harvest prices of wheat, rice and pulses were 282, 364 and 670 in 1988-89 assuming 1970-71 as the base year. It is obvious that shifts in relative prices could not provide impetus for a breakthrough in the country because they were not accompanied by inputs that contribute to an increase in the relative profitability. Further, the expectation that the hike in minimum support price for pulses would boost production may be belied if past trends are any indication. Since 1979-80, the support price for pulses has risen by 60 per cent, however, the production has hovered around 12 million tonnes. Thus, the Government policy of minimum support price as a safety net and the strategy for pulse crops development through the implementation of Pulse Production Programme has failed to enhance production in the country. This paper is an attempt to find out the weak linkages in the chain.

The scope of the paper is confined to major Rabi and Kharif pulses in Haryana. Two districts (Jind and Mahendergarh) have been selected purposively for in-depth study from the State on the basis of the maximum shift and the minimum shift in the acreage of pulse crops during the last two decades. The primary and secondary sources (1964-65 to 1983-84) of data are used to analyse the important issues related to pulse growth in the selected area in the light of the questions raised and conclusions reached in the various studies. A survey of 100 households growing pulses was carried out during 1984 in Jind and Mahendergarh districts of Haryana.

This paper has been divided into five sections for the convenience of the analysis. These are: (i) Acreage under Pulse Crops in Haryana, (ii) Production and productivity of pulse crops, (iii) Comparative profitability of pulse crops vis-a-vis competing crops, (iv) Constraints to Pulse growth in the State, and (v) Summary of conclusions and policy implications.

I

Acreage Under Pulse Crops in Haryana

It is well-known that physical, social and economic factors determine the crop pattern in a region. But, the physical environments having been given economic considerations ultimately dictate as to where, when and how much cereals, pulses, fibres, forages and other crops should be grown. This is due to variations in yield and profitability of different crops. Pulse crops are given relatively low preference despite their nitrogen fixing capacity and other advantages. An attempt has, therefore, been made in this section to highlight the factors affecting acreage under pulse crops in the selected region.

Table 14.1 gives proportion of gross cropped area under pulses and other important crops in Haryana during 1964-65 to 1983-84. The data are analysed for three points of time i.e. triennium ending 1964-65, 1974-75 and 1983-84.

TABLE 14.1: Percentage of GCA, area irrigated and area under high yielding variety seeds of principal crops in Haryana during 1964-65 to 1983-84

	Triennium ending								Compound growth rates of area		
	1964-65		1974-75			1983-84			1964-65 to 1974-75	1974-75 to 1983-84	1964-65 to 1983-84
Crops	GCA	Irrigated	GCA	Irrigated	HYV	GCA	Irrigated	HYV			
(1)	(2)	(3)	(4)	(5)	(6)	(7)	(8)	(9)	(10)	(11)	(12)
Pulses:											
Gram	28.82	20.39	18.69	23.20	—	11.39	26.25	—	−3.68	−3.27	−3.67
Mash	0.20	@	0.24	@	—	0.13	—	—	3.73	−4.16	−1.09
Moong	0.40	@	0.14	@	—	0.13	—	—	−10.81	0.08	−4.57
Massar	0.74	@	0.72	@	—	0.34	—	—	1.31	−5.29	−2.82
Other Pulses	—	—	0.56	N.A.	—	0.67	71.24	—	—	2.14	—
Total Pulses	30.16	20.14	20.36	22.91	—	12.66	27.36	—	−3.20	−3.15	−3.37

(*Contd.*)

TABLE 14.1 (*Contd.*)

(1)	(2)	(3)	(4)	(5)	(6)	(7)	(8)	(9)	(10)	(11)	(12)
Other Important Crops:											
Rice	4.04	58.92	5.62	90.25	31.57	9.86	90.80	77.60	5.88	5.59	6.01
Jowar	6.34	18.97	3.64	27.53	—	2.67	26.32	—	−5.23	−1.83	−3.37
Bajra	17.29	3.03	17.43	9.18	24.22	14.76	14.42	61.90	1.65	−0.59	0.31
Maize	2.03	9.68	2.16	28.52	12.48	0.95	27.78	46.30	2.41	−5.92	−2.82
Wheat	15.80	53.94	24.49	83.50	78.70	31.52	94.03	93.40	7.32	2.90	4.90
Barley	1.97	36.67	2.77	43.18	—	1.33	64.81	—	6.05	−5.14	−0.90
Mustard	3.37	@	4.07	@	—	3.44	N.A.	—	4.01	0.59	1.27
Sugarcane	3.15	81.25	2.42	84.46	—	2.33	91.18	—	−1.72	0.49	−0.43
Cotton	3.82	95.43	4.97	98.56	—	7.13	96.47	—	4.93	3.83	4.53
GCA	100.00	31.14	100.00	47.74	25.34	100.00	63.20	46.68	1.02	0.80	1.14

Notes: (i) The above mentioned compound growth rates are computed by using the following formula: $Y_t = Y_o(1 + r/100)^t$
where Y_o = the magnitude of a variable in the base period;
Y_t = the magnitude of a variable in the final period;
t = time lag in year for the growth of Y_o into Y_t : r = the compound rate of growth.
(ii) @ = Less than 50 Hectares.
(iii) N.A. = Not available.

Source: Statistical Abstracts of Haryana, 1966-67, 1973-74, 1974-75, 1975-76 and 1983-84, 1984-85, 1985-86, Government of Haryana, Chandigarh.

It is observed that pulse crops occupied an important place in the cropping pattern of Haryana as 30.16 per cent of gross cropped area was under pulses at first point of time. Gram was the most important pulse crop with 28.82 per cent of GCA. However, only 20.39 per cent of area under gram and 20.14 per cent of area under total pulses was irrigated. Mash, Moong and Massar were rarely grown. Besides pulses, bajra (17.29%) of GCA, wheat (15.00%) and jowar (6.34%) were significant. The corresponding irrigated area for these crops was 3.03 per cent, 53.94 per cent and 18.97 per cent respectively.

With the increase in irrigation facilities and with the introduction of high yielding variety cereals, the proportion of GCA under pulses was reduced to 20.36 per cent in Haryana at second point of time. The share of minor pulses remained almost same whereas wheat gained the maximum in terms of acreage.

The process of decline in pulse acreage did not stop in early seventies as it reduced further to only 12.66% of GCA at second point of time. However, irrigation facilities improved and 27.36 per cent of area under pulses became irrigated. The information about the adoption of improved technology in the case of pulses, however, is not available.

It is further observed that pulse acreage except mash has recorded negative compound growth rate (–3.37 % per annum) in Haryana during 1964-65 to 1983-84. The rate of decline in terms of area was the maximum for moong, i.e. –4.57 per cent per annum. During the period (1964-65 to 1974-75) mash and massur have shown appreciable growth. However, acreage of pulses as a whole has indicated a negative growth of 3.20 per cent per annum. Particularly, maximum decline took place in the case of moong with regard to its area. During the second period (1974-75 to 1983-84), rate of decline in the acreage of pulse crops became 3.15 per cent per annum. "Other pulses" have shown creditable growth but area under gram declined at the rate of 3.27 per cent per annum.

In comparison to acreage under pulse crops in Haryana, area under rice, wheat, mustard and cotton has registered appreciable increase of 6.01 per cent, 4.90 per cent, 1.27 per cent and 4.53 per cent per annum during 1964-65 to 1983-84, respectively. During the first period, except sugarcane and jowar, all other crops gained and the maximum increase was observed in the case of wheat. However, acreage under rice, cotton and wheat has risen at the rate of 5.59 per cent, 3.83 per cent and 2.90 per cent per annum respectively during the second period.

Tables 14.2 and 14.3 provide the details of acreage, irrigated area and area under high yielding varieties for different crops in Jind and Mahendergarh during 1972-73 to 1983-84. Since these districts were reorganised during 1972-73, comparable data are not available for the first point of time.

TABLE 14.2: **Percentage of GCA, area irrigated and area under high yielding varieties of principal crops in Jind during 1972-73 to 1983-84**

Crops	Triennium ending						
	1974-75			1983-84			Compound Growth Rates
	GCA	Irrigated	H.Y.V.	GCA	Irrigated	H.Y.V.	
(1)	(2)	(3)	(4)	(5)	(6)	(7)	(8)
Pulses:							
Gram	19.98	48.32	–	11.81	56.95	N.A.	–3.84
Mash	0.12	–	–	0.02	N.A.	N.A.	–12.91
Moong	0.06	–	–	0.02	N.A.	N.A.	–7.94
Massar	1.06	–	–	0.87	N.A.	N.A.	–1.53
Other pulses	0.09	–	–	.10	40.00	N.A.	10.46
Total pulses	21.25	45.32	–	12.82	59.55	N.A.	–3.90
Other Important Crops:							
Rice	2.94	97.74	15.04	8.46	100.00	85.80	9.37
Jowar	6.20	50.31	–	2.26	61.64	@	–7.94
Bajra	19.70	34.52	26.19	18.13	43.28	62.60	0.50
Wheat	23.36	93.02	62.33	33.41	97.94	92.30	3.20
Barley	2.70	40.00	–	1.13	63.83	–	–6.96
Mustard	3.03	–	–	1.92	@	–	1.50
Sugarcane	3.86	98.04	–	3.12	97.56	–	–1.63
Cotton	4.54	100.00	–	6.37	96.49	–	3.03
GCA	100.00	66.02	19.70	100.00	68.99	41.12	0.50

@ = Negligible
Source: Statistical Abstracts of Haryana, 1973-74, 1974-75, 1975-76, and 1983-84, 1984-85, 1985-86, Government of Haryana, Chandigarh.

According to Table 14.3, 23.21 per cent of GGA was under pulses in Mahendergarh at second point of time and 3.17 per cent of the area under pulses was irrigated. Again gram was the main pulse crop. Among other crops, bajra (38.63%), wheat (10.27%), barley (8.71%)

and mustard (8.00%) were significant. Particularly, 95.57 per cent of the cultivated area under wheat was irrigated and 88.58 per cent was under the improved technology.

TABLE 14.3: **Percentage of GCA, area irrigated and area under high yielding varieties of principal crops in Mahendergarh during 1972-73 to 1983-84.**

Crops	Triennium ending						
	1974-75			1983-84			Compound Growth Rates
	GCA	Irrigated	H.Y.V.	GCA	Irrigated	H.Y.V.	
(1)	(2)	(3)	(4)	(5)	(6)	(7)	(8)
Pulses:							
Gram	22.83	3.24	—	15.54	9.41	—	−2.56
Mash	—	—	—	—	—	—	—
Moong	0.22	—	—	0.06	—	—	−10.05
Massar	—	—	—	—	—	—	—
Other pulses	0.16	—	—	0.03	—	—	−11.84
Total pluses	23.21	3.17	—	15.63	9.36	—	−2.69
Other Important Crops:							
Rice	@	@	—	—	—	—	—
Jowar	1.37	@	—	1.08	@	@	−1.43
Bajra	38.63	0.69	12.49	39.71	0.67	60.10	0.41
Wheat	10.27	95.57	88.58	18.78	99.20	89.80	5.77
Barley	8.71	73.25	—	4.39	98.48	—	−5.00
Mustard	8.00	—	—	8.56	@	—	1.17
Sugarcane	0.04	@	—	—	—	—	—
Cotton	@	—	—	0.03	—	—	—
GCA	100.00	20.17	16.14	100.00	35.68	17.57	0.85

@ = Negligible

Source: Statistical Abstracts of Haryana, 1973-74, 1974-75, 1975-76 and 1983-84, 1984-85, 1985-86, Government of Haryana, Chandigarh.

Area under pulses came down to 15.63 per cent in Mahendergarh during this period (Table 14.3). The maximum decrease is observed fon gram, while minor pulse crops maintained the same position. The proportion of irrigated area also improved. Wheat gained almost 5 per cent in terms of cultivated area. The rate of decline in area during this period was the highest for 'other pulses' (4.84% per annum). Among other crops, the highest decline in area was observed for barley. The area under wheat increased at the rate of 5.77 per cent per annum which is the maximum among important crops.

So far we have analysed the acreage under different crops during 1964-65 to 1984-85 in the region and in the selected districts. Now, we would like to analyse acreage under pulses on the sampled farms.

It is noticed (Tables 14.4 and 14.5) that the proportion of cultivated area under pulses on the sampled households was higher than the entire state as well as the Districts. The acreage under pulses was as high as 34.85 per cent of GCA in Jind and 23.27 per cent of GCA in Mahendergarh. Out of the total area under pulses, 37.98 per cent in Jind and 26.85 per cent in Mahendergarh was irrigated. Particularly, 15.60 per cent of the pulse area in Jind and 30.41 per cent in Mahendergarh was under the improved variety seeds. Further, gram followed by moong was the dominant pulse crop in both the Districts. Minor pulses comprise mash, massar and moth.

Among other crops, wheat and bajra were very important. Particularly, the area under wheat in Jind was fully irrigated and completely under the improved technology. Bajra has also shown good results with 61.96 per cent of irrigated area and 84.19 per cent of the area under improved technology. On the contrary, the performance of wheat and bajra in terms of irrigated area and the adoption of improved technology had been relatively poor in Mahendergarh.

Class variations were evident in the acreage, irrigated area and in the adoption of improved technology for pulses. In Jind, marginal class variations were found in the proportion of cultivated area under pulses. However, very high variations were observed in irrigated area and the area under HYV seeds. For example, 60.23 per cent of the area under pulses on the small farms was irrigated against 30.29 per cent on the big farms. But, vice-versa was true for the area under improved technology as 30.29 per cent of the cultivated area under pulses on the big farms was under the improved variety seeds against 6.32 per cent on the small farms. Apart from total pulses, the most important i.e. gram, has also indicated marginal class variations in the cultivated area and large variations in the irrigated area. Nearly 58 per cent of the area under gram was irrigated on the small farms against 30.09 per cent on the big farms. The pattern of cultivated area under moong and massar is almost the same among different classes. However, class variations are equally glaring in the irrigated area and in the adoption of improved technology.

TABLE 14.4: Percentage of GCA, area irrigated and area under high yielding varieties of principal crops on the sampled farms in Jind during 1983-84

Farmers/Category	Pulses					Other Important Crops					
	Gram	Moong	Massar	Mash	Total Pulses	Wheat	Bajra	Jowar	Cotton	Mustard	Gowar
Small:											
Percentage of GCA	31.12	4.17	0.95	—	36.24	13.06	30.17	4.75	3.33	0.71	1.66
Percentage irrigated	58.02	67.65	100.00	—	60.23	100.00	61.96	100.00	100.00	100.00	85.71
Percentage under HYV	—	54.90	—	—	6.32	100.00	84.19	—	28.56	—	—
Medium:											
Percentage of GCA	28.29	4.62	0.93	—	33.84	13.68	26.19	3.38	4.28	4.78	4.96
Percentage of Irrigated	46.48	51.67	100.00	—	48.66	100.00	64.92	82.57	100.00	44.44	66.85
Percentage under HYV	—	76.99	—	—	10.50	100.00	86.82	—	100.00	—	—
Big:											
Percentage of GCA	29.78	4.49	0.53	0.13	34.93	17.05	27.39	1.58	6.46	1.41	2.20
Percentage of Irrigated	30.09	25.34	65.67	100.00	30.29	96.14	52.92	77.78	86.39	100.00	60.00
Percentage Under HYV	9.44	86.32	—	—	19.16	100.00	96.15	—	36.05	—	—
All:											
Percentage of GCA	29.61	4.43	0.67	0.09	34.85	15.83	27.47	2.45	5.62	2.06	2.73
Percentage of Irrigated	37.08	36.05	82.61	100.00	37.98	97.85	56.63	84.65	89.69	70.15	65.01
Percentage under HYV	6.26	80.67	—	—	15.60	100.00	92.61	—	32.06	—	—

TABLE 14.5: Percentage of GCA, area irrigated and area under high yielding varieties of principal crops on the sampled farms in Mahendergarh during 1983-84

Category	Pulses				Other Important Crops					
	Gram	Moong	Moth	Total Pulses	Wheat	Bajra	Jowar	Mustard	Barley	Gowar
(1)	(2)	(3)	(4)	(5)	(6)	(7)	(8)	(9)	(10)	(11)
Small:										
Percentage of GCA	25.00	5.21	—	30.21	17.33	33.12	0.67	2.67	3.00	13.00
Percentage irrigated	17.33	21.99	—	18.14	84.62	39.81	100.00	100.00	55.56	53.85
Percentage under HYV	—	61.14	—	10.64	100.00	70.81	—	—	—	—
Medium:										
Percentage of GCA	25.37	4.24	0.44	30.05	16.52	33.96	—	2.95	2.65	13.86
Percentage irrigated	13.95	20.89	—	14.37	92.86	33.87	—	80.00	100.00	53.19
Percentage under HYV	—	43.18	—	6.09	100.00	82.36	—	40.00	—	—
Big:										
Percentage of GCA	18.67	2.22	0.46	21.35	16.69	31.57	0.23	6.69	6.18	16.60
Percentage irrigated	29.23	42.89	56.25	31.23	74.53	42.04	—	70.82	66.51	50.17
Percentage under HYV	41.23	20.02	—	39.49	97.93	75.08	—	24.89	7.44	—
All:										
Percentage of GCA	20.12	2.72	0.43	23.27	16.71	32.04	0.22	5.85	5.43	15.95
Percentage irrigated	25.31	35.00	47.37	26.85	77.99	40.57	20.00	72.41	68.60	50.77
Percentage of HYV	29.88	30.85	52.63	30.41	98.39	76.00	—	25.29	6.61	—

In Mahendergarh, inequalities among the different categories of farmers in the pulse acreage, irrigated area and the area under improved varieties were equally evident. First of all, high variations were found in the cultivated area under pulses such as pulse cultivated area on the small farms was 30.21 per cent of GCA against 21.35 per cent on the big farms. Secondly, only 18.10 per cent of the cultivated area under pulses was irrigated on the small farms against 31.33 per cent on the big farms. Thirdly, the adoption of improved technology has also indicated variations as only 10.64 per cent of the cultivated area under pulses was under the improved seeds on the small farms against 39.49 per cent on the big farms.

After highlighting the significant decline in pulse acreage at the state level as well as the District level, it would be pertinent to find out whether any shift has taken place in the pulse acreage on the sampled farm during last five years.

Percentage of Farmers shifted from Pulses to Other Crops

	Small	*Medium*	*Big*	*All*
Jind	37.50	18.75	22.22	26.00
Mahendergarh	9.00	18.25	30.43	20.00

Jind indicates comparatively higher shift than Mahendergarh due to the availability of irrigation facilities by canal to the majority of the farmers. Mostly, farmers have shifted in favour of wheat on irrigated land, and in favour of bajra and mustard on unirrigated land due to the higher profitability (Section III). In both the districts, proportion of the big farmers reporting to area shift is relatively high due to their sound resource base. In addition, the possibility of growing pulses on inferior quality land was investigated from technical point of view as well as from the angle of economic viability. Most of the farmers opine that pulses can be grown on inferior quality and with little investment on inputs like fertilisers and irrigation in the absence of rainfall.

Finally, it would be logical to analyse the factors which affected pulse acreage on the sampled farms. A survey of literature reveals that a few studies, based on the secondary data, have been carried out on this aspect. In a recent study, current year gram acreage is regressed on lag year income, lag year harvest prices and pre-sowing rainfall. The Nerlovian type of model is fitted. In the case of Mahendergarh, the regression coefficient for weather was 40.35 and was found to be significant at 5 per cent level. This has shown that with one milimetre increase in pre-sowing rainfall, the acreage under gram is expected to increase by 352.9 hectares. Other variables did not show any significant impact. The results for Jind were almost similar and one milimetre increase in rainfall is expected to increase 433.7 hectares under gram cultivation. Let us test this hypothesis (based on the secondary data) for our sampled households. An investigation is made to highlight the factors which affect decision-making of the farmers.

In Jind, timely rainfall emerges the most important factor followed by unirrigated land, home, requirement, economic viability, animal consumption and last year's production. According to small farmers, unirrigated land is the crucial factor, while medium and big farmers express timely rain as the most important factor (Table 14.6).

The situation is quite similar in Mahendergarh where timely rain is the leading factor for all the categories of farmers. Other important criteria are home consumption, unirrigated land and suitability of the soil (Table 14.6).

TABLE 14.6: **Relative importance of factors influencing acreage of pulse crops in Jind and Mahendergarh during 1983-84**

Category	Economic viability	Suitability of soil	Home consumption	Animal consumption	Inferior quality land	Unirrigated land	Timely Rain	Last years' Production
(1)	(2)	(3)	(4)	(5)	(6)	(7)	(8)	(9)
Jind:								
Small	31.25	0.00	37.50	25.00	0.00	62.50	37.50	27.27
Medium	25.00	6.25	37.50	31.25	6.25	12.50	75.00	25.00
Big	11.11	5.55	22.22	11.11	5.55	27.78	77.78	8.69
All	22.00	4.00	32.00	22.00	4.00	34.00	64.00	18.00
Mahendergarh:								
Small	0.00	36.36	54.54	36.36	27.27	18.18	63.64	6.25
Medium	9.09	25.00	31.25	6.25	31.25	31.25	56.25	6.25
Big	8.69	34.78	39.13	17.39	26.09	43.48	59.56	14.29
All	6.00	32.00	40.00	18.00	28.00	34.00	64.00	8.00

* Percentage of farmers.

II

Production and Productivity of Pulse Crops in Haryana

The main objective of this section is to analyse growth of production and productivity of pulse crops in the selected region.

Table 14.7 indicates that the production of pulses during 1964-65 to 1983-84 has declined continuously. During the first time period, production of total pulses declined at the rate of –5.79 per cent per annum. Gram followed by massar recorded the maximum decline (6.14% and 2.70% per annum respectively). The situation was almost same during the second time period but the rate of decline for total production of pulses was relatively lower (–3.72% per annum). Moreover, the production of 'other pulses' has grown at the rate of 6.14 per cent per annum.

In comparison to pulse crops, production of other important crops increased significantly during 1964-65 to 1983-84. During the first time period, wheat, rice, bajra and cotton have shown creditable increase in their production. On the contrary, production of mustard has declined at the rate of 2.32 per cent per annum. During the second time period, there was appreciable increase in the production of rice (11.99%), wheat (8.48%), mustard (3.97%) and cotton (5.51% per annum). In nutshell, except pulse crops, production of all important crops has increased during 1964-65 to 1983-84.

The production of any crop is the function of acreage and productivity. Table 14.8 provides data on yield and its compound growth rate for different pulses and other important crops during 1964-65 to 1983-84 in Haryana. It is noticed that pulses suffered heavily on this front also. The per hectare productivity of pulses has shown stagnation during 1964-65 to 1983-84. During the first time period, productivity of total pulses declined at the rate of –0.40 per cent per annum. Gram was at the top with the highest decrease of –1.95 per cent per annum followed by mash (1.95%) and massar (–3.63%). During the second period, results were encouraging as the productivity of total pulses increased at the rate of 1.65 per cent per annum. The performance of 'other pulses' (6.42%), moong (5.21%) and mash (5.36% per annum) was impressive.

TABLE 14.7: **Production of pulses and other important crops in Haryana during 1964-65 to 1983-84**

Crops	Triennium ending					
	Production in '000 Tonnes			Compound Growth Rates		
	1964-65	1974-75	1983-84	1964-65 to 1974-75	1974-75 to 1983-84	1964-65 to 1983-84
(1)	(2)	(3)	(4)	(5)	(6)	(7)
Pulses:						
1. Gram	806.00	447.00	302.00	−6.14	−4.18	−5.37
2. Mash	5.00	4.87	5.50	−0.29	1.37	0.50
3. Moong	6.20	4.97	4.17	−2.20	−1.91	−2.08
4. Massar	16.87	12.80	11.13	−2.70	−1.53	−2.15
5. Other pulses	—	10.63	18.20		6.14	—
6. Total pulses	874.40	480.27	341.00	−5.79	−3.72	−4.83
Other Important Crops:						
1. Rice	213	465	1287	8.10	11.99	9.93
2. Jowar	50	49	26	−0.19	−6.80	−3.37
3. Bajra	267	458	517	5.57	1.37	3.55
4. Maize	97	123	64	2.42	−7.00	−2.15
5. Wheat	853	1999	4164	8.87	8.48	8.70
6. Barley	79	175	126	8.40	−3.57	2.47
7. Mustard	121	96	136	−2.32	3.97	1.60
8. Sugarcane	548	581	571	0.58	−0.29	0.21
9. Cotton	255	430	697	5.39	5.51	5.43

Source: Statistical Abstracts of Haryana, 1973-74, 1974-75, 1975-76 and 1983-84, 1984-85, 1985-86 Government of Haryana, Chandigarh.

Again other important crops have shown higher growth of productivity as compared to pulses. During the first time period, jowar, bajra, wheat and rice indicated the growth of 5.26 per cent, 2.97 per cent, 3.20 per cent and 2.58 per cent per annum respectively. Commercial crops did not show any promising results. During the second time period, growth of productivity in the case of rice (5.07%), wheat (4.29%), bajra (2.78%), mustard (3.22%) and cotton (1.54% per annum) was appreciable. Commercial crops did relatively better.

TABLE 14.8: Productivity of pulses and other important crops in Haryana during 1964-65 to 1983-84

Crops	Productivity (Kgs. Per Hect.)			Compound Growth Rates		
	1964-65	1974-75	1983-84	1964-65 to 1974-75	1974-75 to 1983-84	1964-65 to 1983-84
(1)	(2)	(3)	(4)	(5)	(6)	(7)
Pulses:						
Gram	610	502	446	–1.95	–1.28	–1.63
Mash	486	397	636	–1.95	5.36	1.43
Moong	383	460	728	1.84	5.21	3.44
Massar	568	391	489	–3.63	2.51	–0.78
Other pulses	—	380	664	—	6.42	—
Total pulses	512	493	524	–0.04	0.65	0.10
Other Important Crops:						
Rice	1254	1621	2521	2.58	5.07	3.66
Jowar	164	274	212	5.26	–2.85	–1.09
Bajra	367	491	628	2.97	2.78	3.11
Maize	914	1050	1074	1.41	0.22	1.31
Wheat	1228	1681	2460	3.20	4.29	3.82
Barley	921	997	1367	0.77	3.56	1.86
Mustard	499	492	654	–0.09	3.22	2.74
Sugarcane	4708	4038	4056	–1.49	0.00	–1.49
Cotton	295	312	328	0.58	1.54	1.05

Source: Statistical Abstracts of Haryana, 1966-67, 1973-74, 1974-75, 1975-76, and 1983-84, 1984-85, 1985-86 Government of Haryana, Chandigarh.

It is evident from Table 14.9 that production of pulses has declined continuously in Jind as well as in Mahendergarh. The rate of decline per annum was 8.59 per cent in Jind against –0.7 per cent in Mahendergarh. In Jind, moong suffered maximum decline while 'other pulses' suffered maximum in Mahendergarh.

In comparison to pulses, production of other crops gained substantially. Rice followed by cotton and wheat were important in Jind, while wheat, bajra and mustard were significant in Mahendergarh. On the contrary, jowar, maize and sugarcane suffered in Jind. None of the major non-pulse crops has shown declining trend in Mahendergarh.

TABLE 14.9: **Production of pulses and other important crops in Jind and Mahendergarh during 1972-73 to 1983-84**

Crops	Jind			Mahendergarh		
	Production in '000 tonnes		Compound Growth Rate	Production in '000 tonnes		Compound Growth Rate
	Triennium 1974-75	Ending 1983-84		Triennium 1974-75	Ending 1983-84	
Pulses:						
Gram	55.67	1733	−9.29	30.33	29.00	−0.08
Mash	0.11	10	0.00	—	N.A	N.A.
Moong	0.84	0.10	−16.19	0.20	0.10	−3.52
Massar	2.23	2.47	0.87	—	—	—
Other pulses	0.20	0.30	3.44	0.20	0.10	−5.60
Total pulses	59.02	20.23	−8.59	30.73	29.20	−0.07
Other Important Crops:						
Rice	16	83	14.47	N.A.	N.A.	@
Jowar	7	3	−5.92	1	1	0.00
Bajra	58	73	1.94	69	65.33	0.42
Maize	2.33	1	−6.78	N.A.	N.A.	N.A.
Wheat	172	336	5.72	71	173	7.75
Barley	8	5	−3.32	51	34	−3.27
Mustard	6	6.67	0.87	18	19	0.33
Sugarcane	70	47.67	−3.15	N.A	N.A.	N.A.
Cotton	23	45	5.77	N.A.	N.A.	N.A

NA = Not Available @ = Negligible

Source: Statistical Abstracts of Haryana, 1973-74, 1974-75, 1975-76 and 1983-84 1984-85, 1986-86, Government of Haryana, Chandigarh.

On the productivity front, the performance of pulses has shown mixed trends in Jind as well as in Mahendergarh (Table 14.10). The per hectare productivity of pulses during 1972-73 to 1983-84 has increased at the rate of 8.41 per cent per annum in Jind and 1.17 per cent per annum in Mahendergarh. Particularly moong and massar have shown appreciable results in Jind. In Mahendergarh, productivity of the pulses i.e. gram, moong and total pulses has increased at the rate of 2.47 per cent, 1.53 per cent and 1.17 per cent per annum respectively. Hence, decline in acreage has been responsible for the reduction in the production of pulses in Jind as well as in Mahendergarh.

TABLE 14.10: **Productivity of pulses and other important crops in Jind and Mahendergarh during 1972-73 to 1983-84**

Crops	Jind			Mahendergarh		
	Triennium 1974-75	Ending 1983-84	Compound Growth Rate	Triennium 1974-75	Ending 1983-84	Compound Growth Rate
Pulses:						
Gram	631	604	.41	369	496	2.47
Mash	486	591	1.67	N.A.	N.A.	—
Moong	449	735	4.60	448	537	1.53
Massar	449	658	3.31	—	485	—
Other pulses	460	550	0.25	460	440	−0.33
Total pulses	567	627	+0.87	406	467	1.17
Other Important Crops:						
Rice	1269	2168	4.57	N.A.	N.A	N.A.
Jowar	254	322	2.01	183	223	64
Bajra	667	903	2.53	370	435	1.39
Maize	623	877	2.90	N.A	N.A.	—
Wheat	1676	2247	2.47	1860	2333	1.88
Barley	618	1035	4.37	1613	1975	1.67
Mustard	433	767	4.87	581	772	2.40
Sugarcane	4183	3535	−1.56	N.A.	N.A.	N.A.
Cotton	341	616	5.07	N.A.	N.A.	N.A.

NA = Not Available

Source: Statistical Abstracts of Haryana, 1973-74, 1974-75, 1975-76 and 1983-84, 1984-85, 1985-86, Government of Haryana, Chandigarh.

In comparison to the productivity of pulses, growth of productivity of other important crops during 1972-73 to 1980-83 had been quite impressive in Jind. In Mahendergarh appreciable increase in the productivity has been recorded by wheat (1.88%), barley (1.67%), bajra (1.39%) and mustard (2.40%).

We have collected and analysed data on productivity of different crops on small, medium and big farms during 1984 in the selected area. The productivity of important crops is shown in Table 14.11.

The productivity of gram in Jind on the sampled farms is lower than the average productivity in the State as well as in the District. For example, the yield of desi varieties of gram on the sampled farms in Jind had been 410 kgs. per hectare against 487 kgs. per hectare in Haryana, 604 kgs. per hectare in the District. In Mahendergarh, situation is quite different and per hectare productivity of gram on the sampled households is above the State as well as the District. The situation of moong, massar and moth is highly discouraging.

TABLE 14.11: **Per hectare yield of important crops on the sampled households during 1983-84**

(Kg.)

Crops	Small Farmers		Medium Farmers		Big Farmers		Total Farmers	
	Improved variety	Desi Variety	Improved variety	Desi variety	Improved variety	Desi variety	Improved variety	Desi variety
(1)	(2)	(3)	(4)	(5)	(6)	(7)	(8)	(9)
				Jind				
Pulses:								
Gram	—	437	—	492	210	375	210	410
Moong	99	200	175	69	101	67	121	109
Massar	—	—	—	279	—	296	—	237
Other Important Crops:								
Wheat	1852	—	1729	—	1729	—	1729	—
Bajra	566	988	716	659	1112	247	679	472
Mustard	—	395	—	329	—	247	—	309
Cotton	618	—	494	—	618	—	593	—
				Mahendergarh				
Pulses:								
Gram	—	316	—	571	531	553	531	534
Moong	12	180	37	146	52	259	40	240
Moth	—	—	—	—	148	326	148	326
Other Important Crops:								
Wheat	1729	—	1482	—	1976	988	1729	988
Bajra	1124	963	753	689	853	319	748	400
Mustard	—	—	—	99	946	351	946	341
Barley	—	—	—	—	—	1400	—	1400
Gowar	—	—	—	247	—	329	—	309

Further, the yield of the improved varieties of pulses on the sampled farms is much below the expected yield which ranges between 12 Qtls. to 20 Qtls. per hectare. Particularly, yield of the improved varieties of gram is lesser than desi varieties in Jind as well as in

Mahendergarh. Only the improved varieties of moong has indicated an increase of 12 kgs. per hectare over desi varieties in Jind, while Mahendergarh has shown extreme failure. Hardly any farmer has adopted improved varieties of massar and moth.

Class variations are found in the productivity of all the pulse crops grown on the sampled farms. The productivity of desi varieties of gram is maximum on the medium farms in Jind as well as in Mahendergarh. The improved varieties of gram were only grown by the big farmers and hence, class comparison is not feasible. The yield of desi varieties of moong in Jind is the highest on small farm, while the improved varieties give the maximum yield on the medium farms. On the contrary, productivity of desi varieties of moong in Mahendergarh is the highest on big farms followed by small and medium farms.

If the yield of other important crops on the sampled households is compared to the State and the districts, the productivity of wheat is substantially below the state. Bajra has done well in both the Districts as the yield of bajra on the sampled farms is above the state. The same is true for cotton in Jind. In Mahendergarh per hectare yield of hybrid bajra, mustard and barley is above the state. Only marginal variations are found in the productivity if compared to the district.

In fact, the yield of high yielding variety cereals and commercial crops on the sampled households is comparable to the state as well as the districts. Unfortunately, the productivity of pulse crops is generally much below the average productivity in the state as well as the districts.

Causes of Low Productivity of Pulses

The poor performance of pulses on the sampled farms may be attributed to a variety of factors such as lack of awareness of improved variety seeds, low level of input use, non-adoption of recommended package of practices, uncertainity of rainfall and lack of irrigation facilities, lack of plant protection and weed control measures.

Lack of Awareness of Improved Variety Seeds

The following percentage of farmers was aware about the improved variety seeds of pulses:

	Small	*Medium*	*Big*	*All*
Jind	25.00	25.00	61.18	36.00
Mahendergarh	18.18	37.50	26.09	28.00

It is evident that only 36 per cent of the sampled farmers in Jind and 28 per cent farmers in Mahendergarh knew about the existence of improved varieties of gram, moong and moth but they were not fully aware about their characteristics and sources of availability. In Jind, 61.18 per cent of big farmers were aware about the improved varieties of pulses against 25 per cent small and medium farmers. On the contrary, in Mahendergarh, a higher percentage (37.50%) of medium farmers and lower percentage (18.18) of small farmers knew about the existence of the improved varieties of pulses.

Low level of Input Use

The recent experience of using higher doses of inputs for high yielding superior and inferior cereals has shown that gradual increase in the use of crop specific and area specific inputs increases productivity. However, pulses reveal exceptions despite the realisation of the need to raise productivity.

It is well established that seed treatment, use of fertilisers, adequate number of ploughings, irrigation, and use of pesticides is essential for obtaining good yield of pulses. However, the use of inputs by the sampled farmers is much below the recommended doses. It can be seen that the number of ploughings is generally one against two or three recommended ploughings. Secondly, the seed treatment is not practiced by any farmer. Thirdly, the use of fertilisers is below the recommended doses. Particularly, farmers in Jind did not use any fertiliser for moong and massar. In Mahendergarh, nothing was spent on fertilisers for gram and moth. Fourthly, only negligible quantity of pesticides and insecticides was used in both Jind and Mahendragarh for pulses. Fifthly, none of the pulses grown on the sampled farms received adequate number of irrigations. The average number was below 1. As a result of the low input use, the yield of even improved varieties is much lower than the potential yield.

Uncertainity of rainfall and lack of adequate irrigation facilities

Pulses are mostly grown as rainfed crops in the selected area. The poor productivity of pulses is attributed to the low rainfall. This is more obvious in the case of kharif pulses which are mainly rainfed crops. On the contrary, the yield of rabi pulses (gram) is not so low because these are generally grown with limited irrigation. Thus, low rainfall not only affects productivity during the kharif season but also affects, to a greater extent, the subsequent rabi pulses. If winter rains fail, productivity without irrigation is very low. In the case of gram, if the initial soil moisture during kharif season is not sufficient, one pre-sowing irrigation is necessary for the proper germination of the plant. But only 15 per cent of the total farmers could manage two irrigations and others tried to grow gram with one irrigation or no irrigation at all. The following table gives variations in the productivity of gram on irrigated and unirrigated farms.

Per Hectare Productivity of Gram on Irrigated and Unirrigated Farms in the Selected Area

Category	Irrigated		Unirrigated		Percentage Variation	
	Desi	HYV	Desi	HYV	Desi	HYV
(1)	(2)	(3)	(4)	(5)	(6)	(7)
Jind:						
Small	551	—	363	—	(51.79)	—
Medium	526	—	479	—	(9.81)	—
Big	259	—	388	210	(-33.25)	—
All	373	—	305	210	(22.29)	—
Mahendergarh:						
Small	1042	—	464	—	124.57	—
Medium	618	—	563	—	9.77	—
Big	951	556	440	514	116.14	8.17
All	902	556	445	514	102.70	8.17

It is observed that the yield per hectare of desi varieties of gram on irrigated farms is higher than unirrigated farms by 22.29 per cent in Jind and 102.70 per cent in Mahendergarh. The maximum variation is found in the case of small farmers in Jind as well as in Mahendergarh. The yield per hectare in Jind and Mahendergarh on irrigated small farms for desi varieties in 551 Kgs. and 1042 Kgs. per hectare against 363 Kgs. and 464 Kgs. per hectare on unirrigated farms.

Lack of Plant Protection Measures

Pulses crops are prone to diseases and pests, hence, enough care is required to save the quantity as well as quality of the produce. Among diseases, gram wilt and blight are very common. For controlling wilt, deep sowing, cultivation of G-24 and H-208 varieties, seed treatment and soil testing are generally recommended. Surprisingly farmers have hardly taken any measure in this regard. Only 10 per cent of the big farmers used pesticides for gram. The small and medium farmers did not report any adoption of these pesticides.

Among pests attacking the pulse crops are termites, cutworms and gram pod borer. There are recommended practices to control them. However, farmers have equally neglected this aspect.

Lack of Weed Control Practices

There is common belief among farmers that pulse crops do not require any hoeing and weeding as they supress the weeds. According to the recommended practices, weeding is essential for realising good yields. Generally, weeds hinder the soil moisture and nutrients which are limiting factors in realising good returns. It also increases the aeration which helps in increasing yield. At least one weeding after a month is a must. Second weeding is required after first irrigation or before flowering stage. Only 20 per cent of the pulse growers in the sample has done weeding only once during the complete crop season.

III

Profitability of Pulse Crops vis-a-vis Competing Crops

Production of pulse crops must increase through increase in acreage and productivity to meet the rising demand of pulses emanating out of growing population in India. However, the magnitude of change in cultivated area depends on the returns from growing pulses as a substitute for other competing crops. The detailed districtwise data about the cost and returns on major pulse crops (gram and moong) in Haryana are available in package of practices, H.A.U. Hissar. But these data are not sufficient to permit an indepth analysis of the problems of cultivation of pulses in the State mainly because of two reasons. First, minor pulse crops having growth potential in future in the state have been completely ignored. Secondly, improved varieties of pulse crops have not been taken into account. Hence, we thought it pertinent to analyse the cost and returns of cultivation of pulses on the sampled households after incorporating the above mentioned missing aspects.

The main objective of this section is to compute farm profitability per hectare from pulse crops vis-a-vis competing crops in Jind and Mahendergarh districts of Haryana during 1983-84. Among pulses, all the pulse crops (gram, moong, massar and moth) grown in the selected area are covered. The selection of competing crops is based on the allocation of land resources in place of pulses in crop rotation. On the basis of this criterion, competing crops included are bajra among foodgrains and mustard among commercial crops.

The gross income is the value of the main product as well as by product of a crop enterprise whether sold, consumed or stocked. Cost includes the following items: (i) Hired human labour, (ii) Owned and hired bullock labour, (iii) Seeds (both farm produced and purchased), (iv) Fertilisers and pesticides, (v) Irrigation charges, (vi) Interest on working capital, (vii) Land revenue, and (viii) Depreciation and other charges.

Thus cost is the sum of expenses that have been incurred by the farmers in cash (Cost A_1 according to Farm Management Surveys). The rates for human labour, bullock labour, machinery and irrigation have been charged at the prevailing market rates in the selected area. In the case of seeds and fertilisers out of pocket cost is included. Interest on working capital is calculated at the rate of 12 per cent per annum for the full crop season. Miscellaneous costs include land revenue, depreciation and other charges.

Tables 14.12 and 14.13 give detail about the per hectare cost, output and net profitability for improved desi varieties of gram, moong, massar, moth, bajra and mustard in Jind and Mahendergarh.

A comparison of cost and returns per hectare of included crops in Jind reveals that mustard is the most profitable crop as it gives the maximum gross returns (Rs. 1360) as well as the highest profits (Rs. 916) per hectare. The second rank goes to desi gram which is at the top among pulses in terms of profitability (Rs. 778) and gross returns (Rs. 1218) followed by improved bajra with a profitability of Rs. 340 per hectare. The fourth rank goes to desi bajra which gives profits of Rs. 293 per hectare. Next are desi massar and improved gram which show the marginal profit of Rs. 29 and Rs. 16 respectively. Surprisingly, improved as well as desi varieties of moong have shown loss of Rs. 101 and Rs. 206 per hectare respectively.

TABLE 14.12: Per hectare net profitability from pulses vis-a-vis competing crops on the sampled farms in Jind during 1983-84

(Rs.)

Items	Gram		Moong		Massar	Bajra		Mustard
	Improved	Desi	Improved	Desi	Desi*	Improved	Desi	Desi*
(1)	(2)	(3)	(4)	(5)	(6)	(7)	(8)	(9)
I. *Cost*:								
(a) Human Labour	220.45	182.49	289.66	333.89	336.81	304.09	164.95	218.89
(b) Owned & Hired bullock labour	157.66	49.35	149.05	198.16	131.36	140.90	80.54	102.60
(c) Mechanical Labour	9.16	19.08	36.99	21.52	53.89	38.49	—	—
(d) Seeds	135.23	122.83	103.74	73.92	49.62	23.54	16.10	18.99
(e) Fertilisers	19.76	13.71	0.00	0.00	0.00	73.88	63.36	25.34
(f) Pesticides	0.00	8.59	0.00	0.00	0.00	0.00	0.00	0.00
(g) Irrigation charges	15.20	3.20	0.00	0.00	0.00	30.80	20.80	20.40
(h) Interest on working capital	33.45	23.96	34.77	37.71	34.30	34.85	20.75	23.17
(i) Miscellaneous	17.00	17.00	19.00	19.00	17.00	16.00	16.00	34.00
Total Cost	607.91	440.21	633.21	685.20	622.98	631.75	382.50	443.39
II. *Output*:								
Value	567	1107.05	484	436	593	883	614	1236
By-Product: Value	56.69	110.71	48.40	43.60	59.30	88.30	61.40	123.60
Gross Returns	623.69	1217.76	532.40	479.60	652.30	971.30	675.40	1359.60
III. *Net Profitability*	15.78	777.55	−100.81	−205.60	29.32	339.55	292.90	916.21

* Improved Varieties were not grown.

TABLE 14.13: Per hectare net profitability from pulses vis-a-vis competing crops on the sampled farms in Mahendergarh during 1983-84

(*Rs.*)

Items	Gram		Moong		Moth	Bajra		Mustard	
	Improved	Desi	Improved	Desi	Desi*	Improved	Desi	Improved	Desi
(1)	(2)	(3)	(4)	(5)	(6)	(7)	(8)	(9)	(10)
I. *Cost*:									
(a) Human Labour	316.75	361.75	344.61	346.00	131.73	463.54	327.39	444.60	244.25
(b) Owned & Hired bullock labour	102.81	84.25	127.65	116.74	41.17	107.48	69.29	111.15	31.56
(c) Mechanical labour	76.57	54.69	24.06	21.46	43.91	75.76	50.04	263.46	91.48
(d) Seeds	148.20	120.02	46.58	33.49	91.12	27.77	11.86	24.70	19.48
(e) Fertilisers	—	—	53.87	47.35	—	74.62	14.47	74.10	14.25
(f) Pesticides	—	—	0.44	—	—	—	—	—	—
(g) Irrigation charges	3.20	3.20	7.20	—	—	4.80	2.80	26.80	8.80
(h) Interest on working capital	38.85	37.43	36.26	33.90	18.48	43.56	28.55	56.69	24.59
(i) Miscellaneous	25.00	25.00	19.00	19.00	19.00	20.00	20.00	25.00	25.00
Total Cost	711.40	686.34	659.67	617.94	345.41	789.53	524.40	1026.50	459.42
II. *Output*:									
Value	1434	1442	160	960	815	972	520	3784	1364
By-Product:									
Value	143.40	144.2	16.00	96.00	81.50	97.20	52.02	378.40	136.40
Gross Returns	1577.40	1586.2	176.00	1056	896.50	1069.20	572.02	4162.40	1500.40
III. *Net Profitability*	866.00	899.86	−483.67	438.06	551.09	279.67	47.62	3135.90	1040.98

* Improved variety was destroyed.

An analysis of cost and returns data of different crops in Mahendergarh shows that mustard is the most profitable crop. It is observed that improved varieties of mustard give a profit of Rs. 3136 per hectare against Rs. 1041 in the case of desi varieties of mustard. Among pulses desi gram has shown the highest profit (Rs. 900) per hectare. Next is improved gram which gives a profit of Rs. 866 per hectare. Moth and moong give profits of Rs. 551 and Rs. 428 per hectare. Surprisingly, profitability of moong is very low and therefore, it gets the last rank.

The ranking order of different crops in terms of profitability is given below:

Rank	Jind		Mahendergarh	
	Improve Variety	Desi Variety	Improved Variety	Desi Variety
(1)	(2)	(3)	(4)	(5)
I	—	Mustard	Mustard	—
II	—	Gram	—	Mustard
III	Bajra	—	—	Gram
IV	—	Bajra	Gram	—
V	—	Masoor	—	Moth
VI	Gram	—	—	Moong
VII	Moong	—	Bajra	—
VIII	—	Moong	—	Bajra
IX	—	—	Moong	—

This is suggested that farmers would prefer mustard over pulse crops in the selected areas.

So far as we have analysed the net profitability per hectare for pulse crops as well as for competing crops in the selected area. As a next step, it would be worthwhile to look into the net return per rupee which gives comprehensive estimates of returns after investing one rupee on a particular crop. Table 14.14 suggests that mustard gives the highest net returns per Re. in Jind as well as in Mahendergarh. Desi mustard gives Rs. 207 after investing one rupee in Jind. The values of net returns per Re. for improved and desi varieties of mustard in Mahendergarh are worked out to be Rs. 3.05 and Rs. 2.27. The lowest net return per Re. in Jind is calculated for improved gram i.e. only 3 paise per Re. Surprisingly, desi as well as improved varieties of moong have shown losses. In Mahendergarh, the lowest net returns are calculated for desi bajra i.e. 9 paise per Re. Again improved moong has indicated a loss of 73 paise over one Re. However, Mahendergarh has shown better results for gram by indicating net returns of Rs. 1.22 for improved varieties and Rs. 1.31 for desi varieties.

TABLE 14.14: **Net returns per re-form pulses vis-a-vis competing crops on the sampled households 1983**

Crops	Jind	Mahendergarh
1. Gram (Desi Variety)	1.77	1.31
(Improved Variety)	0.03	1.22
2. Moong (Desi Variety)	−0.30	0.71
(Improved Variety)	−0.16	−0.73
3. Moth (Desi Variety)	0.05	—
(Improved Variety)	—	—
4. Massar (Desi Variety)	—	1.60
(Improved Variety)	—	—
5. Bajra (Desi Variety)	0.77	0.09
(Improved Variety)	0.54	0.35
6. Mustard (Desi Variety)	2.07	2.27
(Improved Variety)	—	3.05

Note:

Net Returns per Re = $\frac{X_y}{X_j}$

X_y = Net profitability per hectare by growing a particular crop.

X_j = Value of per hectare inputs used for growing a particular crop.

It is evident that Mahendergarh has shown comparatively higher net returns per Re. than Jind for all the pulse crops except improved varieties of moong. This is equally true for mustard. However, Mahendergarh is lagging behind in the case of bajra. In nutshell, the performance of Mahendergarh is better than Jind due to high potential for growing such crops and due to the implementation of different programmes.

A review of cost data in Jind shows that the maximum per hectare cost (Rs. 685) was incurred in the case of desi moong due to the highest cost incurred as wages paid to the human labour as well as animal labour. Next is improved moong with almost similar position. Further, in the array are improved bajra, desi massar and improved gram.

Mahendergarh presents very different picture as the maximum cost (Rs. 1027) was incurred for improved mustard due to higher cost of human labour, animal labour and mechanical labour. Next is improved bajra followed by improved gram and desi gram.

The structure of cost is the same in both the districts. The largest share of cost has gone to human labour followed by animal labour. Irrigation, fertilisers and pesticides constitute a very small proportion of the total cost. Particularly, farmers have not used any irrigation and pesticides for moong and massar in Jind while the same is true for gram and moth in Mahendergarh. Pesticides were only used for desi gram in Jind and improved moong in Mahendergarh. Generally, very low cost was incurred on fertilisers in Jind, however, Rs. 73.88 for improved bajra and Rs. 63.36 for desi bajra were spent on fertilisers. Mahendergarh has shown variations in this regard as fertilisers were consumed for all the crops except gram and moth.

IV

Constraints to Pulse Growth in Haryana

The production of pulses has continued to decline for the past two decades in Haryana. The 1983-84 pulses production of 363 thousand tonnes is less than the 1964-65 and 1972-73 output of 995 and 585 thousand tonnes respectively. This is due to the negative growth of area and productivity of pulses. Particularly, the advances made in rapid growth of area and productivity of cereals after mid-sixties have been limited to a few crops where returns are almost assured. In respect of rainfed crops, a breakthrough in technology is not yet in sight. Pulses have suffered heavily on this account. In this context, a probe into the constraints may provide insight regarding the needed motivation to accelerate the pace of growth in the case of pulses.

Poor Physical Environment

It is indicated that pulses are generally grown on unirrigated pieces of land. The worse happens when farmers cultivate pulses on the unirrigated inferior quality land under highly adverse conditions partly due to high risk and basically due to their prior thinking that pulse crops can utilise limited soil moisture and nutrients more efficiently. As a result, pulse crops are pushed to marginal land at regional level as well as at individual level. Generally, it is argued that the fertility of marginal land can be improved by soil treatment and further by providing adequate irrigation facilities. Although, at the outset, the scope for these crops appears quite bright, yet the possibilities need a close examination. In practice, provision of these facilities dissuades farmer from growing pulses and favours superior cereals of the high yielding variety, inferior cereals or oilseeds due to their high profitability and low risk. Hence, research is needed towards growing pulses on unirrigated inferior quality marginal land rather than growing them by providing the best conditions through the improvement of resource base. It is important here to evaluate and popularise package of practices helpful in improving productivity of rainfed areas with unfavourable physical potential in terms of soil. Soil treatment in the light of requirements of pulses is worth considering because farmers have pointed out it as a main constraint for crop failure. Basically, soil has lost its strength and particularly, crops like pulses which are easily prone to diseases, suffer maximum loss in terms of quality as well as quantity.

Secondly, owing to the uncertain nature of rainfall in areas like Jind and Mahendergarh, it is necessary to make arrangements to store water as and when it is received and use it for irrigation during critical periods. It is suggested by the farmers that production of pulses can be immediately stabilised if facility of strong rain water in farm ponds is created so that some irrigation is possible for pulse crops during the critical periods.

Thirdly, structure of pulse crops also poses a problem in the direction of research. Although, gram is the dominant pulse crop in this area, however, moong, massar and moth are also grown. It puts serious limitations to the single pulse based growth technology in the area and hinders concentrated research on the low productivity and reasons for stagnation. It is suggested that regional stations should focus research on the dominant pulse crop within the prevailing agro-climatic conditions.

Adoption Constraints for Available Improved Technology

Although pulses are important for self-consumption of farmers, yet they assume subsidiary status in the farming system. It causes low attention as well as poor investment. So far no breakthrough in technology has been achieved in the case of pulses. We have already discussed about the poor physical environment in which pulses are grown. We have also analysed the possibility of improving resource base in accordance with the requirements of the pulses.

The non-adoption of improved technology is the basic cause of stagnation of productivity in the case of pulses in the selected area. According to our survey, most of the farmers are unaware of improved technology despite agronomists claim about the evolution of the high yielding varieties of various pulses. There are two rounds of missing link. First is that farmers do not know about the existing improved technology for which Extension Department is to be blamed. Secondly, whatever extension services are available, farmers have not utilised them. It is due to the absence of area specific and crop specific improved technology which includes the complementary package of practices also, such as certified seeds, rhizabium culture and plant protection measures. These are the primary requirements for getting good yield of pulses under improved technology. This is high time to realise that only bringing large area under the improved varieties of pulses is not enough to increase production. It is essential to provide complementary inputs in time and at a reasonable price to the pulse growers.

Farmers have specified about the non-availability of disease resistant certified seeds and the complementary inputs recommended under package of practices for the improved variety of pulses. It happens due to the limited supply and due to the lack of proper distribution system. Particularly, small farmers complained about the discrimination made against them in the distribution of inputs such as packets of rhizabium culture, fertilisers and pesticides. Hence, efforts are needed to increase the production and to create proper channel for the distribution of recommended inputs for the improved varieties of pulses.

Many farmers on the sampled farms had been facing the problems of gram blight for which soil testing and soil treatment is essential. Hence, free of cost, scientific arrangement should be made for all the pulse growers.

Apart from the proper distribution of complementary inputs, guidance towards their application should be provided. Particularly, proper guidance should be given in the case of fertilisers and pesticides. Pulses require mild doses of fertilisers unlike cereal crops. Farmers have to use fertilisers very cautiously. Judicious use of fertilisers will have to be the rule which will include proper choice of fertilisers, time and method of their application. Placement of fertilisers has an advantage over broadcast application. Unless fertilisers are placed at appropriate depths by using suitable implements, it will be difficult to get maximum benefit from fertilisers. Time of application also assumes importance because of limitation of water supply for crop growth.

Finally, owing to the uncertain nature of rainfall and allocation of marginal unirrigated land for pulses, it is necessary to make arrangements for irrigation at least twice during the pulse crops season.

Marketing Problems

Market mechanism plays vital role by influencing the profit of the traders on the one hand and by affecting the consumer price on the other hand. Generally, commodity markets are imperfect but the case of pulses indicates higher imperfections due to the following reasons :

(i) The demand for pulses is very high and supply is low, hence prices are very high but production is not increasing and high prices are not able to provide incentive to the farmers which is quite peculiar.

(ii) Marketed surplus of pulses is low on the sampled households. Gram and moong are the main pulses which are grown for commercial purpose in the selected area. However, 49.70 per cent and 63.30 per cent of the total output of gram is sold in Jind and Mahendergarh. Medium farmers sell the maximum proportion of the produce.

Market Surplus of Pulses on the Sampled Household During 1983-84

Category	Percentage of total output			
	Gram		Moong	
	Jind	Mahendergarh	Jind	Mahendergarh
Small	46.40	51.47	11.03	26.67
Medium	55.87	67.84	11.20	23.53
Big	48.20	63.16	27.59	43.17
All	49.70	63.30	21.29	37.44

The case of moong is very peculiar because only 21.29 per cent and 37.44 per cent of the total output is marketed by the sampled households in Jind and Mahendergarh. Big farmers sell the maximum proportion of the output. Surprisingly, small and medium farmers in Jind have sold only 11 per cent of the total output during 1983. This indicates that gram is the only pulse which farmers grew from commercial point of view. Other pulses like moong, moth and massar are grown for home consumption.

Price spread is relatively high in the case of pulses. The margin between the prices received by the producers and prices paid by the consumers in Haryana for gram is computed in a study for the year 1981-82 for small, medium and big markets. This includes cost of transporting, storing, grading, sorting, processing and various market charges.

	Net Price Received by the Producer Rs./Qtl.	Sale Price of the retailer/ price paid by the consumer Rs./Qtl.
Small Markets	368.43	422.00
Medium Markets	367.32	420.00
Big Markets	370.99	415.00

This shows that for gram the producer's share in the consumer's rupee was 87 paise in small and medium markets and 89 paise in large markets. The per qtl. price received by the producer was the highest (Rs. 370.99) in large markets followed by the small (Rs. 368.43) and medium markets (Rs. 367.32).

Majority of the farmers prefer to sell their pulse produce in the post-harvest period for two reasons. Firstly, small and medium farmers are generally hard pressed by their immediate financial requirements. Secondly, storing pulses is a risky proposition as they are easily prone to insect infestation, which damages quality as well as quantity of the grain. Although controlling measures can be used for fumigation of pulses, however, very few farmers have adopted them. Only the farmers who are financially sound and do have proper storing facilities, sometimes held their pulse produce for a longer period and sometimes till the arrival of the new crop. This distributes the arrivals over different seasons imbalancedly and causes deep seasonal variations. Traders also manipulate the situation. They are financially strong and have sufficient storage facilities. During post-harvest season when the prices are low, they make bulk of the purchases. Later on, they withhold the stock to cause artificial scarcity in the market and in this way, the prices become higher under restricted supplies. Thus, fluctuations occur in the prices of pulses between post-harvest months and lean months but profit goes to traders and not to farmers.

V

Summary and Conclusions and Policy Implications

Pulse crops enjoyed an important place (30.16% of GCA) in the cropping pattern of Haryana during early sixties. With the increase in irrigation facilities and the introduction of high yielding variety seeds, the proportion of GCA got reduced and pulse acreage recorded a negative compound growth rate of 3.37 per cent per annum during 1964-65 to 1983-84. The same is also true for the districts of Jind and Mahendergarh. On the sampled households, the proportion of cultivated area under pulses was higher than the state as well as the districts. Class variations were evident in the acreage, irrigated area and in the adoption of improved technology. It was found that unirrigated land, timely rainfall, suitability of soil, inferior quality land, home requirement, economic viability, animal consumption and last year's production affect farmer's decision regarding the acreage under pulses. But rainfall emerged as the most important factor.

The production of pulses in Haryana during 1964-65 to 1983-84 has declined continuously at the compound rate of 4.83 per cent per annum. The same is true for Jind and Mahendergarh. The productivity of pulses remained almost stagnant and has not made any significant contribution to arrest the declining production of pulses due to continuous reduction in the acreage. The performance of productivity has shown better results in the districts of Jind and Mahendergarh. However, the yield of pulses on the sampled households in Jind is lower than the average productivity in the state as well as the district, while vice-versa is true in the case of Mahendergarh. The poor performance of yield on the sampled farms may be attributed to a variety of factors such as lack of awareness of improved variety seeds, low level of input use, non-adoption of recommended package of practices, uncertainty of rainfall, lack of irrigation facilities, lack of protection and weed control measures.

A comparison of cost and returns per hectare of pulses and competing crops (Gram, moong, massar, moth, bajra and mustard) on the sampled farms reveals that mustard is the most profitable crop as it gives the highest net return per Re. (Table 14.3). The structure of the cost in the case of pulses is the same in both the districts. The largest share of cost has gone to the human labour followed by animal labour. Irrigation, fertilisers and pesticides constitute a very small proportion of the total cost. Hardly any cost is incurred on mechanical labour, however, limited use is made for preparatory tillage.

The main constraints to the growth of production of pulses are poor physical environment, adoption constraints for improved technology and marketing problems, yet the non-adoption of improved technology remains a major cause of stagnation in yield rates. Most of the pulse growers are unaware of the existence of improved technology. Those who know, they have not adopted due to the non-availability of complementary package of practices. Hence, it is necessary to facilitate provision of complementary inputs in time. Apart from availability of complementary inputs, that contribute to an increase in the relative profitability of pulse crops, guidance towards their application should also be provided.

Many farmers on the sampled farms had been facing the problem of gram blight for which soil testing and soil treatment is essential. Hence, arrangements should be made to provide scientific and technical support for all the pulse growers.

Finally, owing to the uncertain nature of rainfall and allocation of marginal unirrigated land for pulses, availability of irrigation at least twice during the pulse crops season would go a longway in raising yield rates. The following constraints should be removed to popularise the improved technology of pulses in the State:

(i) Easy availability of disease resistant improved seeds and other recommended inputs for different pulses,

(ii) Strengthening the extension services which include proper guidance regarding the use of recommended inputs.

A sustained campaign comparable to the one that launched the 'green revolution' is necessary to lower the element of risk and uncertainty now connected with the growing of pulses.

REFERENCES

Bapna S.L., 1976 : "Production of Coarse Cereals in India: Past Performance and Future Prospects", *Economic Programme*, ICRISAT, Hyderabad.

Barbara Harriss, 1979 : "Coarse Grains, Coarse Interventions", *Overseas Development Studies*, London.

Chopra K. and Gurushree S., 1975 : *An Analysis of Demand and Supply in India, 1951-1971*, Institute for Social and Economic Change, Bangalore, Sterling Publishers, New Delhi.

Chopra K., 1982 : "Pulse Production in India — A Statewise Analysis", *Indian Journal of Agricultural Economics*.

Chopra K., 1984 : "Pulse Production in India — The Basic Issues, 1951-1981", Paper Presented for the Study group in Planning Commission, New Delhi.

Gangwar A.C., et.al, 1983 : "Production and Marketing of Gram in Haryana", Department of Agricultural Economics, HAU.

Jodha N.S., 1980 : Intercropping in Traditional Farming Systems, *Journal of Development Studies*, Vol. 6, No. 4, July.

Jodha, N.S. and Singh R.P., 1982 : "Factors Constraining Growth of Coarse Grain Crops in Semi-Arid Tropical India", *IAJE*, July-Sept.

Kumar B.L., 1978 : "Declining trend in Production of Pulse and Factors Affecting it", Vol. XIII, *Economic and Political Weekly*, July 8.

Narain Dharam, 1972 : "Growth and Imbalances in Indian Agriculture", *Economics and Political Weekly*, March 25.

Ray, S.K., 1983 : "Growth and Instability in Indian Agriculture", *Mimeograph*, I.E.G.

Sharma, D. and Jodha, N.S., 1982 : "Pulses Production in Semi-Arid Regions of India-Constraints and Opportunities", *Economic and Political Weekly*, Dec. 25.

Srinivasan, T.N., 1979 : "Trends in Agriculture in India, 1949-50 to 1977-78", Vol. XIV, Special Number, *Economic and Political Weekly*, August.

15

CROPS AND PEOPLE IN SOUTH EAST ASIA

R. D. HILL

FOR MOST countries in South-East Asia and for much of the past quarter century, agricultural production has more than kept space with population growth. This is generally true for rice production and, especially, for maize, even though the area per person under these crops has fallen significantly, suggesting that intensification of production has been more important than extension of the cultivated area. Detailed studies suggest that there is a wide range of area/and production/population relationships. An examination of data for non-cereal crops (groundnuts, soyabean, copra and oil palm) shows that both production and area have generally kept pace with population growth and, in the case of oil palm, have greatly exceeded it.

At the world level the balance between crops and people has received some recently attention. Early in the 1970s, as world food stocks fell from the equivalent of 90-100 days to around 25 days' supply, observers such as Lester R. Brown (1974) and participants in the 1974 Rome Forum (Sartaj Aziz, 1975) pointed to the imminence of mass starvation and a worsening of apparently general chronic malnutrition in the Third World. The prices of basic cereals doubled or tripled in a short space and, if these neo-Malthusian doomsayers were to be believed, the world food economy was on the verge of collapse. That it did not collapse is because most Third World farmers (including those of South-East Asia) still grow all or most of their own food — and farmers account for between two-thirds and four-fifths of the labour force. In normal years, more than three-quarters of the rice harvested is consumed by the producers themselves with only two to three per cent entering world trade, and the balance entering local trade

(Wilhelmy, 1976).

The familiar argument is, of course, that population growth tends to outstrip production. Implicitly, this applies not only to the Third World as a whole but also to the level of major regions and individual countries. A useful corrective to sweeping generalisations is a study by Thomas Poleman (1977), which, in debunking the myths of imminent global starvation, of the "Tragedy of the Commons" and the food/population "myth", attempts to delineate the true dimensions of the situation.

For South-East Asia, the home of around 335 million people, the facts generally accord with Poleman's view and point to a slow but steady rise in the production of basic foodstuffs as measured for the region as a whole and for most of its constituent countries. This is not to suggest that the amount of food per head may not have been greater had population growth been less. Nor is it to suggest that there are no areas, classes and groups of people within the countries of South-East Asia that continue to be relatively short of food especially if seen in relation to local self-sufficiency. Moreover, it is clear that in some instances, any relatively laggard performance in food crops is more than compensated by a spectacular growth in non-food crops, particularly oil palm.

Sources and Reliability of Data

This paper sets out an analysis of national crop production and area data for the last three decades, examining these in relation to population. The crops considered are the two basic cereals, rice and maize, two annual oil-crops, soyabean and groundnuts, and two perennial oil-crops, copra and oil-palm. The basic sources for this analysis are the FAO *Production Yearbooks* which, for all their defects, give the only comprehensive picture of the region. The FAO itself tends, correctly, to regard the data as estimates. Wherever possible, they have been checked against other sources and the data upon which the figures and arguments are based may be taken as being "correct" to within perhaps plus or minus a fifth.[1] Even if the errors are greater than this the high degree of internal consistency would suggest that errors are indeed consistent. Moreover, as there are no alternative sources, the only other available option is not to attempt a study at all which seems rather like cutting off the nose to spite the face.

In order to reduce the effects of reporting errors, variation in stocks and in production from year to year, three-year moving averages have been computed for all production and area data plotted in Figs. 15.1 to 15.9, 15.13 to 15.21. The same method was applied to population data which also show some unexplained variation.[2]

Cereal Crops

Rice

The importance of rice in the diet of South-East Asian peoples' needs no emphasis. Production increased by 15 per cent between 1952 and 1978 (Fig. 15.1). More seriously, the area of rice land per head declined by 29 per cent between 1952 and 1978, that is, by a simple average of just over one per cent per year, and there is some indication that this rate is steadily increasing. In the quinquennium 1954-58 the area per head rose at a rate of 0.5 per cent yearly, but this trend reversed in the following period. For the years 1969-73 the area per person fell by 1.7 per cent each year and this rose to 1.9 per cent yearly for the final period to 1978. During the period 1952 to 1978 the population rose by a total of about 165.3 million, an overall increase of roughly 97 per cent or simple average of around 3.6 per cent annually. Between the same years, the total area under rice rose from 22.3 million hectares to 32.2 million hectares, an overall increase of 44 per cent or an average of 1.6 per cent per year, less than half the rate of population growth. Yet despite this lag, intensification of rice agriculture managed more than to keep pace with population growth (Appendix-I).

Within South-East Asia, the individual countries fell into three groups: first the major exporters, Thailand and Burma; second, the new lands of Indochina, some of which have been exporters, namely (southern) Vietnam and Cambodia; and third, the deficit countries of insular South-East Asia.

In both Thailand and Burma, the exported proportion of the crop has steadily fallen since the Second World War, though production per head has still consistently exceeded internal consumption. Nevertheless, in Thailand especially, rapid population growth has resulted in a fall of about a third in the rice area per head despite the fact that the actual harvested area almost doubled (Fig. 15.2). In Burma, where post-War reconstruction was so slow that the total area under rice did not exceed the 1934-38 average until 1975, population growth has been slower than in Thailand, despite which the area per person fell by 21 per cent between 1948 and 1978 (Fig. 15.3). Despite the decline in the area per head, production per head has tended to remain about the same, though in Thailand the period from the early 1950s to the early 1960s was marked by a substantial fall, as some expansion onto more marginal lands took place. This was later made good by improved methods of cultivation. In Burma the pattern was the reverse, with some improvement in production per head until the mid-1960s, followed by a decline which corresponded with more rapid population growth since that time.

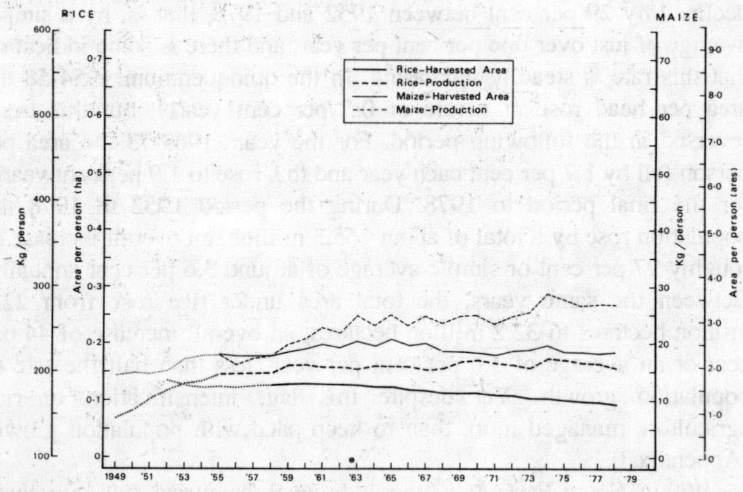

FIG: 15.1: Cereal—South-East Asia.

The socialist countries, Cambodia and Laos (Figs. 15.4 and 15.5) show a substantial year-to-year variations in both area and production, though this could reflect the whimsicalities of local officials as much as realities of a high degree of reliance in rice production upon the vagaries of rainfall. These granted, in the 1960s, the Sihanouk Government made considerable progress in expanding both the rice area and production more rapidly than the population. The data given for the years from 1970 reflect not only the abandonment of land as rural folk fled to escape bombing and other atrocities, and also the fact that the Government controlled increasingly less of the countryside. Vietnam too (Fig. 15.6) was seriously affected by war, though its effects upon production were less marked than in Cambodia. In the south the total area actually rose (by 29%) from 1959 to 1974 resulting in a limited fall, 18 per cent, in the area per person, but this was offset by substantial falls in the north. By the 1970s it would seem that various farm improvements in the south had succeeded in reversing the trend of a falling production per person as well, but again this was marked by an overall downward trend in the country as a whole. In northern Vietnam, in pre-partition days, a rice-deficit area, the exigencies of war and continued population growth led to a fall in the already low levels of both rice production and area per head. The basic problems of food supply in Vietnam have not been solved by socialisation in the countryside, and since the invasion of Cambodia, food supplies in Vietnam have been further reduced as herculean efforts have been made to assist Cambodia as well. The Government has taken steps to diversify and intensify agriculture. These have resulted in some increase in root crops, notably sweet potato, in an endeavour to provide minimally adequate intakes of carbohydrate.

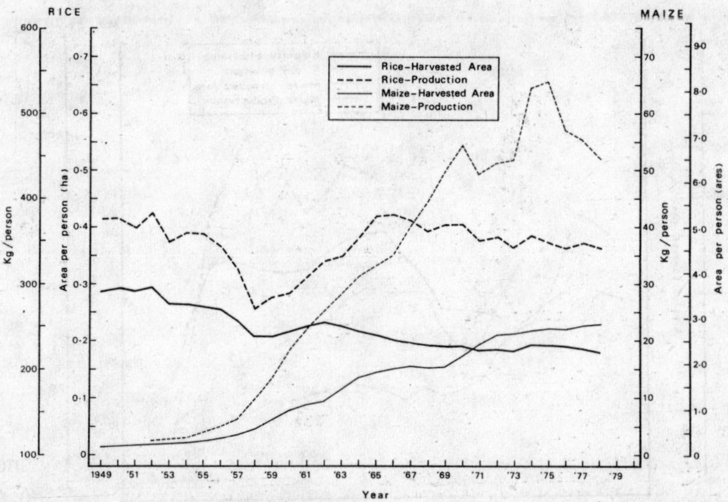

FIG. 15.2: Cereals—Thailand.

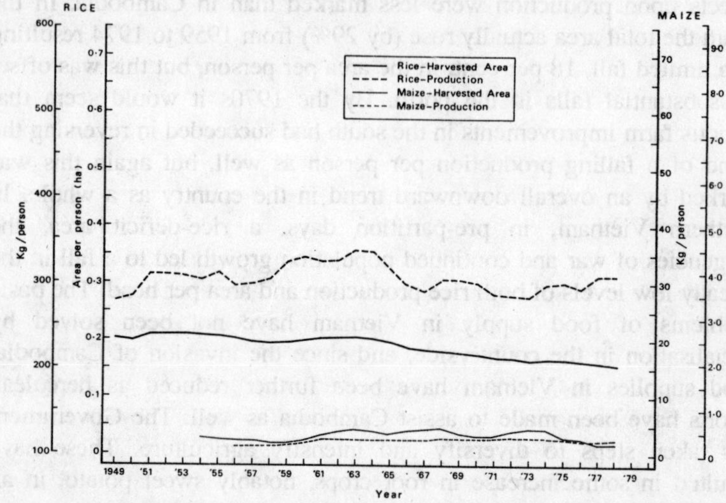

FIG. 15.3: Cereals—Burma.

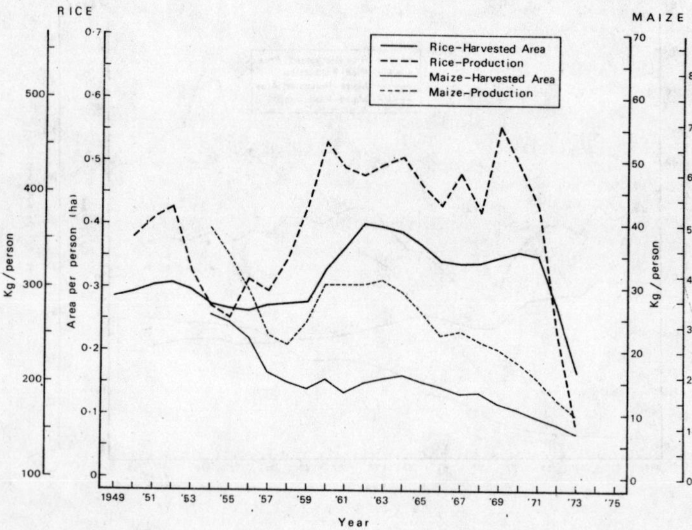

FIG. 15.4: Cereals—Cambodia.

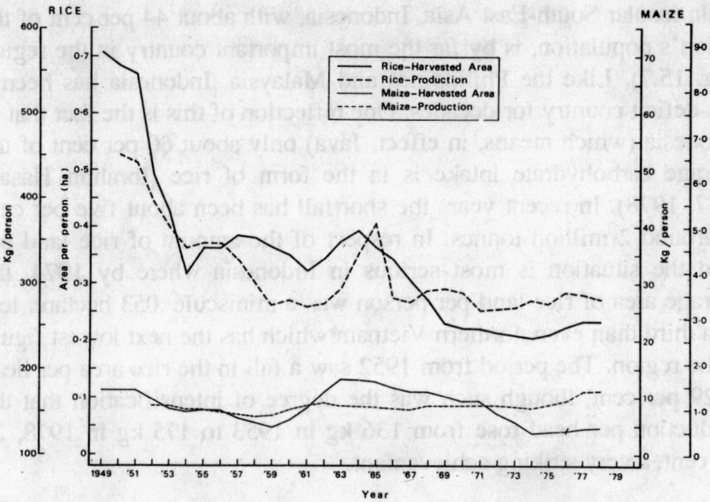

FIG. 15.5: Cereals—Laos.

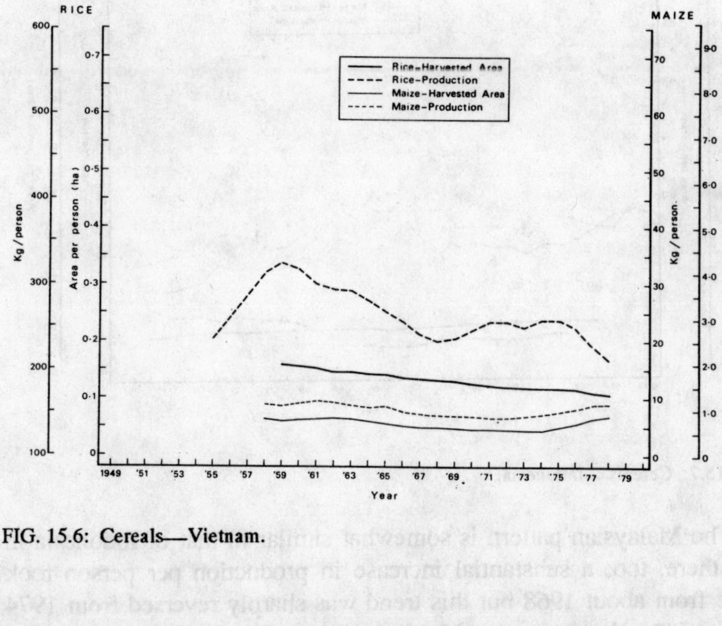

FIG. 15.6: Cereals—Vietnam.

In insular South-East Asia, Indonesia, with about 44 per cent of the region's population, is by far the most important country in the region (Fig. 15.7). Like the Philippines and Malaysia, Indonesia has been a rice-deficit country for decades. One reflection of this is the fact that in Indonesia (which means, in effect, Java) only about 60 per cent of the average carbohydrate intake is in the form of rice (Ibrahim Hasan, 1977, 1978). In recent years the shortfall has been about five per cent or around 2 million tonnes. In respect of the amount of rice land per head the situation is most serious in Indonesia where by 1974, the average area of rice land per person was a miniscule .053 hectare, less by a third than even northern Vietnam which has the next lowest figure in the region. The period from 1952 saw a fall in the rice area per head of 29 per cent, though such was the degree of intensification that the production per head rose from 136 kg in 1953 to 175 kg in 1978, 29 per cent, a very striking achievement.

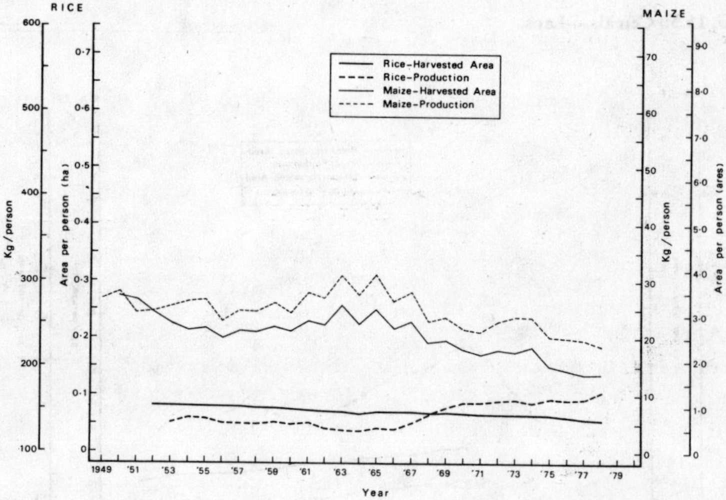

FIG. 15.7: Cereals—Indonesia.

The Malaysian pattern is somewhat similar to that of Indonesia in that there, too, a substantial increase in production per person took place from about 1968 but this trend was sharply reversed from 1974 (Fig. 15.8). However, the Malaysian rice deficit is proportionately larger than that of Indonesia, with about a fifth of consumption requirements being imported. The rice area per person is notably low and has not changed significantly despite an expansion of the harvested area from around 470,000 hectares in 1946,[3] to around 770,000 hectares in 1975. Unlike Indonesia, Malaysia has very large alternative sources of agricultural income, producing some 46 per cent of the world's oil palm products, for example.

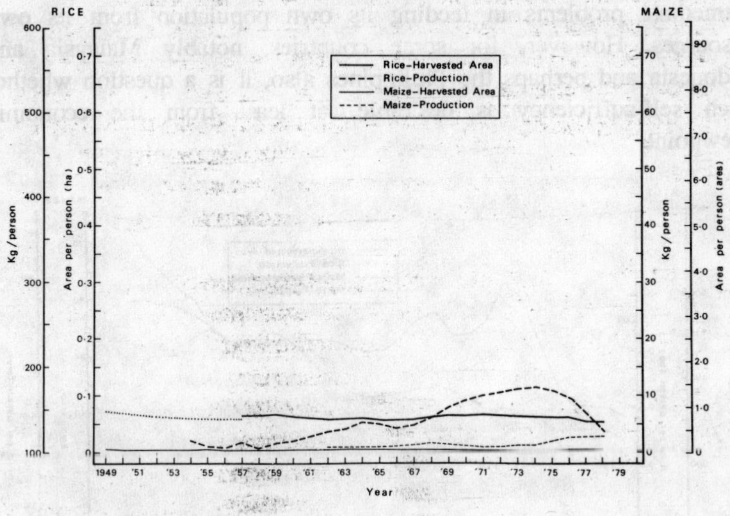

FIG. 15.8: Cereals—Malaysia.

After many years of being a deficit country, the Philippines recently has produced enough rice to feed itself. Unlike any other South-East Asian people, the Philippines, perhaps because of their Spanish/Mexican cultural heritage, are maize-eaters, so that while rice is still the dominant cereal, maize is an important supplement in the diet and many have no other staple. Rice production per head is fairly low (Fig. 15.9). Despite a rise in the rice area per person in the 1950s, as in other South-East Asian countries, the trend has been downward, while production per head has remained more or less static with an upturn from 1973 as high-yielding varieties began to take hold in the countryside.

For the region as a whole some conclusions may be drawn. With the only exceptions of Cambodia, Laos and Northern Vietnam, the production of rice has more than kept pace with population growth.

Where in the 1950s some countries — Thailand, Cambodia, Indonesia and the Philippines — saw rice production falling behind population growth, by the 1960s and especially in the early 70s, this fall had been reversed partly reflecting the adoption of new rice technologies (Asia Society, 1970). In reunified Vietnam, too, there seems every prospect in the long term that the south will regain its position as a rice-bowl. Only Cambodia, now an international "basket-case", faces major and immediate problems in feeding its own population from its own resources. However, for some countries, notably Malaysia and Indonesia and perhaps the Philippines also, it is a question whether such self-sufficiency is desirable, at least from the economic viewpoint.

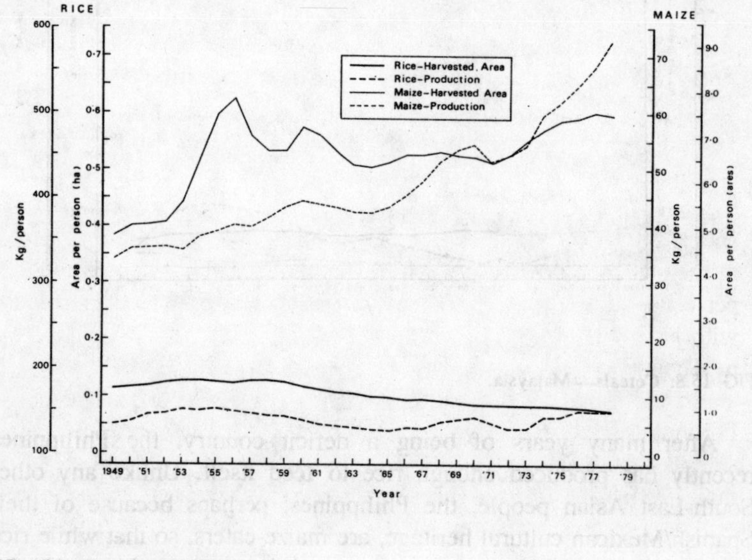

FIG. 15.9: Cereals—Philippines.

Maize

Although maize is of growing importance in the region, it is far from rivalling rice. Where rice production may be measured in hundreds of kilogrammes per person, maize is to be measured in tens, and where the rice area is measured in tenths of a hectare, the maize area is to be measured in hundredths. Maize is of importance for several reasons. First, though rice is and always will be the preferred cereal and the one with the greatest consumer elasticity of demand in the region,[4] maize is an acceptable supplement and in parts of the Philippines, especially the uplands of Mindanao and in Cebu, it may be rather more than a supplement. Second, maize fits into lowland farming where quick-maturing varieties can easily be intercalated with successive crops of wet-season rice. Third, maize (like manioc) fits readily into both shifting agriculture and dry rotational farming on ploughable uplands. Fourth, maize is in demand for feeding pigs and chickens in internal markets and for animal feedstuffs generally, in international markets. Data do not allow estimates of the proportions of production used directly as food or for animal feedstuffs within the countries of production. Only Thailand is a major exporter (89 per cent of production in 1976) although Burma exports about eight per cent of its crop and Indonesia about one per cent (U.N. Statistical Office, *Yearbook of International Trade Statistic*, 1977).

In the region as a whole the area of maize per person has remained more or less constant since 1952 though the total area increased by 3.7 million hectares over the period, that is, by a simple average of 3.3 per cent per year, fractionally slower than the population (Fig. 15.1). Over the same period, production rose from around 15 kg per person to about 27 kg, while the total production increased from about 2.6 million to an estimated 9.7 million tonnes in 1978, an increase of 270 per cent. Of this total, just under one-third was produced in Indonesia with just over one-third contributed by the Philippines and 31 per cent by Thailand.

In the cases of two of these major producers, Thailand and the Philippines, both the area and production per person have increased sharply (Figs. 15.2 and 15.9). In the latter country, the total area under maize increased from just under a million hectares in 1949-50 to 3.5 million hectares in 1978 — a 270 per cent increase in total area but only a 55 per cent increase in the area per person. Thailand has the largest proportionate increase in absolute area (by almost 1.4 million hectares) as well as in area per person (from 0.16 hectare in 1949 to 2.9 hectare in 1978). This contrasts with the rice area per head in these countries which fell steadily. Some intensification of production is indicated for Thailand since the upward slope of the line indicating maize production is steeper than that for area. In the Philippines, the pattern of change has been rather different in that the area expanded sharply in the mid-50s but until the early 70s, production did not keep step, though recently this has risen faster than the area. This presumably represents an initial expansion into somewhat marginal land with low standards of cultivation practice and perhaps also poor-yielding varieties of seed followed by a market improvement in practice and yields.

Although Indonesia produces only a little less than the Philippines, both the quantity and area per person are much lower reflecting the former's much larger population (Fig. 15.7). As in all the other countries of the region, the area per person under maize has tended to fall — in this case by half, while at the same time the total area increased hardly at all in the period 1950 to 1978. In fact, the population grew by 71 per cent overall at this time and, assuming constant yields, it would have been necessary for the maize area to have expanded at an average rate of 38,500 hectares per year just to keep pace with population growth.

The remaining countries are not of particular significance as maize producers. Both Cambodia and Laos (Figs. 15.4 and 15.5) have patterns of considerable variability despite statistical smoothing. In Vietnam, the trend of both production and area per head has been downward, paralleling the pattern for rice. In Vietnam and in Malaysia, maize is of minor importance and both production and area have barely kept pace with population growth (Fig. 15.8).

For the region as a whole it seems fair to suggest that considerable scope still exists for an expansion of both the area and production, since maize fits a number of agricultural niches.[5]

Within-Country Changes

Discussion thus far has centred upon the national level which naturally implies the use of national level data and national averages. The FAO *Production Yearbooks* give no indication of internal patterns of change with time except for the cases of the (former) States of Malaya, Sabah and Sarawak for which data are available for most years down to 1975. These are not important rice-producing areas, however.

The spatial reality behind the national aggregates is not at all easy to establish. While Peninsular Malaysia has excellent time-series statistics of rice areas and production for each State, in no other country is this true. Moreover, Malaysia is scarcely a major producer. Data for provinces or similar administrative units are available only for some countries and then only for particular years as, for example, southern Vietnam in 1970, mapped in Fig. 15.10 and the Philippines in

1971, mapped in Fig. 15.11. Both these indicate that national aggregates hide a very wide range of production and area per person in the provinces, the range being particularly striking in the Philippines. For Indonesia, published data are not available at the provincial *propinsi* or *kabupaten* levels, before the censuses of 1963 and 1973. However, it has been claimed that these censuses, especially the former, are unreliable (Ibrahim Hasan, 1977).

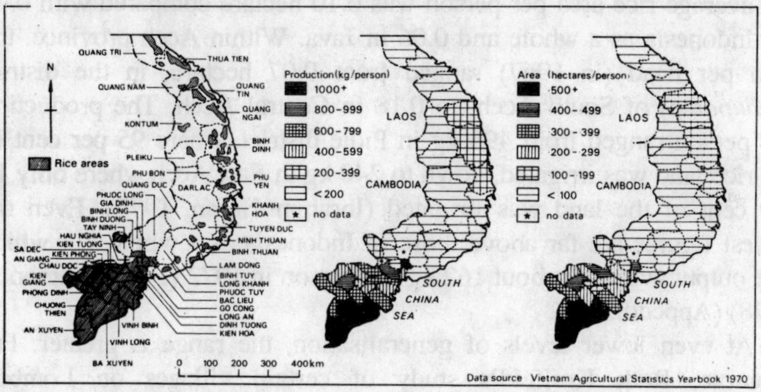

FIG. 15.10: Vietnam rice production, 1970.

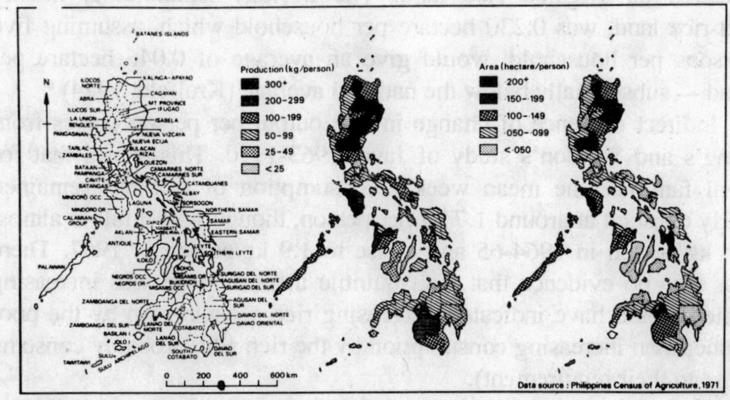

FIG. 15.11: Philippines rice production, 1971.

Further, entirely unsystematic data are to be found in various detailed studies, mainly by economists and anthropologists.

Not unexpectedly, a wide range of figures is found, most being higher than the national averages which may suggest that these averages are an understatement (or that field workers prefer to work in wealthier areas!). For example, Ibrahim Hasan (1977) has shown that in 1972, the average production of rice per head in Aceh province was estimated at 411 kg. more than double the national average. In Aceh, the average rice area per person was 0.10 hectare compared with 0.07 for Indonesia as a whole and 0.06 in Java. Within Aceh province, the area per head (in 1967) ranged from 0.07 hectare, in the district (*kabupaten*) of South Aceh, to 0.18 in Central Aceh. The production per person ranged from 496 kg in Pidie district, where 95 per cent of the rice land was irrigated, down to 240 kg in East Aceh where only 18 per cent of the land was irrigated (Ibrahim Hasan, 1977). Even the lowest figures are far above those of Indonesia as a whole, for which rice output averaged about 162 kg per person in 1972 (175 kg/person in 1978) (Appendix-III).

At even lower levels of generalisation, the range is greater. For example, Ruth Krulfeld's study of certain villages on Lombok indicated that 53 per cent of her sampled households held less than one hectare of land and 39 per cent held between one and two hectares. Compared with these the wealthiest man owned, but did not work, 16 hectares of irrigated rice land. The average landholding, mainly wet-rice land, was 0.230 hectare per household which, assuming five persons per household, would give an average of 0.046 hectare per head — substantially below the national average (Krulfeld, 1974).[6]

Indirect evidence of change in rice output per person comes from King's and Weldon's study of Java, 1963-1970. This shows that for rural families, the mean weekly consumption of rice has remained fairly constant at around 1.7 kg per person, though with a fall to almost 1.5 kg/person in 1964-65 and a rise to 1.9 kg/person in 1967. There was also no evidence that inter-quintile income range was increasing which would have indicated decreasing rice consumption by the poor (rather than increasing consumption by the rich who probably consume close to their requirement).

In contrast, Umehara Hiromitsu's study of a *hacienda barrio* in central Luzon suggested that the average amount of land per person was .22 hectare while Takahashi's study indicated that the average in Barrio Kubukiran was 0.4 hectare per person (Umehara 1974, 18, 30;

Takahashi, 1969, 43).[7] These compare with a national average of around 0.19 hectare at the time of these studies (0.18 hectare in 1978). Other studies, such as that of Purcal (1971) in Province Wellesley, Peninsular Malaysia, could be quoted, but they do little more than help to establish the dimensions of present ranges which are hidden in national averages. The answer to the question of whether the production and area per person (at the district level) decline to a plateau below which people move out and actively seek access to land in other areas is simply not known, though the fact that people may be actively seeking access to land far from their home districts is obvious enough (Supomo Surjohudojo, 1973). The reported increasing incidence of tenant-farmers and share-croppers from outside particular villages suggests that this may indeed be the case, but this and related questions require much more attention than has been given them hitherto.[8]

Food Crops and Diet

One important proposition put forward by the neo-Malthusians is that in the Third World diets are becoming poorer. Lester Brown, for instance, despite the evidence of his own data, has suggested that, "The demand for food will outrun all available indigenous supplies of land and water, creating severe and chronic scarcity. . .the descendants of the present population will be doomed to unbelievably miserable conditions of existence" (Brown, 1974). In fact, food production has, on the whole, kept pace with population growth. It is true that average levels of food intake in the Third World are substantially below those of the wealthy countries of the West and Japan, especially the intake of fats and protein. Yet we need to be reminded that nutritionists still argue about the meaning of "minimum levels".

For South-East Asia, data, especially time series data at the national level, are almost entirely lacking. Evidence from the Philippines (Fig. 15.12) shows that while the per person intakes are below Filipino "sufficiency" levels, they are not much below those levels, which are: total kilocalories per person per day — 2,650, protein — 54.5 gm, fats — 29 gm. In 1972 total food intake was 79 per cent of "sufficiency" down slightly from the figure achieved through most of the 1960s. The corresponding figures for protein and fats were 97 per cent and 96 per cent respectively. Between 1953 and 1972, total intake rose from 1,691 Kcal daily to 2,047 Kcal, a rise of 21 per cent. Over the same period, the intake of protein rose from 41.5 gms per day to 52.9 gms, a rise of 27 per cent, while for fats there was an increase of 46 per cent (NEDA, 1975). These data compare with Burmese national averages in 1958 of a total daily intake of 2,150 Kcal of which 320 Kcal was in the form of fats and oils, the former figure being slightly less than 80 per cent of the Philippine "sufficiency" level, while the latter is nearly four times that level (Nafis Ahmad, 1971). In fact, however, virtually all these figures may represent a significant surplus intake — at least if measured against the current FAO average requirement of 1,900 Kcal/day (Poleman, 1977).

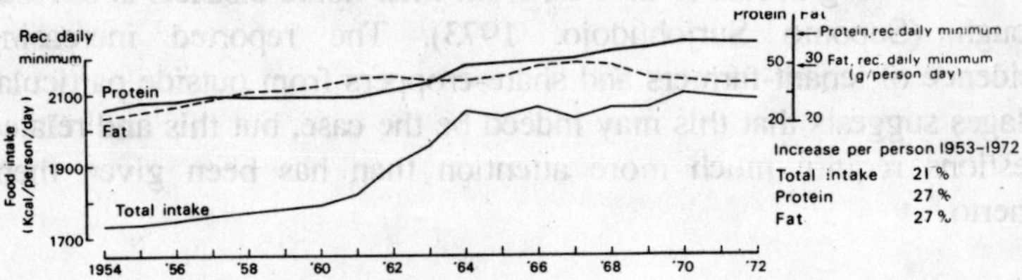

FIG. 15.12: Philippines—food intake per person.

In Indonesia, it is clear that at the national level, rice production per person exceeds the South-East Asian average requirement of 151 kg per year (Wilhelmy, 1976), and contrary to general belief, the proportion of rice in the total carbohydrate intake has tended to rise. In 1960, rice constituted only 55 per cent of total average carbohydrate intake but by 1970 this had risen to 64 per cent (Ibrahim Hasan, 1977). Nevertheless, there is considerable inter-regional variation. According to Bambang Ismawan and Soemartojo (1970), the average weekly consumption per household of milled rice in Javanese villages was 6.98 kg compared with 12.78 kg in villages outside Java, rice forming 50 and 87 per cent of the total starchy food intake respectively. Ibrahim Hasan is undoubtedly correct in suggesting that in Java there is considerable elasticity of demand for rice, but this is not the same as saying that there is an absolute dearth of food. Nor is it true to suggest that there is an absolute dearth of rice since he points to the case of Aceh province as a rice surplus area from which rice is exported with some difficulty because of transportational and other infrastructural rigidities. This raises the question of whether Indonesia and other countries in a similar situation might not be capable of meeting all their own rice requirements if only these rigidities could be removed and a truly national market in rice created.

Annual Oil Crops

While the production of cereals is important, it is the growing of annual and perennial oil crops (and of rubber) which leads to large national incomes from the agricultural sector, as in Malaysia, for instance. Their cultivation is undoubtedly more profitable, though more vulnerable to price fluctuations, than cereals.

The two most important annual oil-crops are soyabeans and groundnuts. They have both been cultivated rather more extensively in recent years than hitherto, both to meet internal demand for blended cooking oils and, to a limited degree, to meet international demand. In the region, the area under soyabeans rose from just over 400,000 hectare in 1950 to 928,000 hectare in 1978 with production increasing from around 270,000 tons to about 700,000 tons (Fig. 15.13). Measured against population growth these figures represent virtually no increase in the area per person and a 21 per cent increase in production, though the actual areas and quantities involved are small — around two kg per person, for quantity. For groundnuts, a rather longer-established crop, the area rose from 680,000 hectare in 1950 to 1.39 million hectare in 1978. Production increased from around 460,000 tons to 1.5 million tons. Discounted for population growth, these figures represent a fractional decline in area but a 40 per cent increase in production. Since around 1963, production per person has tended to increase more rapidly than area. At the same time, the area under soyabeans has increased more rapidly than that under groundnuts.

Annuals such as these fit fairly easily into permanent lowland agriculture in rotation with rice and are also suitable for ploughable upland soils of fair to good nutrient status, both as crops in a rotation and as crops planted beneath coconut or other tall crops which do not cast dense shade. To some degree, the expansion of their cultivation reflects growing commercialisation of peasant agriculture since growers, at least lowland growers, may have to forego part of their rice crop in order to grow them. From the farmer's point of view, their great advantage is that, being annuals, the leadtime to production is short and the possibility is offered of quickly switching to alternative crops should these prove more remunerative. These are advantages the grower of tree-crops lacks.

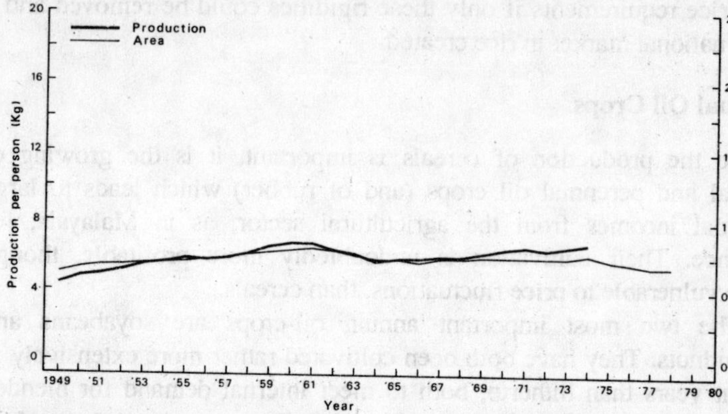

FIG. 15.13: Annual oil crops (groundnuts and soy)—South-East Asia.

Indonesia (Fig. 15.14) is by far the largest producer of annual oil crops in the region with half of both production and area and higher per person figures than any other country in the region except Burma. Despite a decline in both area and production per head during the 1960s, for production this trend has been reversed in the 1970s. By 1978, though the area per person was about 15 per cent lower that of 1950, the production had increased by 13 per cent to 8.3 kg per head.

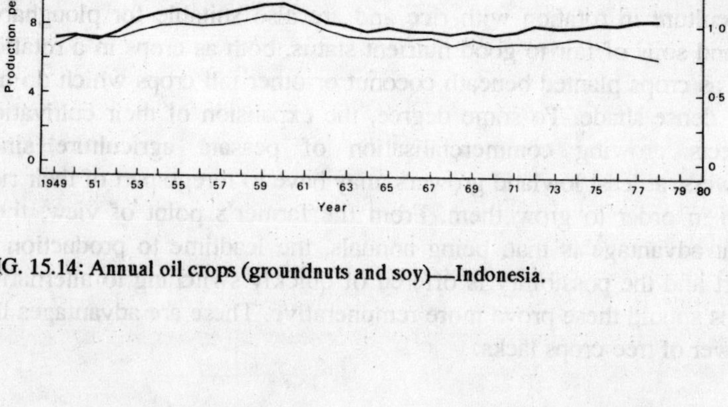

FIG. 15.14: Annual oil crops (groundnuts and soy)—Indonesia.

Burma (Fig. 15.15) produces more annual oil crops (almost all groundnuts) per person and has a greater crop area per person than any other South-East Asian country and accounts for 26 per cent of the area though only 21 per cent of the production. Both production and area have increased sharply since 1950, though with a long-term cyclic pattern probably reflecting rainfall conditions.[9]

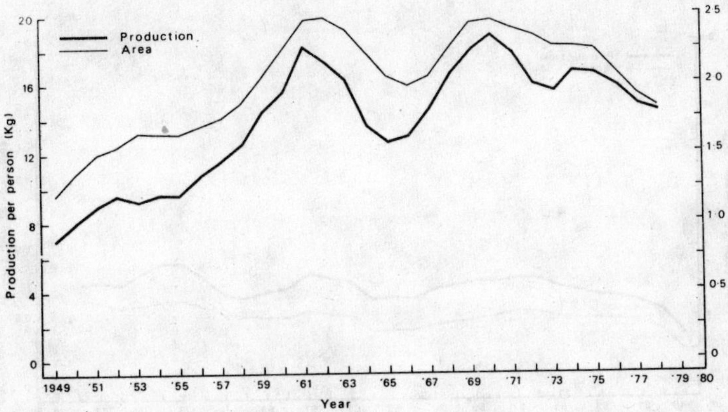

FIG. 15.15: **Annual oil crops (groundnuts and soy)—Burma.**

For the remaining countries there is little of consequence to note except that Thailand, which produces about 13 per cent of the regional total, seemed to be on the edge of a boom in production in the early 1970s but this did not occur (Fig. 15.16). In neighbouring Cambodia, which has similar levels of production per person, though a much smaller total production, war would seem to have had rather less serious effects upon oil crops than upon rice (Figs. 15.17 and 15.4). In the Philippines, the production and area curves are flat, indicating that these have kept pace with population growth (Fig. 15.18). Much the same is true in Vietnam, where in the north and unlike cereals, the annual oil crop production and area have kept pace with population (Figs. 15.19 and 15.6). In the south they have been slightly ahead of population though overall production per person has tended to fall since the early 1970s.

In most countries of South-East Asia the area per head has been either more or less stable or has risen. The prospect would seem to be for little change in area, coupled with production increasing somewhat more rapidly than area. Both crops, especially soyabean, are comparatively new and farmers in many areas are unacquainted with their versatility. From the agronomic viewpoint, soyabeans are especially valuable fixers of nitrogen in the soil. From the dietary viewpoint, any addition of oils and protein is to be welcomed, and in terms of breaking the "subsistence syndrome", these crops have a good deal to offer, at least in areas of better soil.

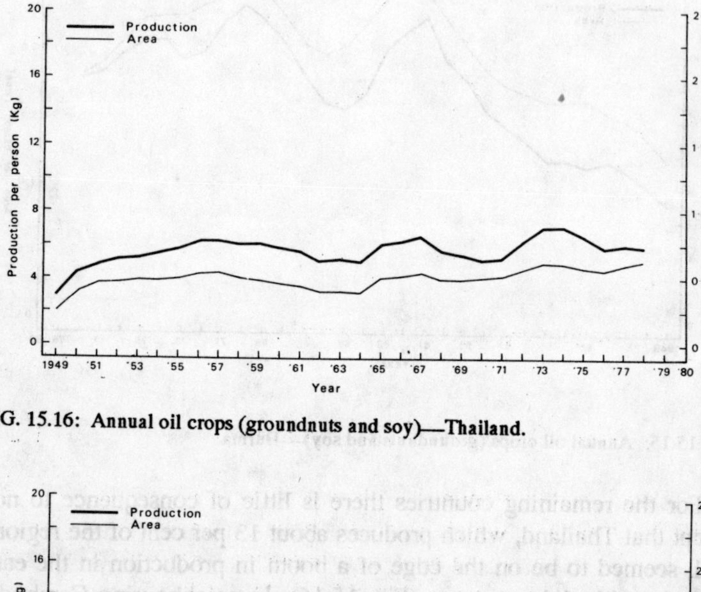

FIG. 15.16: Annual oil crops (groundnuts and soy)—Thailand.

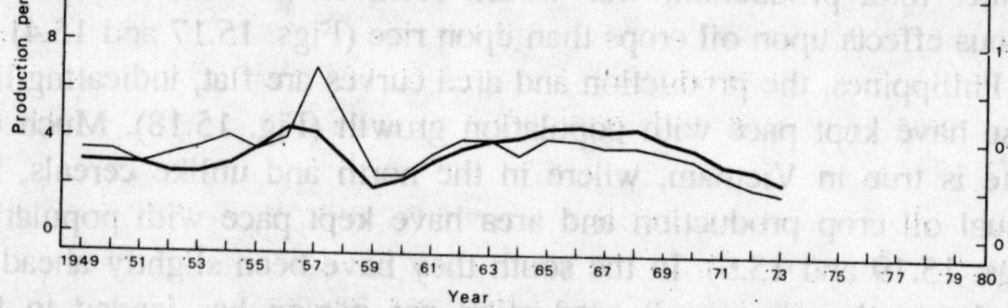

FIG. 15.17: Annual oil crops (groundnuts and soy)—Cambodia.

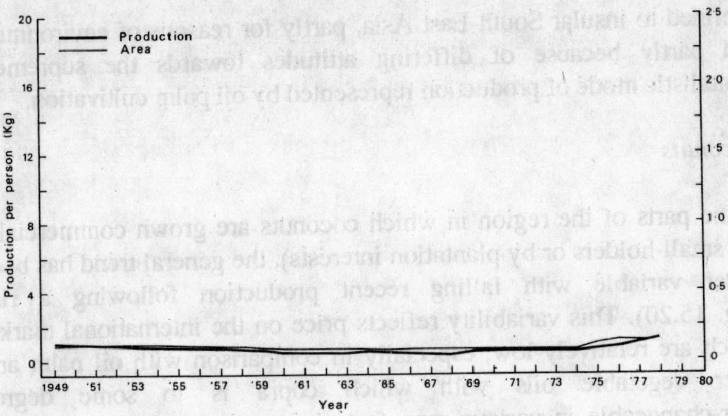

FIG. 15.18: Annual oil crops (groundnuts and soy)—Philippines.

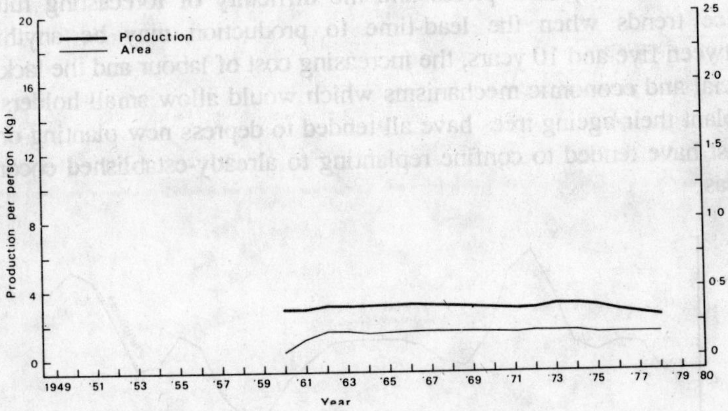

FIG. 15.19: **Annual oil crops (groundnuts and soy)—Vietnam.**

Perennial Oil Crops

The only perennial oil crops of significance in the region are coconut, usually converted into copra, and oil palm, from which both palm oil and palm kernel oil may be extracted. These crops are almost entirely confined to insular South-East Asia, partly for reasons of environment and partly because of differing attitudes towards the supremely capitalistic mode of production represented by oil palm cultivation.

Coconuts

In the parts of the region in which coconuts are grown commercially (by small-holders or by plantation interests), the general trend has been rather variable with falling recent production following a rise (Fig. 15.20). This variability reflects price on the international market which are relatively low, especially in comparison with oil palm and other vegetable oils with which copra is to some degree inter-changeable in various manufacturing processes. (Though actual plantations may be considered "inelastic", production especially from small holdings is in fact elastic as nuts are simply not collected when prices are low.) Low prices and the difficulty of forecasting future price trends when the lead-time to production may be anything between five and 10 years, the increasing cost of labour and the lack of social and economic mechanisms which would allow small holders to replant their ageing trees have all tended to depress new planting or at least have tended to confine replanting to already-established coconut areas.

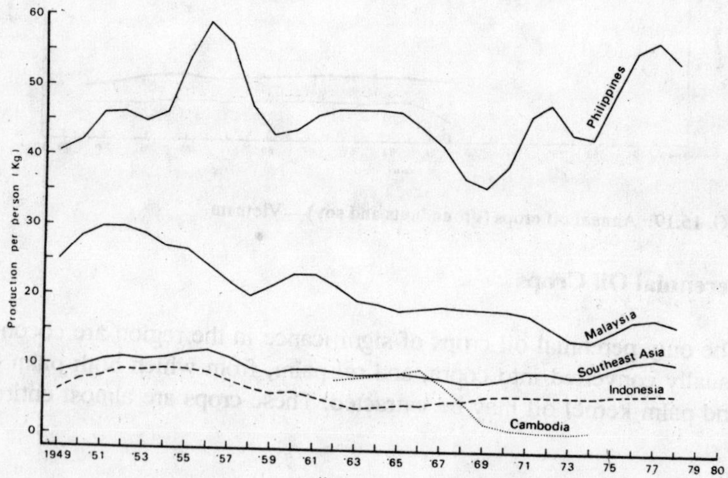

FIG. 15.20: Perennial oil crops—copra production.

Areal data for the period are not generally available so that it is not possible to establish the relationship between production and area. However, data are available for Peninsular Malaysia from 1947 until 1964 (*Malaya Yearbooks*) and these indicate that the area was more or less constant over that period at around 21,000 hectare but the area per person fell from 0.042 hectare in 1947 to 0.026 hectare in 1964, an overall decline of just over a third. This compares with a fall in production per head from 22 kg per person in 1947 to about 15 kg per person in 1964, a fall of exactly one-third. Since there is broad similarity in the conditions under which coconuts are grown, type of planting material and age of trees in the producer countries it may be surmised that the fall in production per head has been paralleled by a fall in the area per head.

Oil Palm

The position with regard to oil palm is rather different (Fig. 15.21). Production is confined to Malaysia, Thailand and the Philippines, in the last two of which production has barely begun. Malaysia now produces close to half of the world total or four-fifths of the regional total. Expansion has been extremely rapid since the late 1950s, with plantings coming into maturity at the rate of around 50,000 hectare per year for the whole country. In 1949 the production per person was only 10 kg, but by 1978 this had reached 166 kg. The output in 1976 increased by about 28 per cent over 1975 to reach 1.45 million tonnes but increases of this magnitude could not be sustained and the 1977-78 increase was only 10 per cent to reach 2.14 million tonnes. Some 35 per cent of the area under oil palm is under the control of two government organisations (*Felda* and *Risda*) which promote production by small-holders on semi-plantation lines. Expansion of the planted area in Sarawak (23,620 hectare in 1976) and in Sabah (planted area, 64,028 hectare in 1976) will continue, and in any case production will increase rapidly as immature palms come into bearing. For the whole of Malaysia, in 1976 the area under oil palm increased by seven per cent over 1975 — three times the rate of population increase which has fallen to about 2.3 per cent annually. Production has more than kept pace with the increases in area. Over the period 1970–75, the average annual rate of increase of the palm oil production was 24 per cent (Ministry of Finance, Malaysia, 1976).

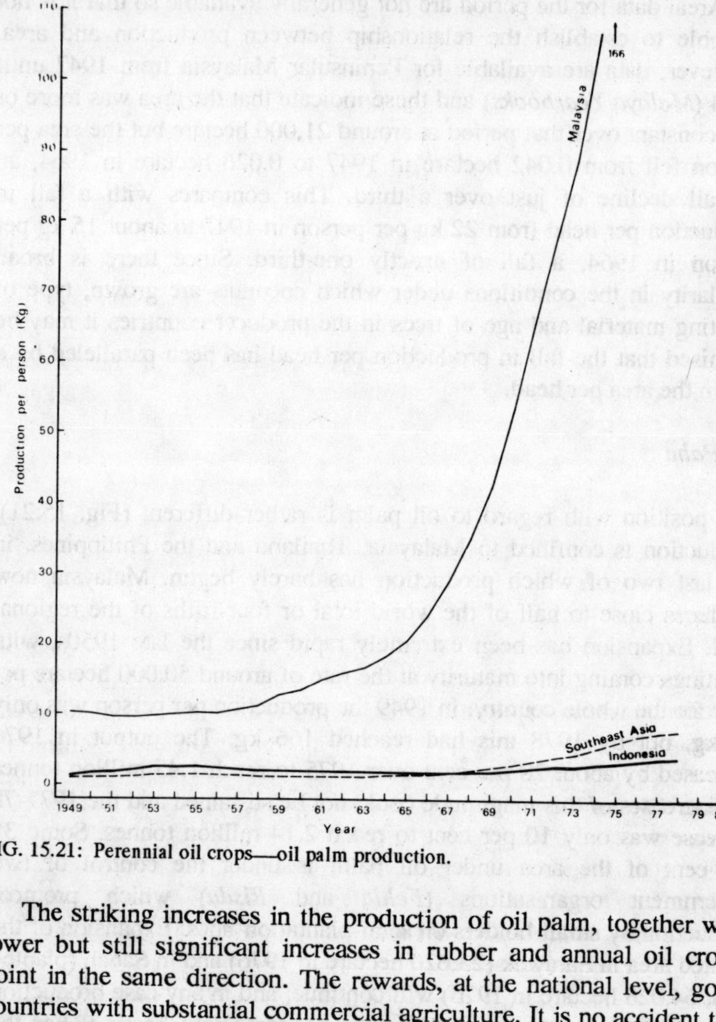

FIG. 15.21: Perennial oil crops—oil palm production.

The striking increases in the production of oil palm, together with lower but still significant increases in rubber and annual oil crops, point in the same direction. The rewards, at the national level, go to countries with substantial commercial agriculture. It is no accident that within the region, Malaysia has a higher GDP per person [about $1,000 (U.S.)] than any other country with a substantial agricultural base. What remains for Malaysia is not merely to continue to increase production of perennial upland crops and to export raw materials, but to make its own some of the forward linkages in the processing of these crops.

For the rest, it is clear that extension of the cultivated area will continue. Great scope still exists for perennial crops, especially in those upland areas of insular South-East Asia where slopes do not exceed 15 degrees and there is no significant dry season. The greatest challenge lies in the uplands of mainland South-East Asia, where there is simply no crop which offers these areas the same potential as do rubber and oil palm in insular South-East Asia. At the same time, there is obvious scope, except perhaps in the crowded plains of Tonkin and northern Annam, for further intensification of production. This process has already begun and will undoubtedly continue if only under the pressure of growing numbers of people. It is clear, then, that with very few exceptions, actual production per head of all the major crops is at least keeping pace with the population growth and, in many cases, even the production of rice, which is lagging behind the others, is, in per head terms, increasing, not decreasing (Appendix-IV). The problem is not to keep pace with population, that is being achieved, but to keep pace with rising expectations. At the same time, the often considerable reduction in the area per person of a number of crops is a cause for concern, but the region is very far from being reduced to the miniscule plots growing manioc (because it gives very high carbohydrate yields per unit of area) which are typical of the most penurious areas of Java.

Conclusion

From the earlier discussion, it is clear that the increases in both production and area over the last quarter of a century or so have been both sustained and large. An increase in the maize area of 7.1 million hectares, between 1952 and 1978 is huge by any standards. But if the question is asked just where, when and by how much did this change or a similar one take place (other than at the national level), the answer must be that this is not known. No country has anything like a complete set of landuse maps or statistics though Malaysia perhaps comes nearest with an aerial photograph-based landuse survey complete with statistics at the State level, but this is confined to Peninsular Malaysia and dates from the mid-1960s to 1970 (Wong, 1966; 1970a, b; Siew, 1970a, b). In Indonesia, landuse mapping of the whole of Java and Bali together with most of the accessible parts of Sulawesi and Sumatra has been completed and maps published by the Landuse Section of the Agrarian Department. These are of varying dates and degrees of accuracy (The mapping has been carried out largely without reference to existing aerial surveys by two men on a motor-cycle stationed in each province). For the rest, so far as I am aware, there is nothing.

The basic problem for any researcher is access to aerial photographs which exist for much of the region but are tightly held usually by the military. In some areas, especially those crucial for land development, successive runs at different times exist and these are invaluable to establish trends. It would seem that governments, apart from that of Malaysia, are unconvinced of the need for landuse mapping or even basic landuse data such as could readily be furnished by local administrations were they ever asked for them, though the agricultural censuses of the early 1970s make some amends.

Change in the structural dimension of the whole economy is easier to document and the data suggest that since 1950 there has been a steady decline in the contribution of agriculture to the gross domestic product in most countries (see Appendix-II). This decline has not everywhere been accompanied by a corresponding fall in the contribution of agriculture to employment, especially if this is measured not merely in terms of those whose livelihood lies within the agricultural sector, but in the labour input terms of actual work done.

There is no reason to believe that the evolutionary mechanisms of peasant societies, as described by Geertz, will not continue to operate "successfully" for a long time yet. Production levels are low in Java and elsewhere but they are rising, if slowly. The data presented here tend to support Boserup's notion that peasant economies contain feedback mechanisms which lead to the intensification of agriculture as a response to population growth. Moreover, the addition to the economy of maize, annual oil crops and especially export crops such as oil palm and rubber, together with vigorous industrialisation employing labour-intensive technology and involving further processing of agricultural raw materials than is now the case do offer some prospect of escape form the "low-level productivity trap", though, in the capitalist-oriented countries, at the cost of vulnerability to the vagaries of the international market place.

APPENDIX I

South-East Asia: population, cereal and annual oil-crop production (000 tonnes) and area (000 ha) estimates, 1948–1978, by country

	Year						
	1948	1953	1958	1963	1968	1973	1978
Burma:							
Population (mill.)	18.1	19.0	20.7	23.7	26.0	28.9	32.1
Rice—production	5153	5741	6590	7783	8023	8602	10500
—area	4624	4004	3998	4789	4763	4879	5200
Maize—production	n.a.	34	26	56	62	80	75
—area	n.a.	70	45	131	154	175	80
Groundnuts—production	124	194	289	337	398	412	450
—area	216	321	397	558	611	663	570
Soyabeans—production	n.a.	n.a.	n.a.	9	10	13	16
—area	n.a.	n.a.	n.a.	14	15	20	23
Cambodia:							
Population (mill.)	3.7	4.0	4.7	5.8	6.6	7.7	n.a.
Rice—production	1270	1523	1153	2622	3251	1050	n.a.
—area	1058	1205	1225	2296	2324	811	n.a.
Maize—production	41	181	69	183	154	73	n.a.
—area	28	135	58	115	113	63	n.a.
Groundnuts—production	4	4	2	7	26	14	n.a.
—area	7	6	3	7	26	13	n.a.
Soyabeans—production	7	5	13	11	4	2	n.a.
—area	9	10	23	14	5	3	n.a.
Indonesia:							
Population (mill.)	72.7	80.1	89.7	100.7	113.3	129.2	147.0
Rice—production	5895 *	10970	11969	11900	17160	21500	25739
—area	3665 *	6387	6990	6738	8021	8404	8812
Maize—production	1600 **	1815	2634	2358	3166	3690	2750
—area	2025 **	1970	2702	2559	3220	3433	2500
Groundnuts—production	270	334	386	392	478	505	687
—area	250	289	331	348	394	407	509
Soyabeans—production	256	306	418	353	420	529	530
—area	360	450	605	540	677	751	671

(Contd.)

(Contd.)

	Year						
	1948	1953	1958	1963	1968	1973	1978
Laos:							
Population (mill.)	1.2	1.3	1.7	2.0	2.8	3.2	2.9
Rice—production	n.a.	530	522	520	771	884	796
—area	n.a.	359	696	753	656	665	690
Maize—production	13	11	12	15	25	27	30
—area	17	12	12	30	38	15	18
Malaysia@:							
Population (mill.)	5.9	6.5	7.6	8.8	9.9	11.3	12.9
Rice—production	527 †	763	872	1188	1428	1979	1590
—area	376 †	362 †	498	524	620	748	630
Maize—production	9 †	n.a.	n.a.	10	14	16	35
—area	7 †	4 †	6 †	9	9	7	11
Groundnuts—production	2	n.a.	n.a.	4	4	23	25
—area	n.a.	n.a.	n.a.	n.a.	n.a.	n.a.	n.a.
Oil palm—production ‡	56.2	62.8	89.4	156.3	345.6	989.2	2141.8
Copra—production	121.0	180.1	139.3	183.2	189.2	158.6	207.0
Philippines:							
Population (mill.)	19.2	21.0	24.4	30.2	34.7	40.1	46.2
Rice—production	2491	3182	3685	3843	4445	5594	6907
—area	2164	2920	3229	3129	3332	3437	3758
Maize—production	618	781	1016	1293	1733	2289	3333
—area	853	1120	1336	1898	2256	2763	3474
Groundnuts—production	21	18	16	14	15	18	47
—area	25	28	20	25	30	33	64
Copra—production	882.8	869.0	1073.9	1511.3	1290.0	1698.4	2600.0
Thailand:							
Population (mill.)	17.7	19.2	23.9	28.7	34.0	39.5	45.0
Rice—production	6835	8239	7186	10029	12410	14898	17000
—area	4930	5931	5108	6730	6500	7743	8288
Maize—production	17	51	186	858	1630	2343	3030
—area	21	47	125	388	692	1050	1400
Groundnuts—production	8	78	121	113	158	152	170
—area	10	72	96	82	101	110	125
Soyabeans—production	7	20	22	33	45	152	125
—area	3	24	20	32	57	139	175
Copra—production	n.a.	15	16	21	21	37	48

(Contd.)

(Contd.)

	Year						
	1948	1953	1958	1963	1968	1973	1978
Vietnam							
Population (mill.)	24.1	25.9	27.9	33.1	37.6	42.7	49.3
Rice—production	7990	2463 [1]	8811	9623	8366	11125	9880
—area	4250	1924 [1]	4491	4497	4894	5030	4889
Maize—production	23	41 [1]	257	254	262	301	460
—area	14	35 [1]	226	273	239	241	450
Groundnuts—production	n.a.	13 [1]	23 [1]	66	74	93	105
—area	n.a.	22 [1]	29 [1]	70	76	89	105
Soyabeans—production	n.a.	n.a.	n.a.	39	49	59	22
—area	14	14	n.a.	30	44	51	45
South-East Asia (totals) [2]							
Population (mill.)	162.6	177.0	200.6	233.0	264.9	302.6	335.4
Rice—production	30161	33411	40788	47508	54472	65632	72412
—area	21067	23092	26335	29456	31110	31717	32267
Maize—production	2218	2880	3971	5017	6878	8818	9713
—area	2965	3393	4510	5403	6721	7747	7933
Groundnuts—production	402	634	833	929	1149	1202	1502
—area	508	738	876	1090	1234	1315	1386
Soyabeans—production	263	326	440	410	493	735	705
—area	386	484	648	628	799	967	928

* Java, Madura, Bali, Lombok only.

** Java, Madura only.

@ As at present constituted.

† Malaya and Sabah.

‡ Oil and kernels.

1 South Vietnam only.

2 Excluding Cambodia in 1978, and with earlier provisos.

APPENDIX II

South-East Asia: proportion of G.D.P. derived from agriculture (incl. forestry, hunting and fishing, at current prices, per cent)

	1950	1953	1956	1959	1962	1965	1968	1971	1974	1977
Burma	40.3	34.5	31.7	32.6	32.2	35.1	39.8	38.3	42.4	47.3
Cambodia				42.9	41.4	41.8	n.a	n.a.	n.a.	n.a.
Indonesia				54.7	57.5	57.0	54.4	46.2	32.5	31.3
Malaysia					33.3	28.6	34.4	33.0	34.6	31.7
Philippines	42.3	42.9	36.8	33.8	33.1	35.1	36.6	31.4	30.9	29.8
Thailand		43.5	40.4	36.7	35.2	36.4	32.8	29.7	32.7	29.7
S. Vietnam					32.8	40.6	45.5	48.7	n.a.	n.a.

Note: The countries concerned changed their system of accounts at various times in the late 1960s. The figures are not, therefore, exactly comparable, though the differences are mostly small (though see Burma) and generally increased the contribution of agriculture by 3-4 per cent. The higher values are given here. The Malaysian figures for 1974 and 1977 are in constant prices of 1970.

Source: U.N. Statistical Office, *Yearbooks of National Accounts Statistics.*

APPENDIX III

Estimated agricultural production per person, 1948–1978*

		Rice (kg)	Maize (kg)	Groundnuts & Soyabean (kg)	Tobacco (kg)	Rubber (kg)	Oil Palm Product (kg)	Copra (kg)
Burma	1948	285	1.78	6.9	n.a.	0.51	—	n.a.
	1958	304	1.26	14.4	1.98	0.58	—	n.a.
	1968	304	2.54	18.0	2.13	0.47	—	n..a
	1978	311	2.34	14.5	1.90	0.62	—	n.a.
Indonesia	1948	n.a.	n.a.	7.24	n.a.	6.04	0.70	5.94
	1958	133	24.5	8.86	0.70	7.96	2.06	7.77
	1968	142	23.0	7.50	0.60	6.52	1.93	4.98
	1978	175	18.7	8.28	0.59	5.92	4.10	6.63
Malaysia	1948	95	1.50	1.62	—	133	9.5	20.5
	1958	107	n.a.	n.a.	—	97	11.2	20.1
	1968	142	1.67	0.40	—	113	35.4	18.5
	1978	123	2.71	1.78	0.66	124	166.0	16.5
Philippines	1948	130	32.2	1.09	0.90	—	—	46.0
	1958	145	41.5	0.68	2.14	—	—	47.1
	1968	137	51.5	0.45	1.66	0.38	—	37.3
	1978	149	72.1	1.39	1.13	0.31	—	54.1
Thailand	1948	360	n.a.	0.85	0.89	5.50	—	n.a
	1958	272	8.9	6.06	2.77	6.24	—	0.65
	1968	363	44.8	5.62	2.39	7.42	—	0.62
	1978	343	52.3	5.92	1.53	9.78	—	1.07
Vietnam	1948	n.a.	n.a.	n.a.	n.a.	n.a.	—	n.a.
	1958	310	9.06	n.a.	n.a.	2.59	—	9.06
	1968	234	6.98	3.38	0.31	0.86	—	5.98
	1978	211	8.80	2.41	0.37	0.88	—	9.43

* Calculated using 3-year running means.

Source: FAO Production Yearbooks, U.N. Demographic Yearbooks.

APPENDIX IV

Population increase and estimated production increase per person 1949–1973 (per cent)

	Population	Rice	Maize	Groundnuts & Soyabeans	Tobacco	Rubber	Oil Palm Products	Copra
urma	75	9	+31	110	1949-78 35	22	—	—
Indonesia	100	1952-78 29	1957-78 −25	14	1951-78 −18	−2	484	12
Malaysia	115	29	81	10	—	−5	1647	−20
Philippines	138	15	124	28	26	—	—	18
Thailand	150	−5	1952-78 2079	565	72	78	—	n.a.
Vietnam	105	1955-78 −11	1958-78 −3	1960-78 −18	1951-78 57	1951-78 −41	—	1955-78 29
South-East Asia (countries listed)	111	1952-78 15	1957-78 55	53	1951-78 −6	1951-78 −10	687	27

NOTES

1. Nevertheless, doubts remain. In most countries, there are few, if any, field checks on yields independent of governments and the methods of estimating crop production, yields and areas are not disclosed. In fact, most countries require a report from local district officials, not always agricultural officers, and it is from these that national figures are complied. *Faut mieux* it has to be assumed that errors tend to cancel each other and that the degree of unreliability has stayed more or less the same through time. Even in the unlikely event of the whole of the past increases in production being attributable to improved reporting, the present situation would not be altered. Since the FAO derives its figures from government publications these are of no value as a check.
2. The published figures for Cambodia have been used only up to 1973 since later data are entirely erroneous.
3. This includes a very dubious figure for Sarawak, which has been omitted from the data given for the period 1949-1957 in the relevant figure.
4. Ibrahim Hasan (1977, 10) has reported income elasticities of demand of 1.7 for low-income groups in rural east Java.
5. In tracing the future progress of this crop one enumerative difficulty is already becoming apparent. At the moment, most maize is grown on land which grows nothing else in the course of a year. If, however, significant proportions of maize come to be grown on land which has already grown another crop in the same year, this could lead to a situation in which it becomes impossible to distinguish this maize from that grown on land which has grown only maize in the course of a year. In other words, it will become impossible to distinguish between intensification and extension of the cultivated area.
6. Since the national averages relate to all people, not just rural people, it could be argued that to achieve strict comparability, the rural figures should be deflated by about one-fifth.
7. The first value has been calculated from Umehara's data for the "municipality of Guimba" and the second is from Takahashi's data for his study-village.
8. In this connection, "longitudinal" studies such as that of a number of villages near Lam Pao, north-east Thailand, a research project by a group from the School of Oriental and African Studies, University of London, would seem essential to establish the nature and direction of these trends at the district level. Incidentally, such studies would also serve to establish the dimensions of what appears to be an emerging agricultural problem (at least in Java). namely, the displacement of rural labour by recent changes in rice-harvesting methods (see Collier, Wiradi and Soentoro, 1973; Boedhisantoso, 1974).
9. I am indebted to Dr. Christopher Dixon for this interpretation.

REFERENCES

Asia Society, 1970 : *Agricultural Revolution in Southeast Asia*, New York, 2 Vols.

Bambang Ismawan and Soemartojo, 1970 : "Masalah pengembangan swadaja petani", *Agro-Ekonomika*, 1(1), 90-102.

Boedisantoso, 1973 : *Rice Harvesting in the Krawang region (west Java) in relation to high yielding varieties*, Centre of Southeast Asian Studies, Monash University, Melbourne.

Boserup, E., 1965 : *The Conditions of Agricultural Growth*, Chicago.

Brown, Lester, 1974 : *By bread alone*, Praeger, New York.

Collier, William L. Gunawan Wiradi and Soentoro, 1973 : "Recent changes in rice harvesting methods, some serious social implications", *Bull. Indon. Econ. Stud*, 9(2), 36-45 (also reprinted by Agricultural Development Council, New York).

Ibrahim Hasan, 1977 : "Rice marketing in Aceh", *Teaching and Research Forum*, 8 , Agricultural Development Council and Singapore University Press, Singapore.

Indonesia, Biro Pusat Statistik 1976a : *Pertanian* (Sensus pertanian 1973) Jakarta, V. 1-4.

———, 1976b : *Jumlah dan luas tanaman menurut jenis tanaman untuk daerah....* (Sensus pertanian 1973), Jakarta.

———, 1976c : *Perkebunan* (Sensus pertanian 1973), Jakarta.

King, D. Y. and Weldon, P. D., 1977 : "Income distribution and levels of living in Java, 1963-1970", *Econ. devel. & cultural change*, 25(4) 699-711.

Krulfeld, R., 1974 : *The Village Economies of the Sasak of Lombok: A Comparison of three Indonesian Peasant Communities*, Ph.D. diss., Yale University.

Malaysia, Ministry of Finance, 1976 : *Economic Report 1976-77*. Ministry of Finance, Kuala Lumpur.

Nafis Ahmad, 1971 : *Economic Resources of the Union of Burma*, Office of Naval Research : Natick, Mass.

Philippines National Economic and Development Authority, 1975 : *Statistical Yearbook of the Philippines*, 1975, Manila.

———, 1974 : *1971 Census of Agriculture*, Manila.

Poleman, Thomas T., 1977 : "World food: myth and reality", *World Development*, 5(5-7) 383-394.

Purcal, J., 1971 : *Rice economy — Employment and Income in Malaysia*, Honolulu.

Sartaj Aziz, (ed.) 1975 : *Hunger, politics and markets, the real issues in the food crisis*, New York University Press, New York.

Siew Kam Yew, 1970a : *The Present Landuse of Pahang*, Ministry of Agriculture and Co-operatives, Kuala Lumpur.

———, 1970b : *The present landuse of Peruk*, Ministry of Agriculture and Co-operatives, Kuala Lumpur.

Supomo Surjohudojo, 1973 : *Life in a Javanese village*, Centre of Southeast Asian Studies, Monash University, Melbourne.

Takahashi, Akira, 1969 : *Land and peasants in central Luzon*, Institute of Developing Economies, Tokyo.

Umehara, Hiromitsu, 1974 : *A hacienda barrio in central Luzon*, Institute of Developing Economies, Tokyo.

United Nations Food and Agricultural Organisation, 1948—1978 : *Production Yearbooks*, Rome.

Vietnam (South) Agricultural Economics and Statistics Service, 1970 : *Agricultural Statistics Yearbook*, Saigon.

Wilhelmy, H., 1976 : "Rice cultivation and alleviation of the food problem in Southeast Asia", *Plant Res. & Devel.*, 3, 29-39.

Wong, I. F. T., 1966 : *The Present Landuse of West Malaysia*, Ministry of Agriculture and Lands, Kuala Lumpur.

———, 1970a : *The Present Landuse of Kedah*, Ministry of Agriculture and Co-operatives, Kuala Lumpur.

———, 1970b : *The Present Landuse of Kelantan*, Ministry of Agriculture and Co-operatives, Kuala Lumpur.

16

GROUNDNUT — CULTIVATION, PRODUCTION AND INDUSTRY IN CHITTOOR DISTRICT

M. V. REDDY

THE CHITTOOR district is the southernmost district of Andhra Pradesh, a largest state in south India. Geometrically, the region lies between 12°37′ and 14°08′ north latitudes and 78°03′ and 79°55′ east longitudes covering an area of 15,136 sq.km. It has a population of 2.75 millions (1981 Census) with a moderate density of 181 people per sq. km. Administratively the district is divided into 15 taluks and 65 *firkas* (Fig. 16.1). The north-south running eastern ghat ranges are in the middle of the district and is the water divide for many seasonal streams. The district slopes down from west to east from an average elevation of about 750 mts. on the west to about 50 mts. on the eastern margin. The major part is covered by red soils (Haplustalfs Rhodustalfs). They include red sands, red loams and red clays. The region in general has a semi-arid climate. But the eastern part receives relatively more rainfall than the western part. According to Koeppen's classification, it comes under 'A_w' type of climate denoting a distinct dry season in winter and marked seasonal rhythm of rainfall. The average rainfall ranges from 555 mm. at the western margin to as high as 1187 mm. at eastern plain.

The Chittoor district is characterised by low industrial and predominantly agricultural economy. This is reflected in the disproportionately large share of population (75.2%) engaged in agricultural sector. Agricultural development cannot go very far unless there is industrial development to absorb surplus manpower. Agro-based industries help to integrate industry with agriculture and promote harmonious and balanced growth of both the sectors. It is this concept that forms the basis for the present study.

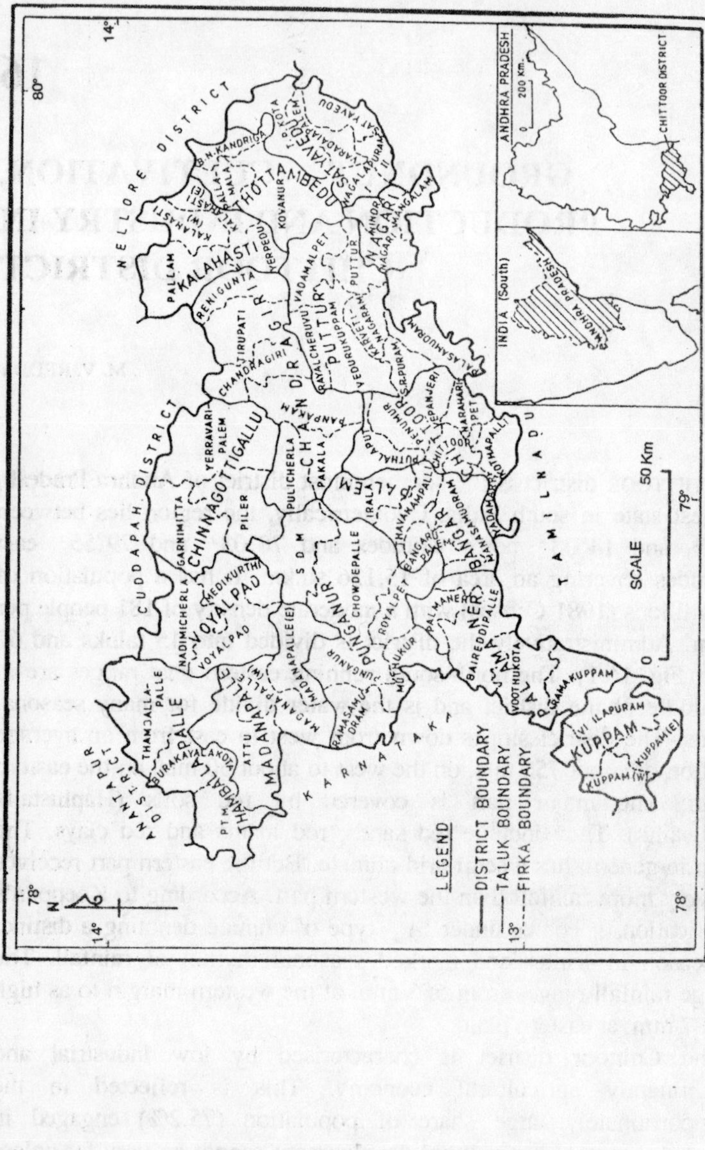

FIG. 16.1: Chittoor district : administrative divisions.

The primary objective of this paper is to analyse the locational aspects of cultivation of groundnut and groundnut based industry and to suggest suitable measures to strengthen the economy of the district. To fulfil this objective, administrative unit namely *firka* is taken as a basic unit and analysis is done for the year 1982-83. To understand the relative position of groundnut crop in the district, the primary crop regions are delimited taking first ranking crop in each *firka*, no matter what percentage it accounts for in the gross cropped area. Locational pattern of groundnut crop and groundnut based industries is studied applying Florence (1959) location quotients. A detailed analysis is made to find out the gaps between groundnut production and groundnut based industries in the district. An attempt is also made to identify new locations for establishing groundnut based industries in the region.

Groundnut Cultivation

Groundnut is by far the most important commercial crop in this region. It accounts for 45.8 per cent of total cultivated area with 6.47 lakh acres. It is a valuable cash crop and plays significant role in the agro-industrial economy of the region.

The spatial distribution of the first ranking crops throws light on the relative position of groundnut cultivation in the district. Groundnut cultivation showed spatial dominance in low rainfall and unirrigated areas of the upland areas, while paddy cultivation showed preference to the high rainfall and plain areas of the eastern part of the district.

TABLE 16.1: Distribution of first ranking crops

Crop	Number of *Firkas*
Groundnut	44
Paddy	14
Ragi	3
Pulses and Grams	4

Groundnut was the first rank crop in 44 *firkas* in 1982-83 (Fig. 16.2). It formed the largest contiguous crop region in the district covering all the upland areas, valley plains, rugged surfaces and also a considerable portion of the plateau surface of the western, northern and southern parts of the district. In these areas the rainfall is low and uncertain. Soils are red sandy loams with low soil depth and poor moisture retentive capacity. They are deficient in organic matter, nitrogen, phosphorous and other plant nutrients. Protective irrigation facilities are scanty. Groundnut cultivation is essentially a kharif crop in the above mentioned areas of the district and widely grown under rainfed conditions.

Groundnut cultivation is favoured by the farmers of the upland areas and low-rainfall areas. Because it is a short growing crop and fetches good income with high price. The cost of cultivation is less and requires less water. It is also a good rotation crop in the paddy fields and can be cultivated under rainfed conditions. Further it is an important commercial oil seed crop.

The spatial distribution of groundnut cultivation in Chittoor district exhibits a great diversity ranging from 3.4 per cent to as high as 76.6 per cent of gross cropped areas (Fig. 16.3).

TABLE 16.2: Distribution of groundnut cultivation

Concentration (Range in per cent)	Number of *Firkas*
Very high (> 60)	8
High (45.1 — 60)	26
Moderate (30.1 — 45)	16
Low (15.1 — 30)	11
Very low (≤ 15)	4

High and very high concentration (45.1 – 60% and > 60%) of groundnut cultivation are found in 34 *firkas* distributed widely in the western, central and south-eastern parts of the district. Here, the crop is largely grown on red sandy and loamy soils where rainfall and irrigation facilities are low. Here it is extensively cultivated in the undulating rugged and sloppy topographical areas. In the high concentration areas of groundnut cultivation, the crop is grown in the kharif season depending on the S.W. monsoon rainfall. Here, the entire gamut of its cultivation depends on the S.W. monsoon.

Low to moderate concentrations (15.1 – 30% and 30.1 –45%) of groundnut cultivation are observed in the eastern and south-western parts of the district. In the eastern part of the district where rainfall is more and less variable and where irrigation facilities are better, groundnut cultivation competes with paddy cultivation. Here,

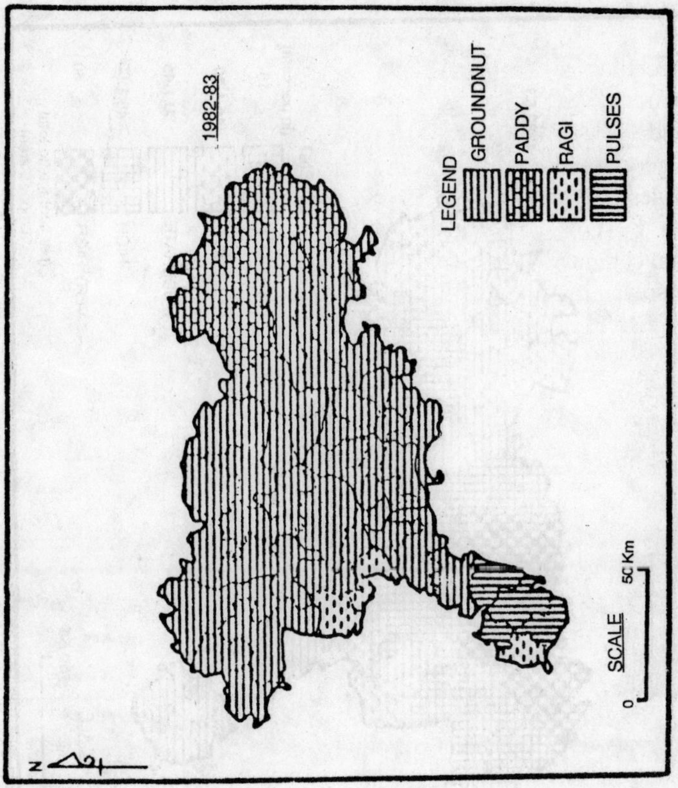

FIG. 16.2: Chitoor district : primary crop regions.

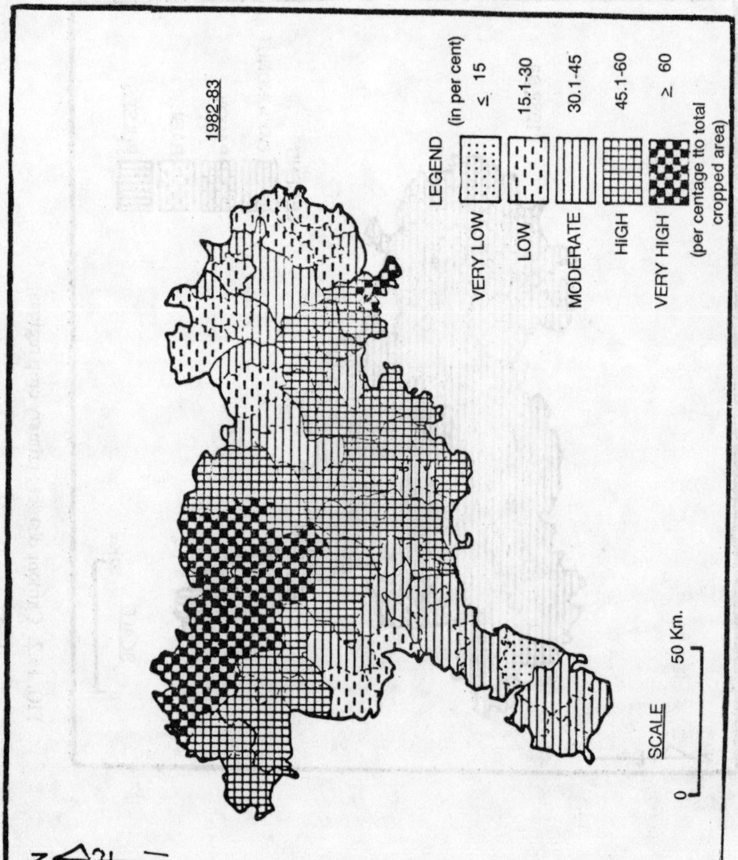

FIG. 16.3: Chittoor district : concentration of groundnut cultivation.

groundnut cultivation is preferred whenever rainfall is less and where the irrigation facilities are inadequate for paddy cultivation. In this region groundnut is cultivated intensively. In the south-western part groundnut is cultivated under rainfed conditions.

In the rabi season, whenever the water supply is sufficient groundnut cultivation is practiced as a rotation crop in the fields where paddy, jowar, bajra, ragi are grown as the first crop. Groundnut in the rabi season is largely cultivated under irrigation and as a result it gives generally high per acre yields than in the kharif season.

Groundnut Based Industries

Chittoor district has 171 groundnut based industries of which 68 are groundnut decorticating mills and 105 are oil mills. They are spread over 37 *firkas* in the district. Though the groundnut is a leading crop in the district, there is no single unit in as many as 28 *firkas* (Figs. 16.4 & 16.5). Most of the oil mills are just serving the local needs. The rest of the production is sent outside the region in raw state.

A study of the spatial distribution pattern (Table 16.3) shows that in as many as 15 *firkas* only 2 or less number of units are present. Only 5 *firkas* are having more than 8 units, many of which are located in the eastern part of the district, where strangely paddy is leading crop. It can be observed that many of the Industries are urban based rather than raw material based. A comparison between distribution of groundnut and groundnut based industries highlights spatial gaps and help to pin-point the potential locations for groundnut based Industries.

TABLE 16.3: **Distribution of groundnut-based industries**

Number of Units	Number of *Firkas*
< 2	15
3 – 4	11
5 – 6	4
7 – 8	2
> 8	5
Total	37

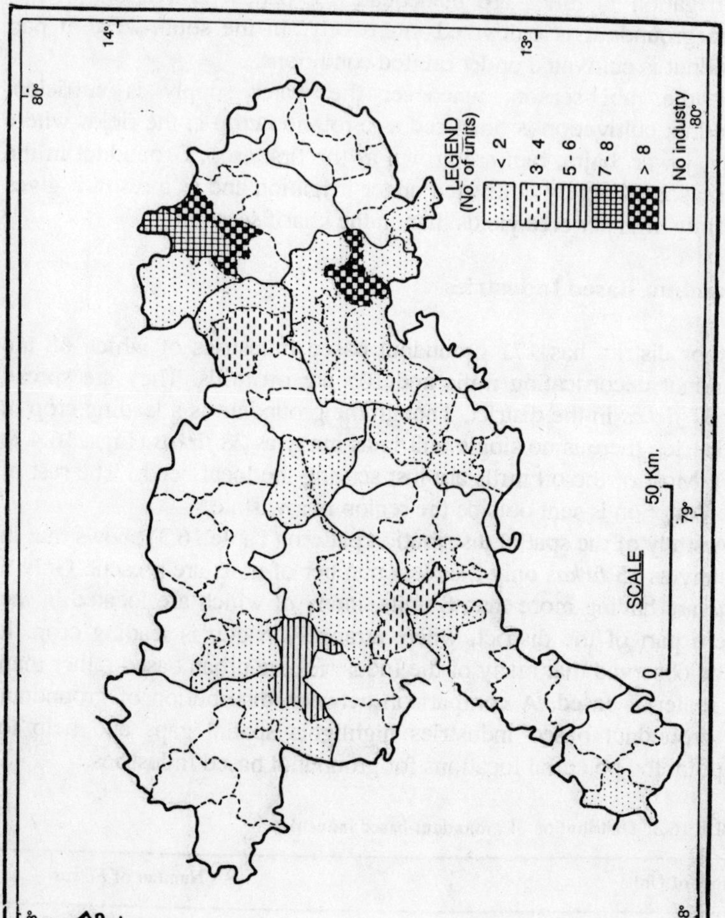

FIG. 16.4: Chittoor district : distribution of groundnut decorticating mills (1982-83).

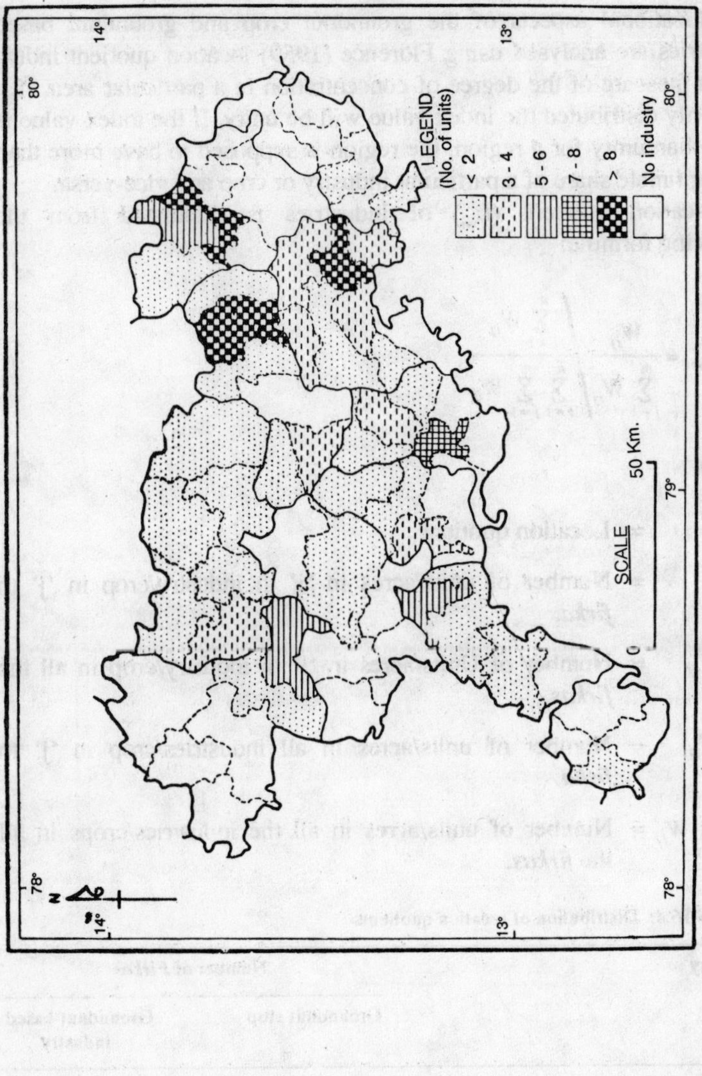

FIG. 16.5: Chittoor district : distribution of groundnut oil mills (1982-83).

Location of Groundnut Crop and Groundnut Based Industries

The locational aspects of the groundnut crop and groundnut based industries are ànalysed using Florence (1959) location quotient index. It is a measure of the degree of concentration in a particular area. If it is evenly distributed the index value will be unity. If the index value is more than unity for a region, the region is supposed to have more than the legitimate share of a particular industry or crop and vice-versa.

Location quotient (L_{wi}) of industries is calculated from the following formula:

$$L_{wi} = \frac{W_{ij}}{\sum_{j=1}^{m} W_{ij}} \Bigg/ \frac{\sum_{i=1}^{n} W_{ij}}{\sum_{i=1}^{n}\sum_{j=1}^{m} W_{ij}}$$

where

L_{wi} = Location quotient

W_{ij} = Number of units/acres in 'i' th industry/crop in 'j' th *firka*.

$\sum_{i=1}^{m} W_{ij}$ = Number of Units/acres in 'i' th industry/crop in all the *firkas*.

$\sum_{i=1}^{n} W_{ij}$ = Number of units/acres in all industries/crop in 'j' th *firka*.

$\sum_{i=1}^{n}\sum_{i=1}^{m} W_{ij}$ = Number of units/acres in all the industries/crops in all the *firkas*.

TABLE 16.4: Distribution of location quotients

Category (L_{wi})	Number of *Firkas*	
	Groundnut crop	Groundnut-based industry
Nil	0	28
< 0.5	11	2
0.5 — 1.0	18	19
1.0 — 1.5	28	15
> 1.5	8	1
Total	65	65

The *firka-wise* location quotients for groundnut crop and groundnut based industry range from 0.07 to 1.67 and 0.2 to 1.5 respectively. Most of the *firkas* are having index values between 0.5 to 1.5. Groundnut cultivation is wide spread and encompasses all the *firkas*. The concentration is very high throughout the district except in few *firkas* in eastern part. On the other hand, groundnut based industries are located only in 37 out of 65 *firkas* in the district. In general oil industry is more important in eastern and central parts of the district. But the industry is not important in groundnut dominant north-western part (Fig. 16.6).

Groundnut Production and Groundnut-Based Industry

Chittoor district has a large surplus of groundnut production. It has a gross cropped area of 6.47 lakh acres and production of 3.3 lakh tonnes. In the following study a detailed analysis of the production pattern, existing installed capacity of both groundnut decorticating and oil mills is attempted. An attempt is also made to identify the potential areas for future location of industries. The total production of groundnut in 1982-83 is 3.3 lakh tonnes, of which 10 per cent is retained for seed purposes. The available groundnut for processing is 2.97 lakh tonnes in the district. The available groundnut for processing showed spatial variation ranging from as low as 250 tonnes to as much as 15,200 tonnes. In 11 *firkas* located in northeast and southwest, the production of groundnut has been less than 2000 tonnes. The 10 *firkas* located in north-west, the production has been over 8000 tonnes. In 20 *firkas* mostly lying in the central part of the district the production has been 4000 to 6000 tonnes (Fig. 16.7).

Groundnut Decortication

Groundnut decortication units are present in only 30 out of the 65 *firkas* (Fig. 16.8). The quantity decorticated during 1982-83 has been 0.39 lakh tonnes only. This means as much as 2.58 lakh tonnes of raw groundnut is sent out of the district. The decorticated groundnut spatially varies from just 360 tonnes in many of the *firkas* to a maximum of 5760 tonnes in Kalahasthi west. In 16 of the 30 *firkas* the quantity decorticated is less than 500 tonnes. The quantity is above 2000 tonnes in 6 *firkas* of them except Kalahasthi west all are having urban settlements. On the whole less than 15 per cent of the groundnut

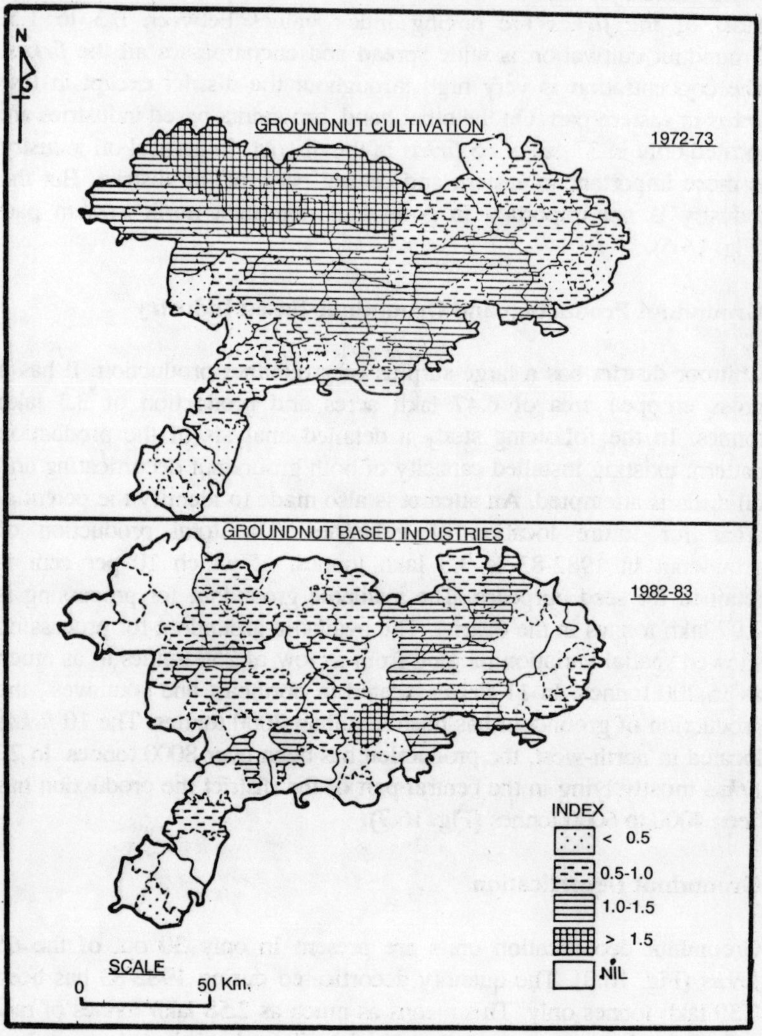

FIG. 16.6: Chittoor district : location quotient index (1982-83).

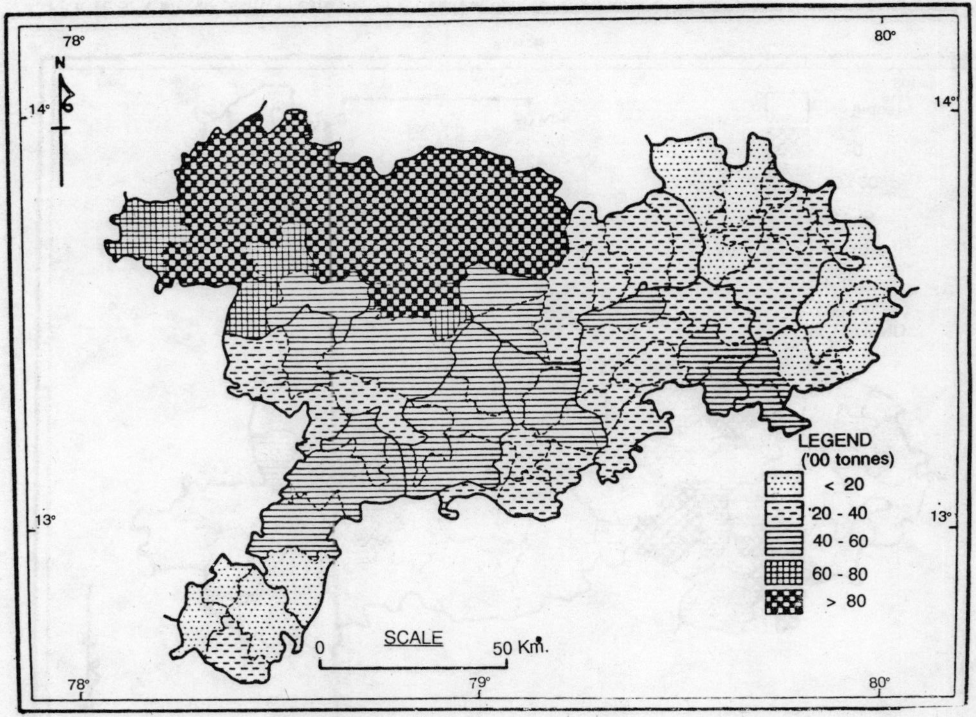

FIG. 16.7: Chittoor district : distribution of groundnut production (1982-83).

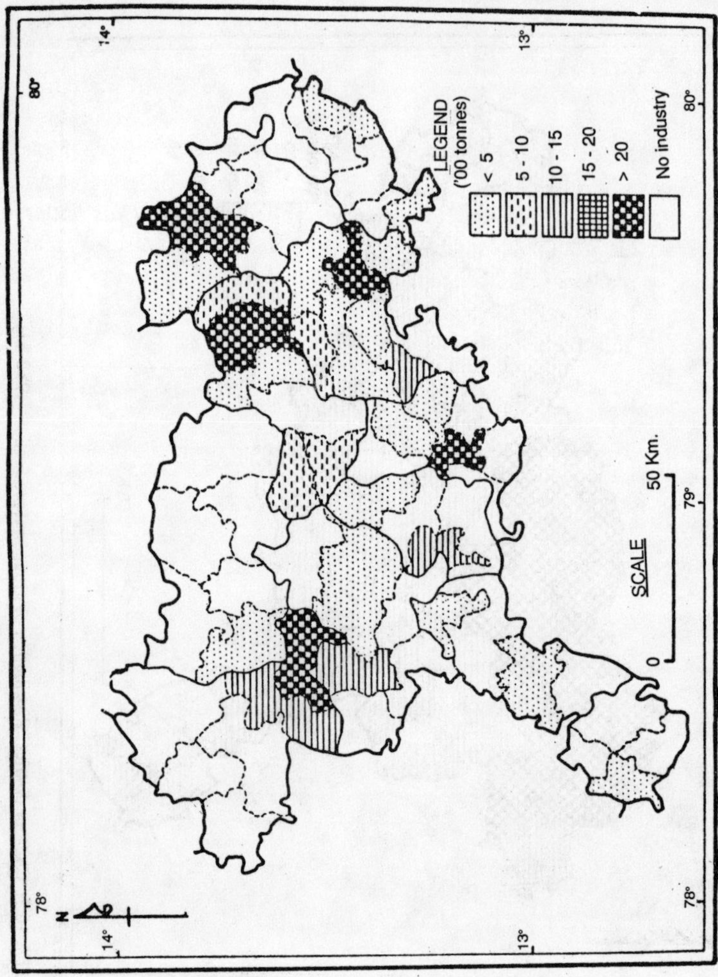

FIG. 16.8: Chittoor district : capacity of groundnut decorticating mills (1982-83).

is decorticated within the district, while over 85 per cent of the production is sent out in raw condition or manually decorticated to some extent.

The surplus groundnut is over 8000 tonnes in 9 *firkas* together mostly lying in the north-western region. A surplus of groundnut production ranging between 4000 and 6000 tonnes in 15 *firkas* which are lying mostly in central and south western parts. The range between 2000 and 4000 in 22 *firkas* mostly lying in the east. The 15 *firkas* under less than 2000 tonnes category are distributed in the east and south-western parts.

TABLE 16.5: Distribution of decortication capacity

Category (tonnes)	Number of *Firkas*
< 500	15
500 –1000	4
1000 –1500	4
1500 –2000	—
> 2000	6
Total	30

TABLE 16.6: Groundnut production, decortication and surplus/deficit

Category (tonnes)	Number of *Firkas*			
	Production	Decortication	Surplus Groundnut	Deficit
< 2000	11	24	15	1
2000 –4000	22	6	22	2
4000 –6000	20	—	15	—
6000 –8000	2	—	1	—
> 8000	10	—	9	—
Total	65	30	62	3

It is estimated that about 10 per cent of the groundnut produced is retained for local consumption and half of it is manually decorticated. It means, still over 2.5 lakh tonnes of undecorticated groundnut remains as surplus (Fig. 16.9).

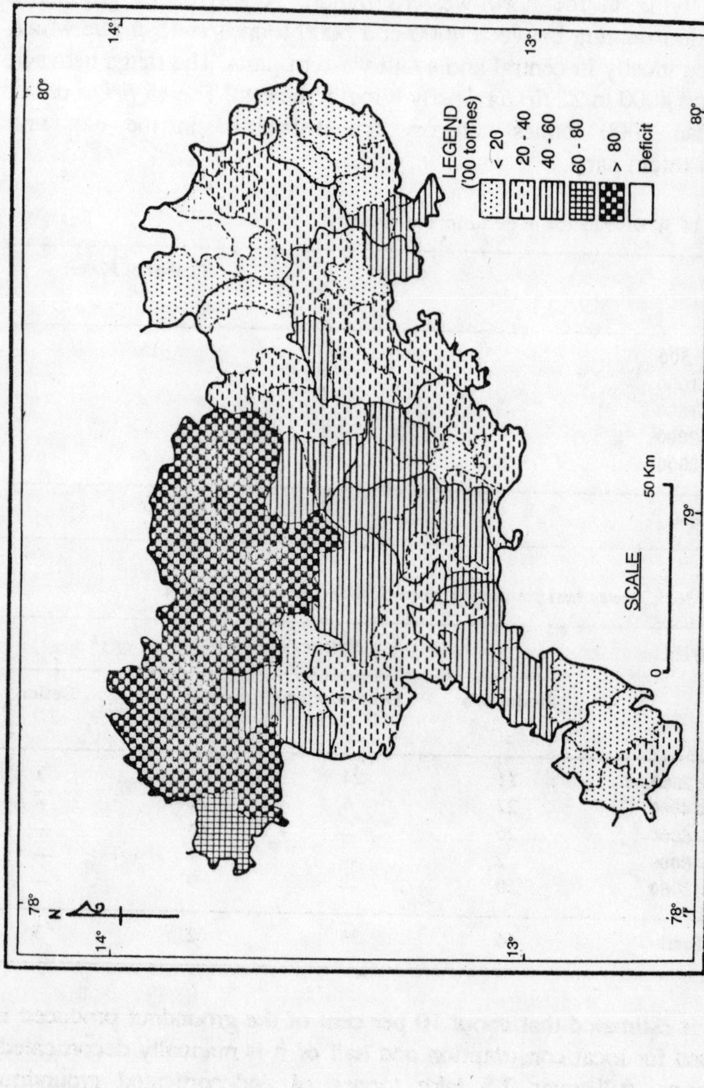

FIG. 16.9: Chitoor district : balance between capacity of decorticating mills and groundnut production (1982-83)

TABLE 16.7: **Available groundnut seed and oil mill capacity**

Category (tonnes)	Number of *Firkas*			
	Groundnut Seed available	Seed utilised for oil extraction	Surplus	Deficit
< 1500	13	33	18	1
1500 – 3000	23	3	29	—
3000 – 4500	17	—	3	—
4500 – 6000	4	—	6	—
> 6000	8	—	8	—
Total	65	36	64	1

Oil Extraction

The total quantity of groundnut seed available in the district is 1.83 lakh tonnes. The local mill capacity for oil extraction is only 0.23 lakh tonnes of seed which leaves a surplus of 1.60 lakh tonnes (Fig. 16.10).

The available groundnut seed is just 170 tonnes in Vijilapuram, while it is as high as 10,130 tonnes in Nadimicherla. The available quantity is over 4500 tonnes in 12 *firkas* mostly located in north-western part. In all these *firkas*, the quantity of locally crushed seed is very little. In 17 *firkas* the available seed varied from 3000 to 4500 tonnes, and in 23 *firkas* from 1500 to 3000 tonnes. In only 13 *firkas* located in eastern and south-western parts, the available seed has been less then 1500 tonnes. The oil extraction mills are located in only 36 *firkas*. The *firka-wise* capacity of the mills varied from 190 tonnes of seed in a large number of *firkas* to a maximum of 2590 tonnes in Kalahasthi east *firka*. In an many as 33 *firkas* the capacity of the oil mills has been less than 1500 tonnes. In only three *firkas* viz., Kalahasthi east, Tirupati and Puttur the capacity of oil mills exceeded 1500 tonnes (Fig. 16.12). In almost all the *firkas* except in Kalahasthi east the available groundnut seed has been far higher than the existing mill capacity and there is plenty of scope for starting new mills (Fig. 16.13).

The total groundnut, if processed within the district yields about one lakh tonnes of groundnut cake and about 80 thousand tonnes of oil. Apart from setting up of industries for decorticating and oil extraction some small scale poultry and animal feed units and oil refining units can also be set-up.

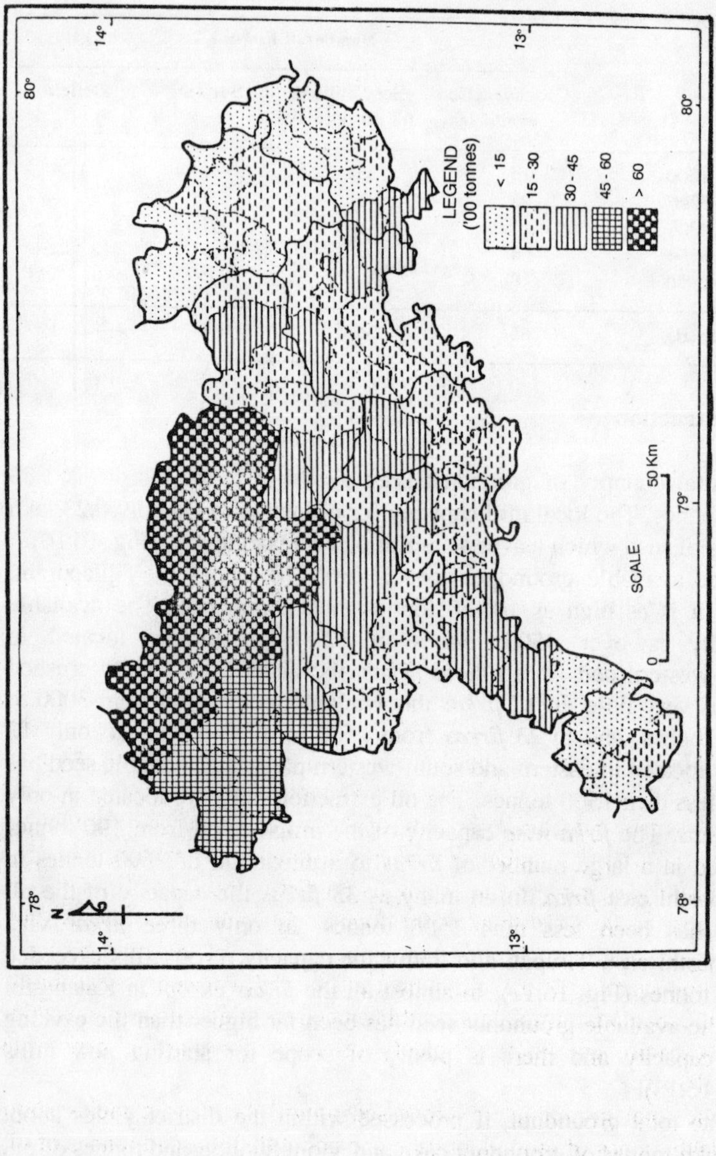

FIG. 16.10: Chittoor district : distribution of available groundnut seed (1982-83).

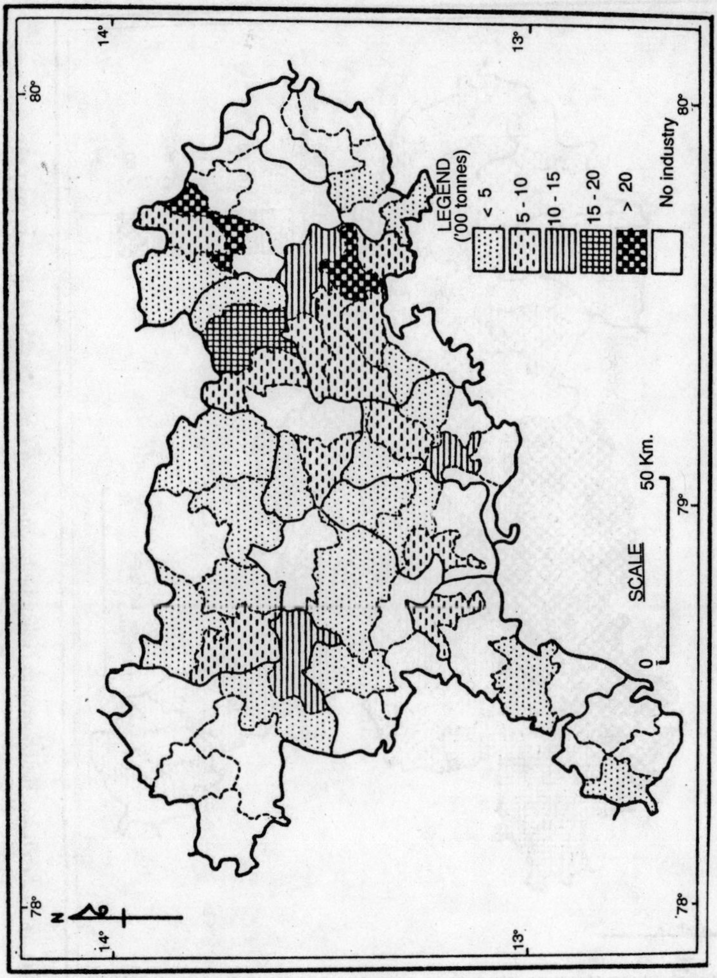

FIG. 16.11: Chittoor district : capacity of groundnut oil mills (1982-83).

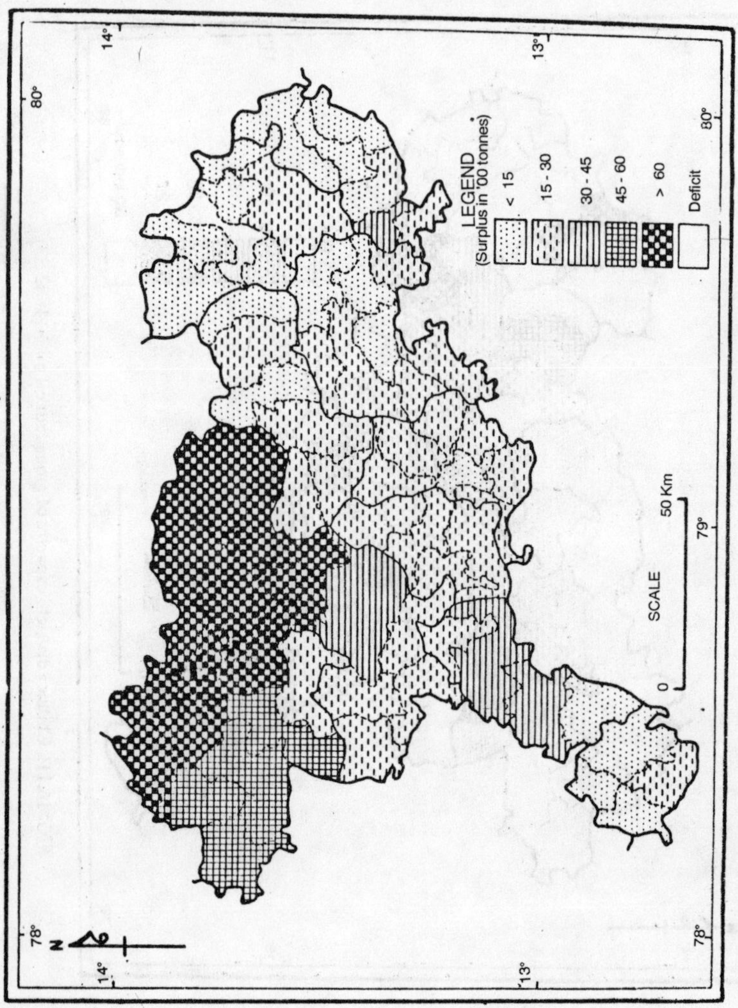

FIG. 16.12: Chittoor district : balance between capacity of oil mills and groundnut seed (1982-83)

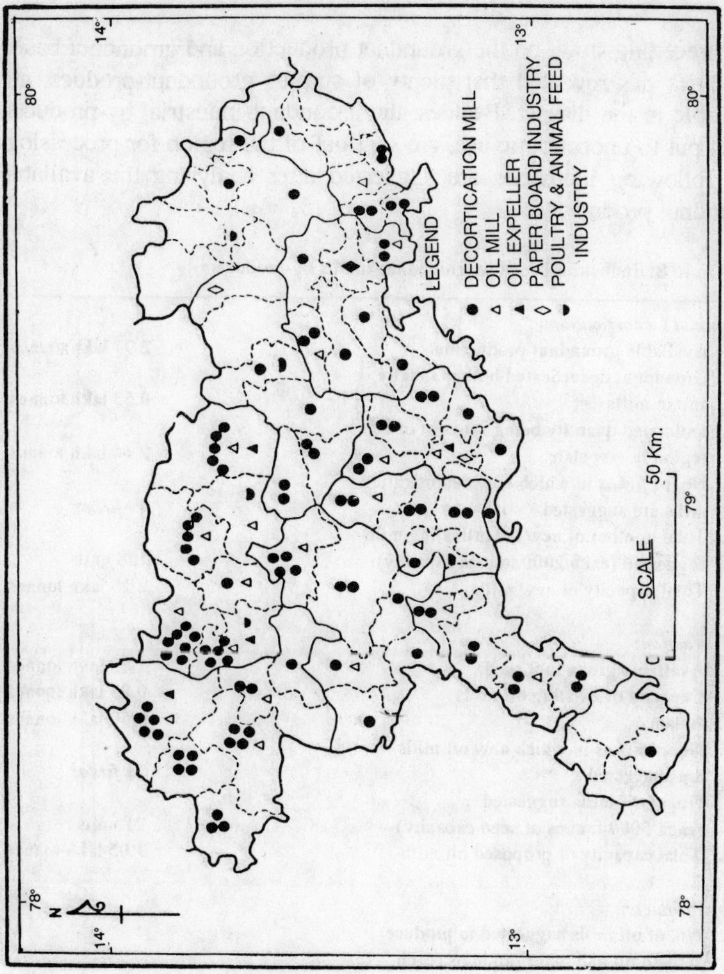

FIG. 16.13: Chittoor district : proposed groundnut based industries.

Potential Groundnut-Based Industries and Locations

The preceding study on the groundnut production and groundnut based industries has revealed that plenty of surplus groundnut-products are available in the district. Besides, the groundnut-industrial by-products, being put to uneconomic use, are sent out of the region for processing. The following industries are suggested after analysing the available groundnut products:

TABLE 16.8: Industries based on groundnut and its by-products

Groundnut Decortication:	
1. Available groundnut production	2.97 lakh tonnes
2. Groundnut decorticated both manually and in mills	0.53 lakh tonnes
3. Estimated quantity being sent out of the region in raw state	2.44 lakh tonnes
4. No. of *firkas* in which new decorticating mills are suggested	49 *firkas*
5. Total number of new decorticating mills suggested (each 2000 tonnes capacity)	108 units
6. Total capacity of new mills	2.16 lakh tonnes
Oil Extraction:	
1. Available groundnut seed	1.83 lakh tonnes
2. Capacity of existing oil mills	0.23 lakh tonnes
3. Balance	1.60 lakh tonnes
4. No. of *firkas* in which new oil mills are suggested	21 *firkas*
5. No. of oil mills suggested (each 5000 tonnes of seed capacity)	21 units
6. Total capacity of proposed oil mills	1.05 lakh tonnes
Other Products:	
1. No. of oil mills suggested to produce refined oil and other products (each 10,000 tonnes oil capacity)	3 units
2. Total capacity of the proposed mills	0.3 lakh tonnes
By-Products:	
1. Total quantity of groundnut shell (33% of the groundnut processed)	1.0 lakh tonnes
2. Groundnut cake	1.0 lakh tonnes
3. No. of proposed poultry and animal feed industries (each 10,000 tonnes capacity)	5 units
4. Total capacity of the mills	0.5 lakh tonnes

Ground shell and groundnut cake could be used to produce cattle and poultry feed as well as fertiliser.

Table 16.9 gives a list of proposed industries in each type based on groundnut and its by-products.

The proposed groundnut industries are supposed to develop an integrated and prosperous agro-industrial economy. It will also contribute to a more stable and balanced economic development of the district.

TABLE 16.9: Proposed groundnut based industries

Type	Number of units
Decorticating Mills	108
Oil Mills	21
Refining Mills	4
Cattle and animal feed and Fertiliser Units	5
Paper Board Mills	2

REFERENCES

Alam, S. M., 1974 : *Planning Atlas of Andhra Pradesh*, Govt. of Andhra Pradesh, Hyderabad.

Alonso, W., 1971 : *The Location of Industry in Development Organisation*, United Nations, New York.

Arora, R. C., 1975 : *Development of Agriculture and Allied Sector*, S. Chand & Co. Ltd., New Delhi.

Bhatia, S. S., 1965 : 'Patterns of Crop Concentration and Diversification in India', *Economic Geography*, Vol. 41, pp.39-56.

Florence, 1959 : *Investment, Location and Size of Plant*, Cambridge University Press, Cambridge.

Karan, P. P. and Jerkins, W. M., 1959 : 'Geography of Manufacturing in India', *Economic Geography*, Vol. 35, 1959, pp.269-278.

Reddy, N. B. K. & Reddy, D. N., 1981 : 'Agriculture in Arid and Semi-arid Regions of India, *Transactions of the Institute of Indian Geographers*, Vol. 3, No. 2, pp.115-135.

Sharma, T. R., 1948 : *Location of Industries in India*, Hind Kitab Ltd., Bombay.

———, 1983 : *Directory of Small Scale Industrial Units, Chittoor District*, Directorate of Industries, Govt. of A.P., Hyderabad.

———, 1983 : *Hand Book of Statistics, Chittoor District*, Asst. Director of Planning and Statistics, Chittoor, Govt. of A.P.

17

THE CROP ECOLOGY

O. P. BISHNOI

THE STUDY of relationships between crop plants and their environment is known as crop ecology. It implies studies of: (a) Climatic factors, such as temperature, moisture and light; (b) Edaphic factors, including parent material and soil; (c) Biotic factors, such as the effects of other organisms; (d) Anthropic, etc., as all these relate to plant adaptation. Any action taken in agricultural development should be based on a thorough understanding of these crop ecological conditions. The appraisal of worldwide crop ecological conditions will facilitate the transfer of plant or animal material and technological experience from one area to another. In the adoption of any species or variety it is necessary that it should appear as population, more or less infinite in size, possessing a range of genetic diversity dependent largely on its origin and mode of production.

Environmental factors in all of their complexity appear to be highly influential in determining the natural distribution of plants. Good (1931) formulated certain basic principles of plant distribution as follows : (i) that plant distribution is primarily controlled by the distribution of climatic conditions, (ii) that plant distribution is secondarily controlled by the distribution of edaphic factors, (iii) that great movements of flora have taken place in the past and are still continuing, (iv) that species movement (plant migration) is brought about by the transport of individual plants during their motile dispersal phases, (v) that there has been great variation and oscillation in climate especially at higher latitudes, and (vi) that at least some, and probably considerable, variation has occurred in the relative distribution and outline of land and sea.

The biotic community has been described by Odum (1959) as a functional unit held together by interdependence of its members. Every ecosystem is composed of four elements (Odum, 1959): (i) abiotic substances, basic inorganic and organic compounds of the environment; (ii) producers, autotrophic organisms, mainly green plants; (iii) consumers, heterophic organisms, mainly animals; (iv) decomposers, heterotrophic organisms, mainly bacteria and fungi. The ratios of the energy flow at different points along the food chain are used as percentages to determine ecological efficiency. The main crop ecological factors have been discussed below.

Environmental Factors

Moisture Factor

The ecological importance of moisture to plants can hardly be overemphasised. The moisture aspects related to the plant growth are atmospheric moistures, precipitation and its effectiveness, adaptation to moisture conditions, transpiration and water use, drought resistance, response to increasing water stress, and effects of excessive moisture. The attempts to correlate climate and vegetation are tentative but that the soil moisture regime and the influence of soil moisture on climax vegetation are extremely important in determining plant distribution.

Warming (1909) made an ecological classification of plants on the basis of their water relations. Three groups were distinguished : hydrophytes, mesophytes and xerophytes. Hydrophytes are those kinds of plants which normally grow in water, but can also grow in soils too wet for ordinary plants. They extend their roots into water or in saturated soil. Stomata are numerous but are located mainly on the upper surface of the leaves. Daubenmire (1959) described a number of ecological groups of hydrophytes : suspended hydrophytes, floating hydrophytes, submerged hydrophytes and emerged hydrophytes. Mesophytes are most common land plants. Stomata are numerous on the underleaf surfaces. Root hairs are abundant. Xerophytes are being described as being capable of enduring prolonged drought without injury. These were classified xerophytes if they wilted permanently only after losing from 50 per cent to 75 per cent of their total water content. These included species of Eragrostis and Danthonia. True Xerophytes were considered those species which could be reduced to a water residum of about 25 per cent without wilting. This group included Agropyron, Phalaris, Oryzopsis, Stipa and Erharta species.

Temperature Factor

Temperature is one of the most familiar ecological factors, and the one which is almost universal in influence, and often becomes limiting for growth and distribution of plants and animals. Various methods have been developed for evaluating and effectiveness of temperature as related to the crop distribution. Most simplest is the determination of length of growing season or the length of optimum range of temperature for a particular species. A positive summation of temperatures above a certain base has been proposed as a measure of thermal efficiency. This involves a summation of temperature in excess of a certain zero point of vital activity, such as 4.4°C. An accumulation of temperatures above a threshold temperature value is the basis of "heat unit systems" in widespread use today. This method, though useful, has the weakness that it does not evaluate the various temperature intensities in accordance with their physiological effects. A refinement was proposed by Decker (1953) by developing a formula in which temperatures were given efficiency ratings, in recognition of the acceleration in the growth rate with an increase in temperature.

Thornthwaite (1931) developed a temperature efficiency index, which was a modification of the temperature summation method, with 0°C taken as the zero point. Later, Thornthwaite (1948), in developing the potential evapotranspiration concept for characterising moisture provinces, suggested that potential evapotranspiration was an index of thermal efficiency as well. It may be considered as a growth index, which expresses growth in terms of the water needed.

Livingston (1916) developed the physiological temperature index based on research on the growth rate of maize. The hourly rates of elongation were divided by the value obtained at 4.4°C as unity. Hopkins (1938) attempted to express the importance of latitude, longitude, and altitude in the distribution and rate of development of plants by means of "Biclimatic Law". This law established bioclimatic zones on the basis of phenological data for all the important crop plants.

Light Factor

Light is the primary source of energy, without which the living organisms could not exist. It also acts as a limiting factor at both the maximum and minimum levels. Light also serves plants in other ways, especially in its stimulating effects upon differentiation of tissues and organs (Daubenmire, 1959). Ecologically, light quality, light intensity and light duration are all important to plants. Day length and intensity probably are the ecological aspects of light and are of greatest importance in plant distribution. It is of basic importance as the primary source of energy for the photosynthetic process. Some plants tolerant to reduced light include spruce, hemlock, beach, sugar maple, spicebush and bloodroot. In contrast, birch, poplar, willow, ponderosa pine, sumac, andropogon grasses, and sunflowers are intolerant of shade. Of the tropical crops, coffee is known to be tolerant to low light intensity. Arabian coffee production was best under conditions of 33 per cent to 50 per cent of full sunlight. Cocoa has been found to produce optimum growth in 25 per cent of full sunlight (Cobley, 1942). It is not easy to explain why shade-tolerant species, sciophytes, do not require more light, nor why the sun-loving species, heliophytes, require intense light for normal development.

Physiographic and Edaphic Factors

The topographic features of elevation and slope, by processes of silting and erosion, and consequently by local geology produce a marked effect on local climates. Plant is dependent on the soil for anchorage, water and nutrients. In extent, the root system may be as great as, or even greater than, the top growth. The surface contact between the soil and the plant roots is great indeed, making the soil truly an intimate part of the plant's environment. Warming (1909) recognised the importance of soil in the natural distribution of plants by attempting to classify them into specific groups based on the kind of soil substratum where they occurred. The soil itself may be considered a paramount factor in the level of productivity of an area. The soil requirements for crop production are :

(i) Suitability for cultural implements required for efficient crop production;
(ii) Effective resistance to destructive erosion or deflation under the management involved in profitable farming;
(iii) Adequate moisture storage to meet normal requirements of the crop, under natural rainfall or irrigation;
(iv) Adequate aeration to a suitable depth to permit development of favourable root system;

(v) Availability of plant nutrients sufficient for profitable yields; and

(vi) Freedom from adverse chemical conditions, such as harmful concentration of soluble constituents, and from other special conditions that favour the development of organisms parasitic to the crop.

Marsden-Jones and Turrill (1945) found marked effects of soil types and development of plants such as the germination ability of seeds, size of plant and degree of errectness, general vigour of vegetative parts, woodiness of the stem, depth of root system, pubescence on stems and leaves, susceptibility to drought, frost and parasites, and the number of flowers and date of flowering.

Biotic Factors

The biotic factors of plant environment are related directly or indirectly to living organisms ranging from soil micro-organisms to man. Every plant is influenced by organisms. The carbon dioxide used by green plants in photosynthesis is made available largely through the respiration of other organisms. The amount of heat, light and nutrients available to a plant is determined in parts by the proximity of other plants, and much of the nitrogen is available in organic form. Both plants and animals serve as environmental factors which influence plant distribution and behaviour.

Crop Distribution on Ecological Basis

An ecological region is an area in which crop ecological conditions follow the same pattern. Usually the region has a uniform climate and distribution of its soil and has a definite pattern. In some cases the soil factor may be more important and the ecological region includes various climates with a definite pattern of soil distribution. In mountainous regions, where climate and soil vary at short distances, an ecological region is an area with a definite pattern of climate-soil-vegetation. Ecological regions may be vast or small, their classification is natural.

Experiments under controlled conditions are adding greatly to our knowledge of the specific requirements for temperature, moisture, light and nutrients but crops must still be grown where the environment is only partially controlled by man, and where the interacting effects of many factors plus their variability are great indeed. Some tropical climates have a monotonous uniformity, whereas the continental intermediate climates are extremely variable. Crops such as wheat and barley have a wide range of tolerance to environmental changes. If data on both weather conditions and varietal yields of crops for long periods are available, an evaluation of the comparative degree of adaptation of specific crop varieties to local environments may be made. Such a method, known as meteorological equivalent method, is described by Azzi (1956) for wheat. The meteorological equivalent may be defined in thermic or pluviometric terms as the degrees of temperature or the millimetres of precipitation which distinguish normal from abnormal conditions. Abnormal conditions may be either the excess or deficiency of an environmental factor. For wheat in Italy the normal rainfall conditions lie between 40 and 116 mm. for the first month preceding heading of the plants. Similarly, the Barger and Thom (1949a, 1949b) method for characterising drought intensity and evaluating drought hazard measures rainfall requirements of an average corn crop. Visher (1955) generalised some relations based on the detailed local studies of environmental conditions and yield of crops.

The ecological concept of productivity by Odum (1959) suggests that the primary productivity of an ecological system may be defined as the rate at which energy is stored by photosynthetic and chemosynthetic activity of producer organisms (green plants) in the form of organic substances which can be used as food material. The rates of energy storage at the tropic levels of consumer and decomposer organisms are known as secondary productivities, and are less and less at successive levels. A summary of estimates of net productivity, on both annual and daily bases, of crops, forests, deserts, and other ecosystems in which it is possible to use a harvest method, has been found to be very useful in demarcating the ecological systems.

Another concept by Klages (1930, 1934) for the evaluation of the relationship between environment and crop yields is the consideration of an idealised growth curve of annual plants with the growth curves of individual years. To be well adapted, a crop should exhibit a vegetation rhythm in harmony with the climatic rhythm. Usually this refers to the moisture available at a given time, or temperature, or occasionally some other factors. Poor adaptation is indicated by a depressed growth curve. The theoretical ecological optimum, as defined by Schimper (1903), consists of the summation of the various harmonic optima. Harmonic optimum refers to the most favourable intensity of any one function in relation to the other functions of the plant. Summation of harmonic optima suggests not only the location of an average point, but also the relative importance of each of the various plant functions in their relation to the growth and behaviour of the entire organism (Klages, 1942). Some of the important crop ecologic conditions have been discussed below.

Crops of the Tropics

The importance of climate as the dominant factor in determining the distribution of world crops has been emphasised. Latitude is the most important general climatic control of the tropical rainy climates. Rising air masses, resulting from high temperatures, tend to foster abundant rainfall. The important crops grown in humid tropics are :

Cacoa

Cacoa is adapted to the tropical rainforest climate. An average annual temperature between 24° and 28°C, with a mean minimum for the coolest month of not less than 15.5°C. High temperature will not limit growth, provided shade is made available. Successful production, in fact, requires shading. Rainfall must be abundant, 80 inches per year, rather well distributed, is needed for good production. An important limiting factor is wind, which tends to increase transpiration and cause the heavy pods to drop from the trees. These conditions promote high yields and facilitate harvest operations when the crop is mature.

Banana

The banana is a staple food for the people of humid tropics. It is adapted to low elevation; well drained lands in the tropics where no frost occurs. Uniformly high temperatures with maximum sunshine are essential to high yields of the fruit. Optimum temperatures range from 24° to 30°C, with a minimum tolerance of 7° to 9°C. The crop is grown commercially in the tropical rainforest climate, with some production in the tropical savanna climate. Large quantities of water are required for successful production, preferably 75 to 100 inches, so distributed that there is no drought. When dry seasons occur supplemental irrigation is used. Soils should be fertile, frible, and have good aggregation characteristics. The top soil is particularly important because the banana plant has a shallow root system. To prevent waterlogging around the roots, clay content should be less than 40 per cent. Good drainage is absolutely necessary.

The banana plant is best adapted to locations having little or no wind. High winds whip the young leaves into ribbons and retard normal development. When the fruit is maturing, wind damage may be disastrous.

Sugarcane

Of the world's production of sugar, approximately 60 per cent comes from sugarcane and 40 per cent from sugarbeets. Sugarcane is a tropical perennial grass, characterised by the high sucrose content of its mature stalks. Sugar extracted from cane comes mainly from tropical climates whereas, sugar from beets comes from the temperate climates. It is grown in subtropics.

Sugarcane grows best in frost-free climates, with plenty of warm to hot sunny weather. The monthly mean temperature ranges from 21° to 27°C throughout the year gives better growth. The so-called "noble" canes originating in Java are known especially for their narrow tolerance or sensitivity to environment. These canes are characterised by their thick, relatively soft stalks and their high sugar content. In Hawaii, Burr et al. (1957) reported that a root temperature of 27°C is optimum for the growth of cane and for nutrient of absorption. Below 24°C root growth is retarded, and at 10°C no growth takes place. Full sunshine is important, yields being reduced practically one-half by cutting the light to one-half full sunshine. Cool nights are harmful, an air temperature of 12°C being about the lowest temperature tolerated without some harmful effects on the plants. In the subtropics, sugarcane is grown commercially at latitudes of approximately 30°. Growth and yield are limited by winter temperatures, however, which often go as low as 0°C on severe winter nights.

Stalk length growth varies directly with temperature. At low elevation in Hawaii, a crop comes to full maturity in two years or less. Without irrigation, a rainfall of 45 inches or more is desirable. Because of the heavy production of dry matter, sugarcane depletes soil fertility, nitrogen is required in practically all fields, and most soils require phosphorus and potassium as well. Cane leaves and parts of stalks may be incorporated into the soil, adding a good deal of organic matter. Torrential rains followed by bright sunshine are ideal. The more the humid conditions, the better is the effect on the rapidity of a growing crop.

Coffee

Although coffee is grown widely throughout the tropics and warmer subtropics, the crop is better adapted to the savanna and particularly to the tropical highland climates than to the rainforest climates.

Coffee is sown from sea level to 7,000 ft. Altitude is important chiefly in its effect on temperature, rainfall, and humidity. Coffee arabica is best adapted to higher altitudes, whereas C. robusta and C. liberica are better adapted to low elevations.

The range of coffee production is limited by the 13°C isotherm for the coldest month and near the equator by the annual mean maximum of 28°C. In most coffee-growing regions, the mean annual temperature is approximately 21°C or slightly higher. The average minimum temperature is 17°C, the average maximum 28°C and the annual mean approximately 20°C. Cold winds and frosts may cause great damage to coffee trees.

At or near the equator, the combination of high temperature and high moisture is undesirable, and direct sunlight at that latitude may be detrimental. Shade is essential for coffee production near the equator.

Successful coffee production has been possible with as little as 30 inches of rainfall or as much as 100 inches, but 45 to 70 inches annually is considered best. Definite wet and dry seasons are desirable. Heavy rains during the flowering season may reduce the crop considerably.

Gentle slopes and adequate soil and air drainage are necessary. Coffee trees soon die in flat areas on heavy clay soils. This adaptation has restricted coffee growing mainly to the hilly or rolling uplands, and in some instances to high, level plateaus with deep porous soils. The flavour of coffee is influenced greatly by climate and soil. In general, coffee grown at high elevations tends to have a mellow or "soft" flavour as contrasted to the "hard" penetrating flavour of lowland coffee.

In addition to the above-mentioned crops, oil palm, rubber, coconut, sisal are some of the other crops grown in the humid tropical climate. The tropical vegetables do not require a cold season and are short day plants. Tomato can grow much better when nights are cool. The most important vegetable crops in the tropics are tomato, pepper, okra, egg plant, melon, water-melon and other cucurbits. Tomato prefers cool nights and dry weather. Frostless highlands are good for tomato. It may be grown in forest climate, and during the humid season of Savanna climates. The most important fruits in tropical locations are citrus, mango, papaw, pineapple, cashewnut, avocado, guava, etc.

Cabbage, cauliflower, lettuce, onion, beet, potato, carrot, etc., can be grown under tropical conditions. The savanna climate is preferred because of cooler nights of the winter. In tropical countries the alternation of short, cloudy days and long warm nights favour leaf and stem growth in cryophilous plants at the expense of roots, tubers, fruits or heads. However, the preference should be given to non-cryophilous vegetables such as egg plant, okra, pepper, cucurbits, etc., that are easier to produce.

In dry tropical climates the low-water requiring millets, sorghum, moong, moth and til are being grown widely.

Millets

Millets are tropical in origin and are heat-loving plants. These are most widely distributed in the sub-humid to semi-arid climates.

For germination, millets require an air temperature of 8° to 10°C. Below it, no germination takes place. When the seeds lie without germination for a long time in soil, they rot.

Millets are heat-loving plants and the temperature of the order of 35° to 40°C is normally tolerated by them. Millets have low moisture requirements and are highly drought resistant. It has been estimated that the transpiration coefficiency of millets is very small and is about one-third that of wheat. They have an exceptionally high capacity of soil-water uptake through the root-system. During lengthy drought period, leaves of millets curl up, growth process is inhibited and plants become dormant. When the drought passes off and humidity has improved, the plant leaves rise again and start their normal growth.

Sorghum

It is an extremely plastic crop. Rainfed sorghum is in fact found in locations where seasonal rainfall as little as 40mm is received. It grows well on a very fine textured soil. It has the ability to withstand high temperatures and also, incidentally, high evapotranspiration rates. The late varieties of sorghum are photoperiodically sensitive, heading time being accurated by shortening days.

Crops of the Subtropics

The subtropics comprise a number of climatic types including subtropical humid, dry-summer mediterranean, coastal marine climates. It follows that a wide variety of crops is grown in these agriculturally favourable climates, including citrus, olives, grapes, vegetables, sugarcane, cotton, cereal grains, and forage crops. Rice and cotton are very important crops of this climate.

Rice

Rice is unique among cereals in its ability to germinate and grows successfully in water or submerged conditions. The plant can transport oxygen from the leaves to the roots and thus live normally in an aquatic environment. It may be considered a facultative hydrophyte. Factors which limit the rice production are unfavourable temperatures, lack of sufficient fresh water for irrigation, and unfavourable soils. For germination, rice needs a minimum temperature of approximately 11° to 12°C; for flowering, 22° to 23°C; and for grain formation, 21° to 22°C (Thomas, 1957). A rather warm temperature of 21°C, or above, is needed for the entire growing period, which may extend from four to six months.

Suitable soils for rice are most important. Clay/silt loam is considered a good soil for rice. These soils do not lose water rapidly by seepage, which makes the use of irrigation water more efficient. In general culture, the land is flooded to a depth of four to six inches, either after the rice seedlings are 6 to 10 inches tall, or just prior to seeding or transplanting, if the transplanting method is used. There are upland varieties of rice, grown on a relatively small scale, which produce a crop without irrigation in regions of a fairly high rainfall. The feet of the plant should remain submerged from the time of sprouting to the milk stage of the grain. At waxy ripeness it needs no water in the field.

Cotton

Cotton is adapted to the humid subtropics where the frost-free season is at least 200-210 days, and the summer isotherm about 25°C or more. In latitude, cotton is grown from 32°S to 37°N. The minimum rainfall limit is considered to be about 20 to 25 inches. Relatively dry, cool conditions at harvest time are much to be preferred. Cotton-planting is delayed in spring until the days and nights are quite warm, with an average temperature of 15° to 16°C.

An abundance of sunshine and a moderate amount of moisture is required. Heavy rains at planting time and during early growth stages are undesirable, for a shallow root system may develop, too much rain may cause excessive shedding of leaves, squares, blooms and balls. Late in the season, rains stimulate top growth, delay maturity, interfere with picking, discolour lint.

Eaton (1950) reported that temperature and day length were important in influencing the type of branching which occurs on the cotton plant. Cool days and cool nights result in mainly fruiting branches. Some varieties respond to day length, requiring short days for normal fruiting.

Under conditions of moisture stress, vegetative growth may decrease by half and ball periods and ball sizes may be reduced, but with no appreciable effect on the relative fruitfulness.

Soils which are formed under grassland or are well-drained, alluvial flood plains are best. Heavy rains tend to delay, and result in greater vegetative growth, which is not desirable from the standpoint of potential insect damage. Level or gentle rolling land is preferred because of the rapid trend toward complete mechanisation of field operations.

Cotton is grown under a wide range of moisture conditions. The minimum rainfall limit is considered to be about 50 to 65 cm unless irrigation is practised. Heavy rain during the planting and early stages is undesirable. At this stage the crop requires an abundance of sunshine and a moderate amount of moisture. Comparatively dry autumn months are desirable for higher production of good quality. Excessive rainfall may cause the shedding of the leaves and balls. High humidity invites many pests and diseases. Low rainfall results in top growth and retards maturity, discolours the lint and interferes with picking. The ideal distribution of rainfall for cotton is the thundershower type with bright sunny days in between rains.

Crops of the Intermediate Climates

Maize

Maize is of the tropical origin and is basically a short day plant, but in its great diversity it is highly variable in response to photoperiod. Most varieties, however, respond to day length. Maize is best adapted where the growing season is long and warm. Varieties of corn are grown successfully over a wide range of latitudes, from 58°N to 40°S, including the tropical, subtropical and intermediate climates. In the tropics the crop is grown from near sea level to 13,000 ft.

The mean summer temperature of 21° to 22°C and mean night temperatures of 14.5°C, and the growing season longer than 140 days are required. The average temperature of 21° to 23°C for the June—August months appears to be favourable for high yields. Night temperatures do not have to be high, but should be above an average of 12.5°C. The temperature requirement for germination and growth of corn is about 10°C. Young seedlings, up to the time they are about six inches tall, can often withstand a light frost without injury, but at older stages corn is quite susceptible to freezing temperatures.

High temperatures, too, may be injurious to crop. Zuber and Decker (1956) in Missouri found that during silking and tasseling period in 1954, the percentage of ears with good seed set dropped from 100 to 32 following three successive days of high temperature (38°C or more).

One of the greatest limiting factors in corn yield is insufficient moisture. The variability in the annual yield of maize is associated with the rainfall distribution pattern. A most critical period for moisture is at the time of flowering and fertilisation. Rainfall most effective was that which occurred three to five weeks before silking and from silking to three weeks afterward (Shaw and Dale, 1959; Runge and Odell, 1958). First, there should be adequate moisture if dry periods develop. May should be slightly warmer than normal, with an average temperature of about 18°C. Precipitation should be approximately 3.5 inches or a little less than normal. June should also be just slightly warmer than the normal of 21°C, with a rainfall of about four inches. For best development of the corn crop, July should have an average of 2° or 3°C below the normal of 22°C, with rainfall about 4.5 inches (which is little above normal). August temperatures are best at normal, 21.5°C, or slightly below normal, with rainfall about one inch above the normal of 5.8 inches. Weather in September should be warmer and drier than the normal of 17.7°C and four inches of rainfall. Another frequent hazard to corn production is early frost. The hazard seems to be the greatest in seasons of high rainfall and below-normal temperatures throughout the summer, for this causes a delay in the normal development of a crop.

Most important climatic types for corn production are humid intermediate climate, subtropical humid. Corn requires well-aerated soils, with an abundance of available plant nutrients. Prairie soils are among the most favourable because of their high level of native fertility.

Wheat

Wheat is a crop widely adapted throughout the world from near the equator to 60°N Latitude and 40°S Latitude. It seems likely that it is well adapted to steppe climate. It grows well in the humid continental, cool summer climate. Wheat is poorly adapted to climates that are continually moist and hot. Disease may cause severe losses under such conditions. It is well adapted in areas with moderate temperatures and sub-humid to semi-arid conditions.

Wheat is grown under a variety of temperature conditions, ranging from extremely low to rather high. It can grow under somewhat high temperatures, provided this condition is not coincident with high atmospheric humidity. Therefore it is widely grown throughout the world. A mean temperature of 15° to 20°C is the optimum for growth and development. A high temperature of about 30° to 35°C has generally detrimental effects on growth. Pollination and fertilisation in the spring wheat are regulated by the temperature and humidity of both soil and air. The optimum range of temperature for the germination of winter wheat and for its vegetative growth is 15° to 20°C. In climates, with higher temperature the winter wheat is generally sown when the air temperature drops to 6° to 10°C. In sub-tropical countries, like north-western India, wheat is sown when the mean air temperature ranges from 15° to 20°C. If the mean temperature surpasses 25°C during the vegetative period, the crop is adversely affected. High temperatures beyond 30°C, when synchronising with low relative humidity at maturity, result in reduced yields.

The most important wheat-producing areas of the world have an annual precipitation of less than 30 inches. For swelling in germination of seeds, the moisture required by the seed is 40 to 60 per cent of the dry weight of the seed. The high moisture content quickens the swelling of the seed. When the soil is dry, seeds fail to germinate. Again the sprouting of the seed is affected by the air and soil humidity. The non-availability of moisture inhibits the formation of side shoots. Dry soil can completely arrest the lateral shoots during the tillering period. When the moisture is available after a long drought, the side tillers begin to increase. These, however, may not be of any significance, as they generally fail to bear grains. Water deficiency at the heading stage results in shrivelled grains and low yield. This effect is especially marked when there are strong winds which hamper normal pollination. The newly evolved high-yielding varieties require much more water than the old ones. These varieties require at least 25 inches of rainfall or an equivalent amount of irrigation applied eight times at suitable intervals.

Potato

It is a crop of temperate climates and its importance is great both for human consumption and industrial use. Potato is planted when the soil temperature at a depth of 10 cm is 8° to 10°C. When it is planted at a higher temperature, the tubers are damaged; but when it is planted at a lower temperature, the sprouting is retarded. It has been estimated that at a mean daily air temperature of 10°C, the length of the period from sowing to sprouting is 27 days. But when the mean daily temperature is 22°C, sprouts appear after 12 days. Experiments have shown that the most favourable temperature for the potato tubers is 18° to 20°C. At 20° to 25°C, the formation of the tubers is retarded and beyond 29°C, it completely stops. A mild frost of up to –1°C partly damages the potato leaves, but when the temperature falls below –2°C, the exposed parts of most of the varieties are destroyed.

The water requirement of the potato crop is not very high during its vegetative period. It has been estimated that a rainfall of 15 to 20 cms or an equivalent amount of irrigation is sufficient for the normal crop. At the time of germination, the crop requires little external moisture, because the potato tubers themselves contain an adequate amount of water. However, as the leaves grow after sprouting, the water requirement of the crop increases. The optimum soil moisture for the potato crop is 65 to 80 per cent of field capacity. Just after sowing and before sprouting, heavy rains adversely affect the crop. Alternate dry and rainy weather leads to an excessive growth of the potato plant and encourages formation of off-shoots on the young tubers.

Potato requires a high light intensity if the crop remains in shade or if it is too densely planted, the yield is adversely affected.

Groundnut

Groundnut is grown under a variety of climatic conditions as compared to many of the other tropical crop plants. In the latitudinal extent, the crop is grown from 45°N to 30°S.

Too high a temperature and frosts adversely affect it. A temperature range of 14° to 16°C is necessary for the seeds to germinate. Lower temperatures retard the germination of the seeds and the growth of the plants. During growth, the temperature requirements of the crop are similar to those of maize and citrus. The crop requires at least 150 frost-free days for successful maturity.

Groundnut comes well with 60 to 125 cms of rainfall. Also good yields have been obtained in areas where the amount of rainfall is 37 to 50 cms, but very well distributed during the crop season. An ideal rainfall for a successful groundnut crop is 7.5 to 12.5 cms during the summer months preceding sowing, 12.5 to 17.5 cms during the fortnight after sowing and 75 to 60 cms distributed during the growth of the crop. The crop needs dry conditions at the time of harvesting. The occurrence of dry sunny weather at harvest facilitates harvesting and drying.

Barley

Barley is mostly used for brewing. Cool nights are more suitable for it. Very fertile soils and irrigation are needed. The crop has similar adaptability quality as wheat, but it can be grown in poorer soils and at lower temperatures in elevated areas. It is tolerant to frosts and salt-affected soils.

Some of the other crops having similar adaptability qualities as barley are jute, tobacco, sunflower, castor, field beans, sesamum, phaseolus, soyabean, cowpea, etc.

REFERENCES

Azzi, G., 1956 : *Ecology*, Constable, London.

Barger, G. L. and Thom, H. C. S., 1949a : "A method for characterising drought intensity in Iowa", *Jour. Amer. Soc. Agron.*, 41, pp.18-19.

———, 1949b : Evaluation of drought hazard, *Jour. Amer. Soc. Agron.*, 41, pp.519-526.

Burr, G. O., Hart C. E., Brodie, H. W., Tanimoto, T., Kortschak, H. P., Tokahashi, D. Ashton, F. and Coleman, R. E., 1957 : "The sugarcane plant", *Ann. rev. Plant Phys.*, 8, pp.275-308.

Cobley, L. S., 1942 : "The effect of shade on the growth rate of cacao cuttings", *Trop. Agr.*, (Trinidad) 19, pp.227-233.

Daubenmire, R. F., 1959 : *Plants and Environment*, Wiley, New York.

Eaton, E. M., 1950 : "Physiology of cotton plant", *Adv. In Agron.* 2, pp.11-25, Academic, New York.

Good, R., 1931 : *A theory of plant geography*, New Phytol, 80 : pp.149-171.

———, 1953 : *Geography of the flowering plants*, Longmans, Green and Company, London.

Hopkins, A. D., 1938 : *Bioclimates, a science of life and climate relations*, U.S. Deptt. Agri. Misc. Publ., 280.

Klages, K. H. W., 1930 : Geographical distribution of variability in the yield of field crops in the States of Mississippi Valley, *Ecol. 11*, pp.293-306.

———, 1934 : "Geographical distribution of variability in the yield of cereal crops in south Dacota, *Ecol. 12*, pp.334-345.

———, 1942 : *Ecological crop geography*, Macmillan, New York.

Livingston, B. E., 1916 : Physiological temperature indices for the study of plant growth in relation to climatic conditions, *Physiol, Res. 1*, pp.399-420.

Mavi, H. S., 1974 : *Agricultural Meteorology*, Punjab Agri. University, Ludhiana.

Odum, E. P., 1959 : *Fundamentals of Ecology*, 2nd Ed., Saunders, Philadelphia.

Rungi, E. C. A. and Odell, R. T., 1958 : "The relation between precipitation, temperature and the yield of corn on the Agronomy, South Farm Urbana, Illinois", *Agron. Jour*, 50, pp.448-454.

Shaw, R. H. and Dale, R. F., 1959 : Climate and corn yields in Iowa, *Weekly weather and crop bulletin*, XLVI (19), pp.7-8.

Schimper, A. F. W., 1903 : *Plant Geography on a Physiological Basis*, Clarendon, Oxford.

Thomas, J. E., 1957 : Rice in Spain. *World Crops*, 9, pp.247-250.

Thornthwaite, C. W., 1931 : "The climates of North America according to a new classification", *Geog. Rev.* 21, pp.633-655.

———, 1948 : "An approach towards a rational classification of climate", *Geog. Rev.*, 38, pp.55-94.

Visher, S. S., 1955 : "Comparative agricultural potential of the world's regions", *Econ. Geog.*, 81, pp.82-86.

Warming, E., 1909 : *Oecological plant geography*, translated by Grooni and Balfour, Clarendon, Oxford.

Zuber, M. S. and Decker, W. L., 1956 : *Effects of 1954 drought on corn. Mo Agr. Exp. Sta. Res. Bul.*, 604.

18

ROTATION OF CROPS

O. P. BISHNOI

CROP ROTATIONS play an important role in increasing productivity per unit area per unit time, without any ill effects on the productivity of the soil, if proper management practices are followed. With the advancement in farm technology in recent years and the introduction of high-yielding varieties of crops, increased use of chemical fertilisers and pesticides, mechanisation and adoption of many other practices, the system of rotation of crops has changed greatly. These have revolutionised the agricultural industry and have increased the production of important crops manifold. They have also significantly helped the farmers in changing their cropping patterns and rendering the fixed sequence of crops less desirable in the present-day agriculture.

It is well-known that soil fertility is depleted as a result of continual cropping. Different types of crops remove different quantities of plant nutrients from the soil. In order to maintain the soil fertility and productivity, it is, therefore, necessary to have good soil management practices, including the supply of organic matter, use of proper crop rotations and adoption of other recommended package of practices. Out of the many factors that play a vital role in maintaining production and productivity of the soil, suitable crop rotations have a definite place.

Advantages of Crop Rotation

Crop rotation has been defined by many writers in different ways. Some of the important definitions are as under:

"A crop rotation may be defined as the growing of different crops in the same field in a proper sequence for getting the maximum crop production."

"A crop rotation is a planned sequence or an order of crops grown in recurring succession on the same land area."

"The rotation of crops in an order in which the chosen cultivated crops follow one another in a set cycle on the same field over a definite period for their growth."

"Growing different crops in a regular sequence is called crop rotation."

"Relay cropping is a new system of cropping analogous to a relay race where a crop hands over the land to the next crop in quick succession."

"Multiple cropping is a philosophy of maximum crop production per unit of land with a minimum of soil deterioration."

It is evident from the above definitions by different authors that the term "rotation of crops" implies maximum crop production by growing different crops in proper sequence in the same land without any ill effect on the soil.

The importance of crop rotation in the general maintenance of soil fertility is well-known. Growing the same crop year after year in the same field depletes the soil of plant nutrients in an unbalanced manner. Crop rotations, therefore, have a number of benefits which are listed below:

(i) A good crop rotation maintains and invariably improves the productivity of the land by introducing the leguminous crops in the crop sequence. The leguminous crops, when included in a crop rotation, help to fix the atmospheric nitrogen in the soil through bacteria in the root nodules and improve the soil fertility. The soil nutrients are utilised more efficiently as a result of proper crop rotations; the growing of the same crop again and again depletes the soil of certain nutrients. The farmyard manure and chemical fertilisers added to the soil are also more efficiently utilised when a suitable crop rotation is followed.

(ii) Weeds, insect pests and crop diseases are better controlled by using a proper crop sequence. The weed flora is generally different for different crops and so are diseases and insect pests. The growing of the same crop again and again in the same field often leads to a higher infection with diseases and a greater infestation with insect pests, thus leading to much reduction in yield. By adopting a suitable good rotation, all the three evils can be considerably reduced, resulting in better yields of crops.

(iii) The farm labour is more uniformly distributed throughout the year in crop rotations than in the cultivation of long-duration crops.

(iv) The losses due to soil erosion are reduced. The spreading crops cover the land and reduce such losses considerably. The growing of such crops also reduces nutrient losses through leaching. Through a planned programme of alternating cultivated crops with grasses and legumes, it is possible to check soil erosion caused by water or wind. The crops like arhar also serve as a good wind breaker for checking soil erosion.

(v) Financial risk may be reduced by growing different types of crops. Financial risk through over-production and low prices can also be reduced.

(vi) Crop yields are increased through planned patterns of cropping. It is a well-established fact that continual cropping with the same crop in the same field results in lower yields than those from well-planned crop rotations. The other factors mentioned above also increase the yields of crops grown in the proper sequence.

(vii) The soil is properly managed with regard to its village and preparation for growing crops in sequence. The sequential growing of crops also helps to keep the soil, texture and soil structure in proper condition.

(viii) The soil micro-organisms play their full role in enriching the soil by their action on the soil organic matter and help bring about other changes which take place in the soil during the growth of a crop.

(ix) The available irrigation facilities are better utilised when a suitable crop rotation is followed. Different crops have different irrigation requirements, and a combination of suitable crops in proper sequence certainly results in a more economic use of irrigation water.

(x) The planned programme of growing crops enables the cultivation of deep-rooted crops, followed by shallow-rooted ones, and of exhaustive crops followed by restorative crops.

Cotton is a deep-rooted crop and invariably it is followed by senji *(Melilotus parviflora)* which is a leguminous fodder crop.

Features of Good Crop Rotation

A crop rotation may be good or bad, as measured by its effects on soil productivity, and on the basis of economic returns. It is, therefore, desirable that a good crop rotation should have the following features:

(i) It should include crops which can grow well on the type of the soil and the topography of the land available. It is well established that soil requirements vary from crop to crop. Therefore, the inclusion of crops in a rotation, according to the soil type, is necessary.

(ii) Leguminous crops should be included to maintain or increase nitrogen and organic matter in the soil. Cereals, cotton and sugarcane are considered exhaustive crops, whereas leguminous crops like pulses, groundnut, guara and berseem, are restorative crops.

(iii) A crop rotation should give the maximum economic returns per unit area.

(iv) It should include crops which result in a better control of weeds, insect pests and diseases.

(v) It should include crops which leave behind good residues in the soil after harvesting for maintaining the organic content of the soil and thus keeping up its fertility.

(vi) It should include a large area under the most profitable crops and should include both deep-rooted and shallow-rooted crops and the crops which give sufficient green fodder for the livestock.

(vii) It should facilitate soil conservation and water management practices, particularly in the rainfed areas.

(viii) It should increase the efficiency of farming with minimum tillage.

Planning a Crop Rotation

The proper and timely planning of a crop rotation for a given area is very necessary in the present-day agriculture. At the time of planning of crop rotation the knowledge of soil type, topography, climate including the rainfall patterns availability of input facilities, plant protection measures, maximum economic gains per unit area, management of soil and water resources, maintenance of soil fertility, etc., is very essential to formulate the various crop rotations.

Rotation of Crops

The delineation of country into homogeneous soil-climatic zones in relation to cropping patterns is very important to seek the feasibility of various crops. Krishnan and Mukhtiar Singh (1972) divided India into 63 homogeneous soil-climatic zones in relation to their suitability for cropping patterns based on moisture index, thermal index and the soil types. Therefore based on climatic and soil features suitable crop rotations may be adopted for the various situations. Based on moisture and thermal conditions, India can be demarcated into 20 zones which can be further sub-divided into 63 zones when the soil types are superimposed over it (Krishnan and Mukhtiar Singh, 1972). Some of the important crop rotations are given in Table 18.1 in the 38 important soil-climatic zones of India.

TABLE 18.1: **Important crop rotations in various soil-climatic zones of India**

S. No.	Soil-climatic zone (Moisture-thermal belt, soil type)	Districts covered in the zone	Feasible crop rotations
1	2	3	4
1.	Extremely dry-hot, sandy soil	Bikaner, Jaisalmer, Jodhpur, Barmer	Bajra/moong/moong-moth/grasses.
2.	Extremely dry mild, sandy soil	Ganganagar gram/mustard.	Bajra/moong/grasses-
3.	Extremely dry-cold, skeletal soil	Ladakh	Ragi-wheat/barley.
4.	Semi-dry-very hot, mixed red black soil	Kurnool, Raichur, Mahboobnagar cotton-moong/jowar.	Bajra/jowar/groundnut/ cotton-moong/bajra/jowar,
5.	Semi-dry-hot, alluvial soil	Churu, Jalore, Banas Kantha, Jaipur, Sikar, Pali, Naguar, Kutch, Jamnagar, Junagarh, Bhavnagar, Rajkot,	Bajra/moong/moth-wheat/ gram/barley/mustard.
6.	Semi-dry-hot, medium and deep black soil	Amreli, Bijapur, Gulbarga	Rice-rice, rice-rice-ragi/ moong/groundnut.
7.	Semi-dry-mild, alluvial soil	Bhatinda, Hissar, Mahindergarh, Ferozepur, Ajmer, Sirohi	Bajra/jowar/moong/ wheat/gram/mustard/ barley, cotton-wheat.

TABLE 18.1 *(Contd.)*

1	2	3	4
8.	Dry-very-hot, red soil	Nalgonda, Salem, Tiruchirapalli	Rice-rice/groundnut/jowar/wheat-rice/moong/maize/jawar/rice-rice/wheat.
9.	Dry-very-hot, mixed red and black soil	Khammam, West Godavari, Krishna, Cudapah, Nellore, Madurai, Ramanatha-puram, Triunelveli, Warangal, Guntur	Rice-rice/moong/rice/moong/ragi/groundnut/urd/gingelly.
10.	Dry-hot, red soil	Hyderabad, Anantpur	Rice-rice/groundnut/jowar/wheat/cotton-rice/maize/sugarcane.
11.	Dry-hot, alluvial soil	Gurgaon, Alwar, Tonk, Bharatpur, Mathura, Agra, Mainpuri, Kanpur, Forukkabad, Etawah	Bajra/jowar/moong/maize-wheat/gram/mustard-summer moong/fodder.
12.	Dry-hot, grey brown soil	Mehsana, Surendranagar, Kaira, Ahmedabad	Cotton-wheat, bajra/maize/groundnut/rice-wheat. Tobacco+Soyabean, Cotton+Soyabean –bajra, Soyabean – Jowar fodder-sugarbeet/ragi.
13.	Dry-hot, medium and deep black soil	Jalaun, Hamirpur, Ratlam, Mandsaur, Medak, Bellary, Chitra-durga, Coimbatore, Bhind, Datia, Shivpuri, Kotah, Sabarkantha, Panch Mahals, Jhabua, Dhar, Nimar, Baroda, Dhulia, Jalgaon, Buldana, Akola, Amravati, Parbhani, Aurangabad, Nasik, Ahmednagar, Bir, Sholapur, Osmanabad, Bidar.	Jowar/bajra/rice/ground-nut-wheat/maize/moong/cotton/gram-moong/castor/fodder.
14.	Dry-mild, red soil	Mysore, Udaipur, Dungarpur, Banswara	Rice/maize/jowar/cotton/wheat/gram/barley/mustard.
15.	Dry-mild, alluvial soil	Gurdaspur, Amritsar, Kapurthala, Jullundur, Ludhiana, Sangrur, Patiala, Karnal, Delhi, Meerut, Bulandshahar, Etah, Aligarh	Maize/bajra, jowar/groundnut/arhar-wheat/gram/mustard/barley.

TABLE 18.1 *(Contd.)*

1	2	3	4
16.	Dry-mild, mixed red and deep black soil	Chittorgarh, Dharwar, Poona, Satara, Sangli, Belgaum	Jowar/cotton/goundnut/ maize/arhar-wheat/gram.
17.	Slightly dry-very hot red soil	Visakhapatnam, Srikakulam, Chittor, Ganjam, North Arcot, South Arcot, Chingleput, East Godavari, Karimnagar	Rice-wheat/rice/ groundnut-jowar/rice/ moong/maize/ragi.
18.	Slightly dry-hot, mixed red and alluvial soil	Kolar, Tikamgarh, Adilabad, Hardoi, Sitapur, Unnao, Lucknow, Barabanki, Fatehpur, Raibareli, Allahabad, Pratapgarh, Faizabad, Sultanpur, Varanasi, Ghazipur, Ballia, Azamgarh, Shahabad, Patna, Mongyr, Gaya	Ragi/jowar/rice/arhar/ cotton/bajra/groundnut-wheat/gram/barley.
19.	Slightly dry-hot, medium and deep black soil	Betul, Jhalawar, Wardha, Yeotmal, Nizamabad, Nanded, Baroda, Broach	Rice/wheat/maize/moong/ cotton-moong/castor/ jowar/arhar/groundnut/ rice-wheat/gram.
20.	Slightly dry-mild, red soil	Chikmagalur, Hassen, Mandya, Tumkur, Bangalore	Rice-rice/wheat/ groundnut/jowar/rice/ moong/maize/ragi.
21.	Slightly dry-mild, alluvial soil	Ambala, Saharanpur, Muzaffarpurnagar, Badaun, Moradabad, Shahjahanpur, Jammu, Kathua	Rice/maize/jowar/ groundnut/arhar/ soyabeen-wheat/barley/ gram/peas.
22.	Slightly dry-mild medium and black soil	Banda, Ujjain, Shajapur, Dewas, Indore	Cotton/jowar/groundnut/ arhar/soyabean/rice-wheat/gram/peas.
23.	Slightly moist-hot, red soils	Bolangir, Purulia, Bankura, Kalahandi, Baudh, Khondmals, Puri, Bhubaneswar, Keonjhar, Dhenkanal, Midnapore, Birbhum, Bilaspur, Raipur, Santal Parganas, Hazaribagh, Palamau, Dhanbad, Murshidabad, Burdwan, Malda	Rice/maize/jute/wheat/ gram/rice, rice/jute-potato/rice/maize/rice/ groundnut/green gram.
24.	Slightly moist-hot alluvial soil	Nadia, Hooghly, Bhagalpur	Rice/maize/jute/wheat/ gram/linseed.

TABLE 18.1 *(Contd.)*

1	2	3	4
25.	Slightly moist-hot, mixed red and black soil	Singhbhum, Chanda, Bhandara, Shahdol, Sidhi, Nagpur, Hoshangabad	Rice/jowar/maize/cotton/arhar-wheat/gram/barley/peas.
26.	Slightly moist-mild, alluvial soil	Hoshiarpur, Bijnor, Rampur, Bareilly, Pilibhit, Kheri, Bahraich, Gonda, Basti, Gorakhpur, Deoria, Muzaffarpur, Dharbhanga, Shaharsa	Rice/maize/groundnut/arhar/sugarcane-wheat/gram/mustard.
27.	Slightly moist-mild, mixed red and yellow/black soil	Raigarh, Surguja, Mirzapur, Chhattarpur, Panna, Satna, Rewa, Sehore, Vidisha, Sagar, Guna, Damoh	Rice/maize/jowar/arhar cotton-wheat/linseed/gram/barley/mustard.
28.	Moist-hot, red soils	Durg, Bastar, Sundargarh, Sambalpur, Balasor	Rice/maize-wheat/potato.
29.	Moist-hot, alluvial soil	24 Parganas, Howrah	Rice/jute-rice/mustard.
30.	Moist-hot, mixed red and black soil	Balaghat	Jute/rice/maize-rice/wheat/barley.
31.	Moist-mild, alluvial soil	Nowgang, Champaran, Purnea	Rice/maize/jute-rice/wheat/barley.
32.	Moist-mild, mixed red & black soil	Ranchi, Seoni, Mandla	Rice/maize/jowar/arhar-rice/wheat/gram.
33.	Moist-mild, medium and deep black soil	Chhindwara, Narsingpur, Jabalpur	Rice/maize/arhar/jowar-wheat/gram/peas.
34.	Wet-hot, alluvial soil	Thana, Colaba, Ratnagiri, N, Kansara, S. Kanara, Cannanore, Trichur, Kozhikode, Ernakulam, Kottayam, Alleppey, Quilon, Trivandrum, Darrang	Rice-rice-groundnut/maize/ragi/bajra.
35.	Wet-mild, red and yellow tarai soils	Dehradun, Tripura	Rice/maize/sugarcane/arhar-wheat/jute/mustard/gram.
36.	Extremely wet-mild, alluvial soil	Lakhimpur, Kamrup, Goalpara, Coochbehar	Rice/jute-rice/mustard.
37.	Extremely wet-cold, brown hill soil and tarai soil	Mandi, Mahasu, Almora, Garhwal, Tehri, Garhwal	Rice/maize-wheat/barley/mustard.
38.	Extremely wet-cold, sub-montane soil	Udhampur, Chamba, Kangra	Maize/rice-wheat/barley.

In special situations, as an example, crop rotations of 200 to 400 per cent intensity of cropping in Haryana State are given in Table 18.2.

TABLE 18.2: Multiple cropping systems for Haryana State

Intensity of Cropping	Jan.	Feb.	Mar.	Apr.	May	June	July	Aug.	Sept.	Oct.	Nov.	Dec.
400%	Wheat			Moong/ Cowpeas 68			Maize Bajra		Toria/ Potato/ Radish/			
300%				Juar Fodder/ Bajra Fodder/ Cowpeas Fodder/grain Guara Fodder/ G.M. Moong (grain) Cheena (grain) Maize Sathi (grain)			Rice/ Maize/ Bajra/ Moong/ Cowpeas/ Groundnut Sesamum				Wheat Peas Barley Raya Sarson	
				Cotton							Wheat Peas Barley	
200%				Cotton+ Moong/ Cowpeas							Sainji Gram Potato	
200% (Sugarcane areas)		Sugarcane Sugarcane+ Cowpeas/ Moong/ Bhindi										Sugarcane Peas/ Onion Wheat Wheat
						Sugarcane (sprcuted buds)						

Crop Rotations in Some Specialised Farming Areas

Around big cities and towns with the availability of special facilities like inputs, marketing and remunerative prices, there are some specialised areas for growing fodders, vegetables sugarcane, etc.

Dry and green fodder are in great demand for meeting the requirements of the dairy farms and so are the vegetables. Sullage water is the main irrigation source which meets the fertiliser requirements of fodder crops to a great extent. With the availability of input facilities, farmers can grow a number of both Kharif (maize, jowar, bajra and cowpeas) and Rabi (berseem, oats, senji and lucerne) fodders round the year. Green fodder rotation is followed as early Kharif fodder — main season Kharif fodder — Rabi fodder.

Vegetable growing is also very popular around cities. Very high intensity of cropping varying from 200 to 600 per cent (Table 18.3) can be followed when the irrigation facilities are adequate and the inputs like fertilisers, seeds, labour and plant protection measures are readily available. The timely marketing of the produce is a must, potato cultivation is mainly mechanised. Planting, earthing up and digging are performed with tractor-drawn implements. The growing of dhaincha, sannhemp and green-manuring them before potato is very popular in the potato-growing areas in the submontane zone. The cultivation of ponda sugarcane is also practised in some areas as an intercrop in summer vegetables. The vegetable crops can also be followed with the main crop or fodder rotations. Some of these are chillies-wheat/peas, chillies-Rabi fodder (berseem), cabbage-Kharif fodder, Bhindi-wheat, Kharif fodder-peas-wheat, peas+radish-wheat, chillies+onion-wheat/oilseeds, Brinjal-wheat, sugarcane+brinjal/bhindi-sugarcane, chillies-senji-sugarcane, onion+cucurbits-sugarcane, sugarcane, sugarcane celery+tinda, cotton-chillies.

The seed production of vegetables, like radish, bhindi and carrot has also gained popularity with the progressive farmers.

Sugarcane is a long-duration crop but is important near the sugarmills in the area. Among autumn intercropping, combinations like cane+potato/lentil/peas/tomato/wheat/lahi/ and in spring cane+spring tomato/onion are very remunerative. The intercropping with vegetables have been found to be more remunerative than with the crops like maize.

Intercropping of banana with toria, radish, cucumber, turmeric, ginger, coriander, maize, field beans, peas, squashes have given economic yields in the banana-growing areas.

TABLE 18.3: Multiple cropping systems for vegetable crops in Haryana State

Intensity of Cropping	Jan.	Feb.	Mar.	Apr.	May	June	July	Aug.	Sept.	Oct.	Nov.	Dec.
600%	Spring Potatoes & Cucurbits (Tar Kakri)	—	Cucurbits	Leafy vegetables	—	—	Radish (desi)	—	Palak	—	Potato	—
500%	Spring Potato	—	—	Chauli	—	—	Early Cauliflower	—	Mid-Cauliflower	—	Autumn Potato	—
	—	Onion for green vegetables	—	Bhindi	—	—	Radish (desi)	—	Mid-Cauliflower+	—	Methi	—
		Bhindi	—	Chauli	—	—	Tori	—	Palak	—	Palak	—
	—	—	Onion for green vegetables	—	leafy veg. (Kulfa)	—	early cauliflower	—	Carrot Potato	—	Late Potato	—
	Tomato early (under cover)	—	Cowpeas	—	—	—	Bhindi	—	Palak	—	Potato	—
	Brinjal (Ratoon)	—	Cowpeas	—	—	—	Chillies (green)	—	+Radish	—	Palak	—
	Spring Potato	—	—	Leafy vegetable (Kulfa or Chaulai)		Early Cauliflower		Palak	Potato			

TABLE 18.3 *(Contd.)*

Intensity of Cropping	Jan.	Feb.	Mar.	Apr.	May	June	July	Aug.	Sept.	Oct.	Nov.	Dec.
400%	Spring Potato	—	—	Leafy Vegeta-tables	—	—	Early Cauli-flower	—	—	Potato	—	—
	Spring Potato	—	—	Cowpeas	—	—	Early Cauli-flower	—	—	Potato	—	Potato Potato
300%	--	—	—	Cowpeas	—	—	Early Cauli-flower	—	—	Lettuce	—	—
	Spring Potato	—	—	—	Leafy vegetable cucurbits			—	—	Cauliflower		—
	—	Tomato				Cauliflower				Potato		
300%	Onion	—	—	—	—	—	Cauliflower		—	Potato	—	—
	—	Capsicum	—	—	—	—	Leafy vegetable		—	Cauliflower		—
	—	Water-melon	—	—	—	—	Bhindi	—	—	Potato	—	—
	—	Cucumber	—	—	—	—	Bhindi	–	—	Potato (Radish)		—
	Spring potato	—	—	Leafy vegetables	—	—	Tomato	—	—	—	—	—
	—	Musk-melon	—	—	—	—	Bhindi	—	—	Potato	—	—

TABLE 18.3 *(Contd.)*

Intensity of Cropping	Jan.	Feb.	Mar.	Apr.	May	June	July	Aug.	Sept.	Oct.	Nov.	Dec.
	—	Bhindi		—	—	—	Luffa	—	—	Potato	—	—
	—	Musk-melon	—	—	—	—	Bhindi	—	—	Potato+ Radish	—	—
	—	Chillies	—	—	—	—	Carrot (late July)	—	—	Potato	—	—
	—	—	—	—	Leafy vegetables	—	Cucurbits	—	—	Pea	—	—
	—	—	—	—	Leafy vegetables		Carrot (late July)		—	Broad Bean	—	—
	—	Cucurbits (in the mid of Potato)	—	—			Chillies	—	—	Potato	—	—
	—	—	—	—	Leafy veg.	—	—	Cauli-flower	—	—	Vicia faba	—
	—	Tomato	—	—	—	—	Bhindi	—	—	Potato+ Radish	—	—
	—	Onion	—	—	—	—	Chillies	—	—	—	Lettuce	—
200%	Cucurbits	—	—	—	—	—	Brinjal	—	—	—	—	—
	Bottle gourd	—	—	—	—	—	Tomato	—	—	—	—	—

Tobacco is an important cash crop. Maize-tobacco+green onion-maize has given higher economic returns as compared with the best cereal rotation paddy-wheat-maize in north Bihar. The other rotations are urad-tobacco+union-maize, sanhemp-tobacco+green onion-maize/moong/onion.

Topica, yams, amorphophallus, etc., can be grown as inter-crops in coconut and arecanut in most of the uplands specially in Kerala. Topica and yams can be grown during the first two to four years in rubber, coffee, tea and cashew plantations depending upon the coverage of plantation crops in hilly tracts. Topica can also be grown in rotation with banana once in two years. Also legumes like cowpea, groundnut, black gram, green gram, sannhemp, soyabean, etc., and vegetables like bhindi, coleus, etc., can be grown successfully as inter-crops in cassava and amorphophallus fields. In low-lying areas sweet potato, coleus, colocasia pulses, vegetables, etc., can be grown as rotation crops. Two crops of paddy followed by one crop of sweet potato or colocasia can also be taken. Thus there is good possibility of growing these tuber crops in multiple cropping.

The growing of fodder crops is becoming very popular around towns. It can be grown from 200 to 400 per cent cropping intensity. The rotation with 200 per cent cropping intensity are berseem + Japan rape intercropped during winter in Napier X bajra hybrid, lucerne intercropped with Napier X bajra hybrid, berseem+Japan rape-Jowar+cowpea, oats + berseem/senji-jowar + guar. The rotation with 300 per cent cropping intensity are Berseem + Japan rape-cowpeas/maize/jowar/bajra/jowar/cowpeas/maize, oats/sanji/metha-guar/bajra/cowpea-jowar-maize/cowpea. The 400 per cent intensity of cropping possible are oats-bajra+cowpea-maize + cowpeas-maize + copweas.

Crop Rotation in the Rain-fed Areas

Water is the major limiting factor in the dryland agriculture. The adoption of good management practices, like the selection of proper choice of crops and varieties, the optimum plant population, the judicious use of fertilisers and control of insect and pest diseases, etc., are very important in this connection. The amount of soil water available to plants depends upon the depth to which their roots can penetrate and density of the root system. To induce deep rooting means to increase water use from the subsoil layers resulting in higher yields under the restricted availability of soil moisture.

Water harvesting, as a promising technique, can be used in areas of low rainfall with the increase of more available water in proportionally lesser areas. Contour-bunding, land-shaping and conservation bench-terracing can help to retain rainwater and store it in the land profile. Suitable farm implements are helpful in checking evaporation from the soil profile. The use of straw mulches has proved very beneficial in boosting crop production under dryland agriculture. The yields under rainfed conditions can be increased by appropriate package of practices. Control of weeds is very essential in dry farming in order to remove their competition with the crops for space, moisture and nutrients. The deep placement of fertilisers can induce deep rooting and lead to a more efficient use of soil moisture. In the low rainfall areas (less than 500 mm annual) only one crop is possible. In medium to high rainfall areas with well distribution of rainfall and proper conservation practices can be adopted for double cropping. The crops which are hardy and can tolerate drought are most suited to be grown in such cases. In medium rainfall areas (500 to 750 mm annual) the crop rotations of the form bajra/jowar/guar/gram/rabi oilseeds, moong/cowpea/til-gram/barley can be followed. In high rainfall areas (above 750 mm annual) maize/mash/arhar lentil/rabi oilseeds/gram/ barley/wheat can be adopted. Paddy-wheat rotation is very popular in high rainfall areas with the better soil and water conservation practices.

Intercropping and Mixed Cropping

This is a special practice being followed in situations mainly as a safeguard against the failure of a single crop. The inter-row space in the widely spaced crops, like sugarcane, cotton and maize, can be better utilised by growing another crop as an intercrop. Similarly, mixtures can be grown where row spacing in the crop is less. In low rainfall areas, the sowing of wheat+gram, raya intercropped in gram and wheat, and sarson as an intercrop in wheat are very old practices. Its importance decreased with the introduction of high yielding varieties of wheat. Intercropping of vegetables, like brinjal, bhindi and onions and even sunflower in the inter-row spacing of sugarcane is being practised.

The intercropping of wheat in potato crop; sesamum in mash crop; sesamum, moong and guara in groundnut; summer moong in cotton; barley in lentil; mash in cotton and maize; gram and taramira in desi cotton are a few important examples being followed by the farmers.

Vegetables also can be grown in mixtures. These examples have been already given Tables 18.1 to 18.3 in detail.

Economic Aspects of Crop Rotations

With the introduction of high-yielding varieties of crops, particularly of rice and wheat, there has been the maximum expansion of area under the high-yielding varieties since 1966. The choice of crops to be grown in a crop sequence depends mainly on the economic returns the farmers can get per unit area per unit time. The availability of physical facilities like irrigation, inputs, labour and mechanisation have also an important bearing on the cropping patterns and the crop rotations to be followed in any area. The important crop rotations being followed in northern India are paddy-wheat, maize-wheat, groundnut-wheat, cotton desi-wheat, bajra-wheat/gram+rabi oilseeds, manuring-potato-wheat, cotton American fallow/wheat/barley, toria-wheat, sugarcane-sugarcane, Kharif vegetables-rabi green vegetables. Out of all these rotations paddy-wheat in the northern regions and paddy-paddy rotation in the southern tract are very common in the rice-growing areas. Maize/groundnut/arhar-wheat rotation is very popular in the maize-growing areas. In low rainfall areas bajra/jowar followed by wheat/gram/Rabi oilseeds are normally followed.

In the cotton belt, cotton-wheat is very popular and remunerative. Growing of sugarcane with intercrops of vegetables and short duration crops are also very paying. Potato-wheat is also more paying.

It may be worthwhile to mention that no single crop rotation can be paying all over the region. Its adoption depends upon the agro-climatic conditions of the region, the availability of input facilities and remunerative prices. It the National Demonstration trials paddy-wheat gave the highest production and net income per unit area per unit time. This rotation was followed by the maize-wheat, bajra-wheat and groundnut-wheat rotations.

The three crops rotations of cotton-wheat-fodder, paddy-wheat-sathi moong, paddy-wheat-fodder, paddy-wheat-sathi maize, bajra-wheat-sathi moong, bajra-wheat-fodder, groundnut-wheat-sathi moong give good economic returns.

Thus keeping in view the climatic factors, oil type, availability of irrigation and other inputs, prevalent market rates and other guidelines indicated above, the farmers may choose such crop rotations as give them the maximum production and income per unit area per unit time in the present day agriculture.

REFERENCES

"Cropping patterns in India", 1972 : *Proceedings of the Symposium* held at Delhi, Jan. 27-31, 1968, under the Indian Council of Agricultural Research, New Delhi.

"Multiple Cropping" : *Proceedings of a Symposium* held at Haryana Agricultural. University, Hissar, Under the Indian Society of Agronomy and ICAR, Oct. 7-8, 1972.

Singh, K. and Bishnoi, O. P., 1976 : *Cropping Patterns in Haryana*, Research Publication No. DR 2/76, Haryana Agricultural University, Hissar.

Singh, K., 1975 : *Crop Rotations*, published by Punjab Agricultural University Ludhiana, 1975.

"Water Management" : *Proceedings of a Symposium* held in May, 1966, at the University of Udaipur, Udaipur, under the Indian Society of Agronomy and ICAR.

19

CROP-COMBINATIONAL ANALYSIS

A Review of Methodology

ABDUL MAJEED

AGRICULTURE IS by far an important single use of land in most of the countries of the world, and the description and analysis of various problems relating to agricultural activities have contributed a great deal to the content of the regional and economic geography of these countries. Such studies have, however, confined themselves to the description of dominant individual crops rather than complex agricultural system, with the result that they usually suffer from overgeneralisations, because the individual crops, except in relatively rare circumstances of extreme mono-culture, are not grown alone but characteristically they appear in combinations (Weaver, 1954). Moreover, it is significant to note that any process of development has an inherited property of pushing the economic system from simpler to more complex. This complexity as manifested by diversification of places in terms of their economic activities can be measured by the different combinations of their activities.

An intuitive classification based on subjective judgements like commercial towns, industrial towns or wheat zones, rice zones, etc., have no statistical basis. All the places, in fact, have multiple functions, but all of them may not be equally important at every place. Some of them may be very significant, while others might be quite inconsequential. A statistical basis is, therefore, necessary for identifying the group of significant functions, which represent the functional personality of an area (Mahmood, 1977).

Recently attention has been focused on combinational analysis of different elements in geographical studies which give a scientific treatment to the geographical problems. Geographers are particularly interested in classifying the places into different combinations of dominant activities including those which pertain to agricultural production and landuse. This method ultimately minimises the chances of oversimplified generalisations too.

A general understanding of the particular combination of crops and the relative importance of each in an area can be very helpful in interpreting some aspects of social and economic geography of the region. J.C. Weaver has pointed out three different lines in which crop-combinational studies may have more significance. Firstly, they are essential to an adequate understanding of individual crop geography; secondly, crop-combination in itself is an integrated reality that demands definitions and distributional analysis; thirdly, such regions are essential for the construction of the still more complex structure of valid agricultural regions (Weaver, 1954).

Many methods have so far been devised to demarcate the set of elements that play a significant role in any particular system. The present paper does not introduce a new method of forming such element-combinations, but it is only an attempt to review some of the methods already worked out and to assess their accuracy and usefulness.

The Quartile Method

Johnson (1958) applied a mathematical procedure in order to derive primary crop-combinations. In forming the crop-association regions of East Pakistan (now Bangladesh) he used five crops — three main crops — namely *aman* paddy, *aus* paddy and jute — one, a group of dry *rabi* crops; and one, *boro* (paddy) and devised a simple scale of gradations which may be called as Quartile Method.

The percentages of these five crops to net sown area were arranged in decreasing order and using the average percentages for the whole of East Pakistan as the mean point, he determined "Very High (VH)", "High (H)", "Medium (M)", "Low (L)", and "Very Low (VL)" areas for each crop. The scale was calculated by dividing the range between the mean value and the maximum recorded percentage, in the ratio 1 : 2 : 2 upwards, and similarly for the lower end of the scale. Then the crops with a medium or higher percentage were grouped together to give the crop association regions. Thus six distinctive regions were formed for the whole of East Pakistan.

The method is very simple since the scale of gradations is

calculated only once and then the regions are formed by constantly referring to it. Thus practically no further calculations are required; and the author has also explained how he has solved a few minor problems that arose in the adoption of this method. But there is one main defect which must be pointed out here. The combinations obtained depend upon the location of *medium* (or "mean" in the case of East Pakistan); and this in turn depends upon the *mean* chosen. The combinations will be different, if, instead of choosing a large area, only a part of it is chosen (Ayyar, 1969). Thus the Quartile Method devised by Johnson is not suitable for the large areas where too many crops at each level of importance produce an unmanageable number of variants. Again, every crop, however, insignificant it might be in comparison to other dominant crops, has also its Very High, High and Medium grades and, therefore, is included in the combination. Since these grades are based on averages for the whole of the region, it often happens that the pre-dominant crops of the unit of the region is allotted a lowest grade owing to its low percentage in comparison to others, while the insignificant crop of the same unit may get a higher grade owing to its greater importance in comparison to others. The defects introduced by the averages of the whole region may be removed by considering the mean of all the crops in each unit area (Siddiqui, 1967).

Mean Positive Deviation Method

This method was devised by L. L. Pownall (1953). He applied this method in the functional classification of towns in New Zealand, and calculated the means of various functions for all the towns of New Zealand and grouped the functions having a positive deviation from the mean in each town. This method can be put statistically as :

$$X_i - \bar{X} > 0, i = 1, 2, 3, \ldots n$$

(X_i being the individual value and $\bar{X}$ their mean)

This is an ordinary method and requires only grouping of these elements which have a positive deviation from the mean. Thus the combinations based on simple positive deviations from the average are quite reasonable but the average value is very sensitive to variations and is easily influenced by a number of insignificant crops which have a very low percentage (Siddiqui, 1967).

Standard Deviation Method

This is simply an improvement on Pownall's Mean Positive Deviation Method and was introduced by H. J. Nelson (1955). After calculating the Standard Deviation (SD), he allotted three degrees of deviation from the mean to each function. These three degrees of variation were : mean + 1 SD, mean + 2 SD and mean + 3 SD. All the functions having a value more than mean were grouped together and those over + 1SD were rated as 1, over + 2 SD were rated as 2 and over + 3 SD or more were rated as 3. The percentage of a function which was less than its mean + 1 SD was not considered as significant and hence not included in the combination.

Since each function having a positive deviation is included in the classification both by Pownall and Nelson, the combinations are the same. Thus the only improvement brought out by Nelson is in allotting 1, 2, or 3 ratings to various elements according to their degree of variation from the mean. Nelson's ratings, although consuming a lot of time and labour in calculating standard deviation for each unit area, make it more expressive by assigning the degree of individual crop strength to the combination.

The crop-combinations based on this method are more objective, well defined and nearer to truth except the shortcoming described in the case of Pownall's Method (Siddiqui, 1967).

Minimum Deviation Method

This method, advanced by J.C. Weaver in 1954 in his study on the crop-combination regions in the Middle West (USA), is essentially statistical and takes into consideration percentages of crop acreages to total cropped area.

In the statistical method of crop-combination study devised by Weaver, the percentages mentioned above are arranged in decreasing order. The method devised and adopted by Weaver is fully described in his paper and will be only briefly explained here :

A standard theoretical distribution is taken as follows:

100.0% of GCA*	for mono-culture	single crop
50.0% of GCA	for each crop	two-crop

* Gross Cropped Area.

		combination
33.33% of GCA	for each crop	**three-crop combination**
25.0% of GCA	for each crop	**four-crop combination**
20.0% of GCA	for each crop	**five-crop combination**

and so on for each next crop-combination.

The actual percentages of various crops in any area will, of course, generally be different from any of the above theoretical values. For determining combinations, Weaver calculated deviations of real percentages of crops in the unit area considered against a theoretical standard (as given above) and allotted the crop combination to that one where the actual percentages showed the minimum deviation from the theoretical values.

Let $d_1, d_2, d_3 \ldots\ldots\ldots d_n$ be the deviations (plus or minus) of the actual percentages from any one of the theoretical values. These deviations, for every set of crops, are squared and summed. Each sum is divided by the number of crops in the set. The square root of this result will give us a measure of the deviation of the concerned crops-combination from the theoretical distribution. Calling this δ we have :

$$\delta = \sqrt{\frac{\Sigma(d)^2}{n}}$$

where $\Sigma\,(d)^2$ is the sum of the squares of individual deviations and n is the number of crops being considered.[1] That combination with the smallest value for δ will be the combination to be selected.

However, as Weaver has pointed out, "Since we are not directly interested in the actual value of the δs but only in their relative sizes, the square root need be extracted to save work and we might use the square of δ (i.e. variance or δ^2) viz.

$$\delta^2 = \frac{\Sigma(d)^2}{n}$$

To illustrate his technique Weaver gives the example of Keokuk country, where corn occupied 54 per cent, oats 24 per cent, hay 13 per cent, soybeans 5 per cent and wheat 2 per cent. In the example given below.

Mono-culture

$$\frac{(100-54)^2}{1} = 2116$$

2-Crop Combination

$$\frac{(50-54)^2 + (50-24)^2}{2} = 346$$

3-Crop Combination

$$\frac{(33.33-54)^2 + (33.33-24)^2 + (33.33-13)^2}{3} = 309$$

4-Crop Combination

$$\frac{(25-54)^2 + (25-24)^2 + (25-13)^2 + (25-5)^2}{4} = 347$$

5-Crop Combination

$$\frac{(20-54)^2 + (20-24)^2 + (20-13)^2 + (20-5)^2 + (20-2)^2}{5} = 354$$

Since the variance for 3-Crop Combination (309) is the smallest (i.e., minimum deviation from the Standard Theoretical Value), the county is designated with a combination of three crops, i.e. COH (Corn-Oats-Hay) (Weaver, 1954).

Weaver's method to determine the minimum deviations appears to be quite simple, but in practice, it requires much calculation. Occasionally it also tends to produce highly generalised results in areas of large number of variants, or as admitted by Weaver himself, in areas where the first-ranking crop is very large, although not large enough to justify the combination as mono-culture, and the remaining crops are comparatively small. In such circumstances the results tend to "show the lowest deviation for a crop-combination that includes every crop occupying as much as one per cent of total harvested cropland" (Weaver, 1954).

Such a problem of unwieldy and most generalised combinations was met in Bundelkhand (Siddiqui, 1967), Madhya Pradesh (Ayyar, 1969), Malwa Tract (Singh, 1965) and Uttar Pradesh (Husain, 1972) while evaluating the crop-combinations, considering all the crops of the region. This does prove that the method is intrinsically defective and has the weakness of overgeneralisations. In fact, the method only emphasises the variety of crops grown in a region and that it would be admirably suitable if we consider smaller administrative units, or there is considerable crop specialisation.

Least Squared Deviation Method

Thomas (1963) modified Weaver's method by including all the crops with zero per cent theoretical values also in each step of the comparison and applied this method in deriving the crop-combinations in Wales. Instead of restricting the crops to the number involved in the combination, Thomas includes all the crops in the series under discussion. As this method involves all the crops in each step, there is no need of dividing Σd^2 by n. Thus the formula for the least squared deviation method may be expressed as follows :

$$\delta^2 = \Sigma d^2$$

In each step we will get some value of δ^2. The theoretical percentage which gives the minimum value of δ^2 (i.e. least squared deviation) is considered to be matching best with the reality and hence, the area is identified with it.

Thomas's formula, however, makes the calculations unnecessarily tedious (although the use of computer makes it simple) without giving marked differences from the results obtained by Weaver's method.

Maximum Positive Deviation Method

Weaver's method (even after modification), however, in cases of one or two functions is predominantly significant and in all other functions are equally insignificant: it gives undue importance to the functions of very low significance. In many empirical problems, therefore, Weaver's method does not give intuitively correct results. The basic reason of this fallacy in Weaver's method as pointed out by S.M. Rafiullah (1965) is that

(a) it ignores the signs of the deviations, and
(b) for a higher number of crops or functions the theoretical percentages go down and so the deviations caused by least values are underestimated.

While applying Weaver's method for functional classification of towns, Rafiullah suggested another modification in it in order to remove these shortcomings and thus introduced a new deviation formula (Maximum Positive Deviation) which, although suffers from the defects of laborious calculations like those of Weaver, is quite suitable for delineating sharply the primary crop-combinations. The formula introduced by him is :

$$\delta = \sqrt{\frac{\sum D_p^2 - \sum D_n^2}{N^2}}$$

where δ is the deviation, D_p and D_n are the positive and negative differences respectively from the medial value of the theoretical-curve value of the combination and N is the number of functions in the combination.[2] Since only relative rank of the deviation is needed the root sign may be ignored to save calculations and the formula may be used as :

$$\delta^2 = \frac{\sum D_p^2 - \sum D_n^2}{N^2}$$

Here the maximum positive deviation (i.e. maximum positive value of δ^2) gives the critical combination.

The Maximum Positive Deviation Method devised by Rafiullah gets the ascendancy over Weaver's Minimum Deviation Method as it reduces the number of crops in the combination and thus simplifies the crop-association map which can be helpful in the designation of agricultural regions of the area chosen (Husain, 1972).

Modified Minimum Deviation Method

Another modification of Weaver's method has been put forward by Kikukazi Doi (1959). This new method substitutes $\Sigma d^2/n$ of Weaver's with the sum of squared differences (Σd^2).[3] The combination having the smallest Σd^2 will be the combination of the primary crops. But it is not required to calculate Σd^2 for each combination, as it can be found by only consulting a table provided by Doi, which presents critical values for different elements at various accumulated percentages. Thus Doi, avoiding to calculate Σd^2 for each combination, has made the procedure more simpler giving all the critical values in a tabular form. In this case the method requires not the squaring of the differences but only a summation of the percentages of different crops. The crop percentages have to be checked with the table of critical values. If the percentage held by a single crop is lower than the critical value the crop is dropped from the combination; and if the percentage is higher than the critical value it is included in the combination.[4]

On deriving crop-combinations for different regions (Siddiqui, 1975; Mohammad, 1975; Majeed, 1978; Khan 1973), with Doi's method it was found that the results are more realistic in comparison to other methods discussed earlier and that the results are not mere generalisations. Thus the Modified Minimum Deviation Method is a better method for deriving accurate crop-combination regions within no time.

Maximum Distance Method

N.P. Ayyar (1969) has devised a new method in order to determine the number of the first few crops which could be considered as the most important, i.e. the number of the crops beyond which the addition of an extra crop does not significantly add to the previous combination. Since the farthest point on the actual curve for the unit area from the theoretical curve gives the number of crops to be included in the crop-combination, the method is called Maximum Distance Method.

To find out the distances (y's) for different crops of a unit area Ayyar introduced the following formula :

$$y' = y \cdot \cos\theta - x \cdot \sin\theta$$

where y' = distance on the actual curve for each crop from the theoretical curve.

$$y = \frac{\text{accumulated percentage of the crop}}{\text{scale chosen on the } y\text{-axis}}$$

$$\theta = \tan^{-1}\frac{\text{Total units on the } y\text{-axis according to scale}}{\text{Total units on the } x\text{-axis according to scale}}$$

(On y-axis accumulated percentages of the crops and on x-axis the number of crops are shown)

and x = Number of crops included in the accumulated percentage concerned.

The corp number for which y' is highest will be the required crop-combination.

To illustrate his technique Ayyar gives the example of Raipur District of Madhya Pradesh where 26 crops in all are grown — rice occupied 58.4 per cent of Gross Cropped Area, other pulses 24.1 per cent, linseed 6.9 per cent, wheat 0.9 per cent, gram 0.7 per cent, bengal gram 0.6 per cent, sesamum 0.6 per cent, groundnut 0.5 per cent, condiments 0.2 per cent, maize 0.1 per cent, other fibres 0.1 per cent, other non-food crops 0.1 per cent (other crops occupy individually less than 0.1 per cent of GCA).

For the formula $y' = y \cdot \cos\theta - x \cdot \sin\theta$ Ayyar calculated θ (choosing a scale of 1 unit for one crop, i.e. 26 units on the *x*-axis and 1 unit for 10 per cent, i.e. 10 units on the *y*-axis) by $\tan^{-1} 10/26$ which is equal to 21°2′ (thus cos 21°2′ = 0.93337 and sin 21°2′ = 0.35891); *ys* by dividing the accumulated percentages of the crops concerned with 0.1; *xs* being 1 to 26 as the case may be and tabulated the results which are as follows :

Crop	*Accumulated %*	*y*	*y′*
1	58.4	5.84	5.09
2	82.5	8.25	6.98
3	89.4	8.94	7.27
4	95.1	9.51	7.44
5	96.0	9.60	7.17
6	96.9	9.61	6.89
7	97.6	9.76	6.60

and so on for each next crop.

The combination for Raipur is then a Four-Crop Combination as the *y′* for four crops is the highest (7.44).

The two main advantages of this method (as stated by the author himself) are its simplicity and ease of calculation. Further, it is not generally necessary to consider more than 10 crops. Thirdly, there will be only one or at most two maximum points. That is, once the maximum point has been obtained and *y′* has started decreasing the calculations could stop.

Lower Limit Method

A.G. Athawale (1966) has evolved three new methods for crop-combinational analysis which are less time consuming and sometimes give lesser number of crops. These three methods are : (1) Method of Difference, (2) Method of Summation, and (3) Lower Limit Method.

These methods devised by Athawale have been fully described and tried in Andhra Pradesh at district level to derive the crop-combinations in his paper. After describing all the three methods in detail he himself opines that the third method is the best because of its simplicity and flexibility and hence, the Lower Limit Method (which itself is modified one) is being taken into consideration in the present paper.

In this method Athawale has attempted to associate the value of N (the number of crops to be included in the combination) to determine the crop-combination. To attain the lower limit in terms of acreages of crops he introduces the formula given below:

$$A = G/3N$$

where A is the lower limit in terms of acreage;

G = Gross Cropped Area;

N = Number of crops having acreage more than or equal to $G/100$

According to this formula all crops having acreages more than A will find place in crop-combination.

To illustrate this technique Athawale has given an example which is as follows:

Let G = 1,260 and the crop acreages be 800, 200, 160, 50, 25, 10, 10 and 5.

Then $G/100 = 1260/100 = 12.60$

Hence $N = 5$ (since only 5 crops have acreage more than 12.60) and lower limit $A = G/3N = 1260/3 \times 5 = 84$

There are three crops above 84 and hence the combination would be tri-culture, i.e. 3-crop-combination.

Athawale's Lower Limit Method is very simple as it does not require much calculation and also saves the time taken in calculating the percentages of all the crops from Gross Cropped Area (GCA) and hence it appears to be a better and suitable method for deriving accurate combinational regions within no time without any drawbacks of laborious calculations the other methods have.

Successive Quotients Method

This method has been applied by the Polish geographers to determine the crop-combinations (Kostrowicki, 1972). In this method the land occupancy strength of each crop is not studied directly but the numerous crops based on their agro-technical similarities are firstly categorised into three groups : (1) Intensifying Crops requiring more careful cultivation and more fertilisation. Ridged up or row crops, vegetables and most of the industrial crops have been classified into this group; (2) Structure Forming Crops, that do not require much labour or fertilisation. All leguminous crops are included in this group; (3) Extractive Crops are those that exhaust soil the most. Mostly cereals and other crops with similar requirements have been included in this group, and then the dominant elements in each group are considered.

After grouping all the crops growing in these three groups the area under particular groups of crops is divided successively by 1, 2, 3, 4, etc., until the final number of classified places (quotients) is obtained. To get a proper degree of differentiation the six successive quotients have usually been accepted for more detailed studies and four successive quotients for less detailed studies.

Then the crop-combination is expressed in a formula in which by big characters are shown the groups; in small characters the dominant crops occupying the largest hectareage, and by figures the role of particular groups of crops determined by the number of successive quotients.

The following example of Prochna Commune illustrates the determination of crop-combination by the Successive Quotients Method:

Crop-combinations

Taking four successive quotients :

E_2 *av* + I_1 st + S_1 *ti*, i.e. oats with potatoes and clover.

Taking six successive quotients –

E_4 *av* + I_1 *st* + S_1 *ti* i.e. oats with potatoes and clover with the dominant role of extractive crops (cereals).

In this method the calculation is easier as no percentage counting is necessary and the number of crop-combinations is usually less. The method thus seems to be more practical with the only drawback of leaving the crops of greater hectareage than those with lower hectareage due to grouping.

TABLE 19.1: Arable land utilisation in Prochna (total sown area = 936 hectares)

Specification	Symbol*	Hectareage
Extractive Crops	*E*	*495*
Wheat	tv	136
Rye	Sc	117
Barley	hs	42
Oats	av	200
Intensifying Crops	*I*	*236*
Oleaginous	h	21
Fibre Plants	t	3
Maize	zm	20
Sugar beet	bs	15
Fodder root crops	pr	40
Vegetables	lg	6
Potatoes	st	131
Structure Forming Crops	*S*	*232*
Vetch	vs	2
Clover	ti	135
Lucerne	ms	15
Serradella	os	55
Leguminous	px	25

TABLE 19.2: Successive quotients

Divisors	Extractive Hectares	Crops E Quotients	Intensifying Crops* Hects.	Intensifying Crops* Quots.	Structure forming Crops *S* Hects.	Structure forming Crops *S* Quots.
1	495	1	236	3	232	4
2	247	2	118		116	
3	165	5	78		77	
4	123	6	59		58	
5	99		47		46	

* The symbols for particular crops have been based on their Latin (botanical) names.

NOTES

1. This value is obviously the one referred to in statistics as the Standard Deviation. The sum of the squares of the deviation of the individual items from the mean is known as Variance. The SD or Standard Deviation is the square root of Variance.
2. Unlike Weaver the difference of actual values is calculated from the middle value of the theoretical standard.

3. Although Thomas has also substituted Weaver's $\Sigma d^2/n$ with Σd^2 but by including all the crops with zero per cent theoretical values also in each step in deriving the crop-combinations, he makes the calculation unnecessarily tedious consuming a lot of time and labour giving no marked differences from the results obtained by Weaver's Minimum Deviation Method.
4. To illustrate this, an example is given below :

   Individual crop percentages in a unit area are as : wheat 48.06 per cent, maize 16.38 per cent, barley 11.08 per cent, millets 9.94 per cent, sugarcane 6.48 per cent and pulses 5.66 per cent, etc.

   Now according to the table of critical values all those crops are included in the combination whose sum of percentages is below 50, as the critical value for all the elements at 50 is 0. In the above case, the first crop, wheat, has a percentage of 48.06 and the next ranking crop is maize with 16.38 per cent. As 48.06 per cent is below 50 per cent, the next crop is added in it and the sum of both the crops (wheat and maize) is 64.44 per cent. Under the heading, 65, the critical value for the third-ranking element, is given as 8.66. As barley with 11.08 per cent is larger than it, it is also included in the combination. Now the sum of the percentages ranking first to third is 75.52 and the corresponding fourth critical value for 75 is 10.57. Because millets with 9.94 per cent is lower than this critical value, it is excluded from the combination. Thus a three-crop combination (i.e Wheat-Maize-Barley) is designated for the unit area.

REFERENCES

Ahmad, A. and Siddiqui, M. F., 1967 : "Crop Association Patterns in the Luni Basin", *Geographer*, Vol. XIV, pp. 71-80.

Athawale, A. G., 1966 : "Some New Methods of Crop-Combination", *Geographical Review of India*, Vol. XXVIII, No. 4, pp. 29-33.

Ayyar, N. P., 1969 : "Crop Regions of Madhya Pradesh : A Study in Methodology", *Geographical Review of India*, Vol. XXXI, No. 1, p. 17.

Doi, K., 1959 : "The Industrial Structure of Japanese Prefectures", *Proceedings of International Geographical Union, Regional Conference in Japan*, 1957, pp. 310-16.

Husain, M., 1972 : "Crop-Combination Regions of Uttar Pradesh : A Study in Methodology", *Geographical Review of India*, Vol. XXXIV, No. 2, pp. 143-147.

Johnson, B. L. C., 1958 : "Crop-Association Regions in East Pakistan", *Geography*, Vol. XLII, pp. 86-103.

Khan, A. H., 1973 : "Crop Combinations in Rohilkhand", *Geographer*, Vol. XX, No. 2, pp. 153-162.

Kostrowicki, J., 1972 : "Some Methods and Techniques to Determine Crop and Other Landuse Combinations as used in the Polish Landuse Studies", in *Symposium on Landuse in Developing Countries, 21st International Geographical Congress*, 1968, (eds. M. Shafi, M. Anas and M. F. Siddiqui), Department of Geography, A.M.U., Aligarh, pp. 83-88.

Mahmood, A., 1977 : "*Statistical Methods in Geographical Studies*", Rajesh Publications, New Delhi, p. 115.

Majeed, A., 1978 : "*Changing Patterns of Cropland Use in the Union Territory of Delhi*, (Unpublished M. Phil. Dissertation), Aligarh Muslim University, Aligarh, pp. 153-157.

Mohammad, N., 1975 : "Crop Association Regions in Ghaghara-Rapti Doab", *The North Eastern Geographer*, Vol. VII, Nos. 1 and 2, pp. 3-9.

Nelson, H. J., 1955 : "A Service Classification of American Cities", *Economic Geography*, Vol. XXXI, No. 3, pp. 189-210.

Pownall, L. L., 1953 : "The Functions of New Zealand Towns", *Annals of the Association of American Geographers*, Vol. XLIII, pp. 332-50.

Rafiullah, S. M., 1965 : "A New Approach to Functional Classification of Towns", *Geographer*, Vol. XII, pp. 40-53.

Siddiqui, M. F., 1967 : "Combinational Analysis : A Review of Methodology", *Geographer*, Vol. XIV, p. 83.

———, 1967 : "Combinational Analysis : A Review of Methodology", *Geographer*, Vol. XIV, pp. 91-92.

———, 1975 : "Crop-Combinations and Specialisations in India", *Geographer*, Vol. XXI, No. 1, pp. 78-84.

Singh, H., 1965 : "Crop-Combination Regions in the Malwa Tract of Punjab", *Deccan Geographer*, Vol. 3, No. 1, pp. 21-30.

Singh, T., 1977 : "Bihar : A Study of Crop-Combination Regions", *Indian Journal of Regional Science*, Vol. IX, No. 2, pp. 147-152.

Thomas, D., 1963 : "Agriculture in Wales During the Nepoleonic Wars, (Cardiff), pp. 80-81, cf. Coppock, J. T. (1964), *An Agricultural Atlas of England and Wales* (London), pp. 211-212.

Tripathi, V. B. and Agarwal, U., 1968 : "Changing Patterns of Cropland Use in the Lower Ganga-Yamuna Doab", *Geographer*, Vol. XV, pp. 134-136.

Weaver, J. C., 1954 : "Changing Patterns of Cropland Use in the Middle West", *Economic Geography*, Vol. XXX, No. 1, p. 2.

———, 1954 : "Crop-Combination Regions in the Middle West", *Geographical Review* Vol. XLIV, No. 2, p. 176.

[illegible] Dudley, [illegible]: "Agriculture in Sussex During the Napoleonic Wars" [illegible] (1961): The Agricultural Area of England [illegible] (London), pp. 211-217.

INDEX